URBAN AVANT-GARDES

•

Can art or architecture change the world? Is it possible, despite successive failures, to think of a new cultural avant-garde today? What would this mean? *Urban Avant-Gardes* attempts to contribute to the debate on these questions, by looking back to past avant-gardes from the nineteenth and twentieth centuries, by examining the theoretical and critical terrain around avant-garde cultural interventions, and by profiling a range of contemporary cases of radical cultural practices.

The book begins with a reconsideration of the first avant-garde of the nineteenth century, followed by commentaries on the avant-gardes of early Modernist art and architecture. It then engages with the theories as well as cultural practices of the 1960s, and seeks to identify flaws in the concept of an avant-garde that may still disable cultural interventions. Moving on through the 1990s, the book interrogates practices between art, architecture and theory. It does not propose a new avant-garde but does find hope in emerging practices that in various ways engage with the agendas of environmentalism and social justice. At this point the terms art and architecture, as well as avant-garde, cease to be useful; what emerges is a need to re-imagine a public sphere.

Urban Avant-Gardes brings together material from a wide range of disciplines in the arts and social sciences to argue for cultural intervention as a means to radical change, while recognising that most such efforts in the past have not delivered the dreams of their perpetrators.

Malcolm Miles is Reader in Cultural Theory at the University of Plymouth, author of *Art, Space and the City* and co-editor of *The City Cultures Reader*.

URBAN AVANT-GARDES

ART, ARCHITECTURE AND CHANGE

•

Malcolm Miles

LONDON AND NEW YORK

First published 2004
by Routledge
11 New Fetter Lane, London EC4P 4EE

Simultaneously published in the USA and Canada
by Routledge
29 West 35th Street, New York, NY 10001

Routledge is an imprint of the Taylor & Francis Group

Typeset in Sabon and Helvetica Neue by
Florence Production Ltd, Stoodleigh, Devon
Printed and bound in Great Britain by
TJ International Ltd, Padstow, Cornwall

British Library Cataloguing in Publication Data
A catalogue record for this book is available from
the British Library

Library of Congress Cataloging in Publication Data
Miles, Malcolm.
Urban Avant-Gardes: art, architecture
and change / Malcolm Miles.
p. cm.
Includes bibliographical references and index.
1. Art and society. 2. Architecture and society.
3. City planning – Social aspects.
4. City planning – Environmental aspects.
I. Title.

N72.S6M55 2004
720′.1′03–dc22 2003018038

ISBN 0–415–26687–4 (hbk)
ISBN 0–415–26688–2 (pbk)

CONTENTS

•

LIST OF PLATES

•

All photographs by M. Miles unless otherwise stated in the plate captions

ACKNOWLEDGEMENTS

•

Research for this book has been supported in the following ways: the Small Grants scheme of the Arts & Humanities Research Board (AHRB) enabled me to visit north America and continental Europe during 2001–2 to interview artists and visit sites; a visit to Barcelona in 2002 was funded within another AHRB Small Grant in collaboration with Sarah Bennett, John Butler and Antoní Remesar; attendance at a symposium in Potsdam in 2001 was funded by the Schweisfurth Stiftung; a visit to Pittsburgh in 2001 was assisted by the Studio for Creative Inquiry, Carnegie Mellon University; the University of Barcelona and the Portuguese Design Centre assisted my participation in symposia in Barcelona and Lisbon during 2001–3; the European League of Institutes of Art funded various visits within its network; and the University of Plymouth provided time to write and met incidental costs beyond those covered by the above sources. The book's arguments were tested in seminars at Oxford Brookes University, the University of Barcelona, and the University of Plymouth, and initially in doctoral research in the School of Architecture at Oxford Brookes University. I am grateful, too, to individuals with whom I have had face-to-face or electronic conversations, including Cariad Astles, Kevin Atherton, Colin Beardon, Ian Bentley, Franco Bianchini, Iain Biggs, Iain Borden, Daniella Brasil, Jackie Brookner, David Butler, Mario Caeiro, Simon Clarke, Ian Cole, Jeff Collins, Tim Collins, Michael Corris, Diarmuid Costello, David Cross, Vera David, Monica Degen, Deborah Duffin, Peter Dunn, Graeme Evans, Jo Foorde, Sofia Fotinos, Murray Fraser, Raimi Gbaidamosi, John Goto, Reiko Goto, Jean Grant, Dan Gretton, Tim Hall, David Haley, Nabeel Hamdi, Helen and Newton Harrison, Richard Hayward, Peter Hill, Valerie Holman, Kathrin Horschelmann, Mark Jayne, Maria Kaika, Jeff Kastner, Nicola Kirkham, Suzanne Lacy, Anya Lewin, Katy MacLeod, James Marriott, Steven Miles, Lucy Milton, John Molyneux, Joanna Morra, Patricia Phillips, Marjetica Potrc, Robert Powell, Herman Prigann, Tim Putnam, David Reason, Antoní Remesar, Jane Rendell, George Revill, Marion Roberts, Dorothy Rowe, Judith Rugg,

Esther Salomon, Emma Sangster, Kirk Savage, Nick Stanley, John Stevenson, Paul Stickley, Joost Smiers, Heike Strelow, Ben Stringer, Valerie Swales, Erik Swyngedouw, Jane Trowell, Mierle Ukeles, Toshio Watanabe, Jackie West, Sarah Wigglesworth, Elizabeth Wilson and Paul Younger. Sincere thanks are offered all the above and to others I have inadvertently forgotten to include. Finally, I thank Andrew Mould and his team at Routledge for their forbearance and aid – the book took longer than anticipated to write and went through many changes from the original plan, but I am confident that it is the successor to my previous book in this field and it is due to the efforts of the publisher as well as my own that it appears in print.

GENERAL INTRODUCTION

•

I begin with a brief rationale for the book. This needs to go beyond reasons for writing such as the clarification of my own ideas or publication of my own research. Those are both necessary motivations for the writer, but I hope the book will contribute to debates on urban issues during the first decade of the twenty-first century. In particular, I hope it will illuminate what certain kinds of cultural practices contribute, not only reflectively, but in actively shaping the agendas of future urban development and change. The agendas are shaped already by contexts such as climate change and globalisation, yet it seems important that criticism should be not only reactive to such contexts, but also informed by alternatives to the scenarios of the present situation. It seems, too, that much of what has been published in urban studies, cultural and urban geographies, and cultural policy emphasises the role of cultural institutions in urban regeneration while ignoring more radical forms of practice that irritate those institutional structures. From another angle, recent writing in sociology, while taking a cultural turn, tends to define almost anything as a tactic of resistance. This suggests a need for writing that begins from an involvement in practices which enact alternative scenarios – in my case as someone whose practice is theory (which is produced and has its textual forms just like art is produced and has its visual or tactile forms) – but also crosses into surrounding academic territories. The writer needs to get home alive, of course, but on the way to have contested the assumptions that limit present discussion, to have gained a new insight into the home territory by seeing it from outside, and to have articulated something of the values, implicit or explicit but including some of the big ones like freedom, of those practices. But there is a difficulty in that the language we use sometimes articulates concepts and meta-concepts that were developed in modernity and which have lost their currency. The concept of an avant-garde is one such, deeply flawed by elitism and an assumption that the new society is not here or now but located in a utopian future, which becomes a never-never land. One response would be to drop the idea. But this could be to reject the hope it embodies, of which I cannot quite let go. So I am left unpacking the baggage and sifting through the failures, asking what is left but also what is different in the work of radical cultural practices now. I have had to be selective in what I write about, have left out much no doubt, but have tried to make the story interesting. Now I will try to outline the aims, scope and organisation of the book, and its relation to my previous book, *Art, Space and the City* (1997).

As indicated, the first aim is to ask what can be retrieved from the concepts of an avant-garde formulated in the nineteenth and twentieth centuries; and the second is to comment on recent and contemporary practices. These are difficult to categorise but exist between art, architecture and the processes of urban formation. The third is then to introduce readers to the literatures of fields within that triangulation other than their own. I have attempted to integrate various insights and perspectives rather than to set them out like a row of market stalls. It is not that the kinds of literature compete, but that read together they offer more than they do alone.

What happens, then, if I dig up the idea of an avant-garde? First, I find its histories more encouraging to those in power than to resistors. But I do not argue for a new avant-garde. If it seemed in the mid-nineteenth century that artists might lead society towards a future built on social justice, the terms were often one-dimensional (class consciousness) or utopian (a dream of social organisation that is as aesthetic as art). These problems have not gone away; but to class are added the categories of race and gender, and other more local differentiations in the recognition that common interests replace geographical coherence in patterns of urban sociation. Meanwhile utopianism is largely discredited. And yet the dream of a better world does not go away; to ask what can be excavated from the histories of cultural movements for a better world may thus offer insights into a necessary revision of the question and a necessary revision of tactics.

On the second aim: the practices on which I comment are included because I read them as critical interventions in current conditions; and because the practitioners were willing to engage in conversation and to answer questions that were not restricted to appreciation. Many others could have been included; I have followed the needs of viability within the limitations of time and resources. As to whether they should be taken as art, architecture, or something else altogether (like activism), I see no interest in arguing over that – if they are there, the angels continue to dance on the pins regardless of being counted.

On the perhaps more predictable aim to introduce readers to elements of the literatures of other fields than their own: it is also an aim to make connections between ideas and critical frameworks, and between theories and enactments of theory. I have tried to create access to complex material but not at the cost of masking complexity, and would add that the aim includes drawing attention to practices that are outside the main stream, or difficult to categorise, and tend to be less widely known than they should be.

SCOPE

This is an academic book for second- and third-year undergraduates in art, architecture, cultural geography, cultural planning, cultural studies, urban sociology and urban studies; and for post-graduates in trans-disciplinary fields of culture, society and environment. If the book offers something to readers in critical theory I should be delighted, though I do not claim to add to the achievements of the Frankfurt School (only to draw on them).

The book covers a period from Realism in France after 1848 (and initial uses of the term 'avant-garde' to describe a politicised art before that) to the present. The years which introduce the chapter titles run from 1871 to 2001 – the Paris Commune to 9–11. But the book is situated in the present in which it has been produced. This happens to include a millennium with its adrenalin-producing tales of doom; but I see more continuity than sudden change, as the reaction to 9–11 reproduces an emphasis on security and denial of difference already well established through the Cold War. But, if 1989 (Chapter 5) is the mid-point in the book's trajectory, Chapter 4, taking 1967 as a point of departure, is equally pivotal because there seem to me continuities, too, in hope. Although society was not transformed in 1968, the utopianism of the era (in student protest, in art, in dropping out and in philosophy) may offer insights for today even if the tactics failed and/or the utopianism itself was flawed. In face of what appears an abolition of politics, it seems vital to set aside the feeling of helplessness that the present situation engenders.

The book may be perceived as occupying interstitial spaces between fields and disciplines. This reflects my own tendency to work in trans-disciplinary areas (where tolerance is greater because one is less of a threat, but where recognition is compounded with a pejorative sense of non-belonging). If my personal state of psyche draws me to border places, then I should say still that critical theory requires such an approach, and that a trans-disciplinary enquiry is more likely to produce new insights into the social, cultural and political conditions in which the practices discussed intervene than one based in a single discipline.

ORGANISATION

The book is arranged in nine chapters, each designated by a date between 1871 and 2001. This arrangement has two interlocking architectures: one of three sections and the other of three points with links – vaults, as it were – which draw together aspects of the material across the book's chronology. The first section looks back to periods in which different avant-gardes have emerged; the second also looks back, but to a period that stretches from the build-up to the events of 1968 to 1993 (an arbitrary division in some ways, but convenient to introduce a necessary theme); the third begins in 1993 but with a future scope rather than a past scope, and ends in 2001 with the attack on the World Trade Center in New York and its aftermath (which is far from over). Taking the second architecture, Chapters 1, 5 and 9 all concern the public realm and its furnishing with signs of social ordering and disordering. Between Chapters 1 and 5 the fields of art, architecture and theory are examined. Between Chapters 5 and 9 a number of contemporary practices are investigated, many of which enact the collapse of conventional boundaries between practices and fields. Chapter 5 begins in 1989, at the end of the Cold War – a convenient point at which to re-assess and extend ideas from Chapter 1 on the destruction and recoding of monuments.

Each chapter begins anecdotally. The dates and events taken may have an oblique relation to the chapter's main content; but they act also as a frame, or grid, against which the book's material pushes – it is a way of telling a story that leaves a certain amount to the imaginative and deductive powers of the reader. Chapter 1 opens with the destruction of the Vendôme Column during the Paris Commune of 1871 and moves to Realism as a first avant-garde. Chapter 2 begins with Raymond Williams' allusion to Strindberg's birthday procession in Stockholm in 1912 – an oblique perspective until it is noted that the procession was organised by a workers' commune – leading to discussion of a second avant-garde in early twentieth-century art. Chapter 3 begins with an account of Le Corbusier's desecration of a villa by Eileen Gray at Cap-Martin in 1938, and links his Modernism (an architectural avant-garde) to orientalism as well as the political situation of the 1930s. Chapter 4 begins with a question following a lecture by Herbert Marcuse at the Free University, Berlin, in 1967, and asks why the hoped-for transformation seems never to occur. Chapter 5 begins with the fall of the Berlin Wall in 1989 and asks how the genre of the monument may be democratised, subverted, or reclaimed. Chapter 6 starts in 1993 at the opening of the Holocaust Memorial Museum in Washington, D.C. and reconsiders the construction of historical narratives. Chapter 7 begins a few days later with a performance at an art centre in New York, and addresses participation and provocation in recent art and architecture. Chapter 8 looks to issues of sustainability and how cultural interventions address the green agenda, beginning with a meeting of activists in Brazil in January 2001. Chapter 9 takes responses to the attack on the World Trade Center eight months later as point of departure for a reconsideration of the public sphere, setting the current regime of a security-state beside a potential for dynamism and cosmopolitanism in a world reclaimed by its inhabitants.

In an effort to write a clear and succinct main text, various and sometimes copious details, sources and tangents are put into the notes that follow each chapter (put there not at the end for the reader's convenience and because each chapter can be used as a seminar text). The book offers two ways of reading: as a main text alone; or as a text plus notes. The reader will decide which route to take, and in which order to read the chapters. I use the Harvard system for references but to minimise clutter in the text give references only after direct quotations, putting supplementary sources in the notes. I have not given notes for further reading because the end-notes meet that need.

RELATION TO PREVIOUS WRITING

There are two differences between this book and my previous writing: first, I write here in the first person, having previously used the academic third person because it seemed to place greater value on the material than on my view of it. Perhaps now I am relaxed enough to see 'I' as affirming a legitimate presence of the writer in what is discussed. The second difference is a shift in position since I wrote *Art, Space and the City*: a move away from public art – which I now see as a departure subsumed back into a main stream that has itself become more fractured and interesting – towards cultural practices which are critical regardless of category or site, and which in many cases collapse the boundaries of production and reception. If it all looks like work at the edges, this only suggests the obsolescence of the categories used hitherto. Manfredo Tafuri argues that 'It is useless to propose purely architectural alternatives' (Tafuri, 1976: 181); and Iain Borden notes the death of an architectural avant-garde that he defines as an 'elitist group, small in number, somehow apart yet ahead of the rest of society and prescient of its future direction', seeing radicalism now as no longer oppositional but working 'ironically and irritatingly against the dominant systems of capitalism, colonialism and patriarchy' (Borden, 2003: 117–18) – which could almost be a summary of my argument except that I still hold on to hope.

1

1871
SPITTING ON BONAPARTE

•

In this opening chapter I attempt to set a scene of rapid social change during the Paris Commune of 1871 and to establish within it the role of cultural processes, including in this case the destruction of a public monument. Through discussion of Gustave Courbet's art and his involvement in the destruction of the Vendôme Column, I sketch what I take to be a first avant-garde, which is epitomised by French Realism. This avant-garde, which is politicised through a link to French utopian socialism in the mid-nineteenth century, is not entirely extinguished by the fall of the Commune. It contrasts with the anti-art avant-garde of early twentieth-century art discussed in Chapter 2, yet has some relation to the utopianism of the Modernist project in architecture discussed in Chapter 3. The problem of what, apart from public monuments like the Vendôme Column, constitutes a public sphere is taken up in Chapter 9. Setting the pattern for the book, I begin with an anecdote:

> the impulse to attack and destroy public works of art is part of the general attack on the continued presence of signs of the *ancien régime*. It is confirmation also that in moments of 'madness', publics will treat these monuments almost as if they were the actual leaders themselves . . . For instance in a report from 1871 on the destruction of the Vendôme column, the *London Illustrated News* gave this account of what happened after the column was felled: '[The crowd] treated the statue . . . as the emperor himself, spitting on his face, while members of the National Guard hit his nose with rifles.'
>
> (Lewis, 1991: 3, quoted in Mulvey, 1999: 220)

I PLACE VENDÔME

Anecdotes are not documentation. Nonetheless, they provide useful insights into histories. There is another, too strange to be a trick of memory or invention, that the Communards went through Paris shooting the public clocks, acting not like rat-catchers but as executioners.

In the first story, Bonaparte's effigy stands in for the person of Napoleon III, and is treated as the Communards would have wished to treat that person (by then elsewhere). Perhaps some of those present remembered the revolution of 1848 and the election of Napoleon as Emperor in 1851 by a conservative provincial vote, a vote against Paris, which sealed its failure. Napoleon III presided over a bourgeois state, an economic boom in the 1850s, the making of many fortunes, and the remodelling of Paris under Baron Haussmann which carved wide streets through the working-class quarters, redistributing the poor to the peripheries. On August 15th, 1870 the Emperor had planned to unveil a statue at Place de Clichy – *Monument to the 1814 Defence of the Barrier at Clichy* by Amédée Doublemard[1] – but instead he rode out to his armies to be defeated at Sedan on September 1st, with which his currency became worthless. In the second story, the face of a clock with its regularly spaced numerals stands, a more dispersed and abstract sign than a statue, for another regime, that of the routines of labour on which modern industrial production depends.[2] In a more direct expression of hate for the toppled regime, the Communards shot two generals.[3] In this context, the toppling of the Vendôme Column, bringing the bronze statue of Bonaparte down to street level where it could be spat on, is not an ephemeral act of destructiveness, or a prank, but a purposeful re-enactment of the abolition of a regime through the destruction of one of its monuments. The re-enactment replays the shift of power as public spectacle, affirms in the freedom to do it that a change of power has taken place, and reclaims public space from the previous regime.[4] Similarly, when the Berlin Wall was opened in 1989 people hacked it to pieces, taking them home as material evidence of having been there at its destruction.

The Vendôme Column commemorated Napoleon Bonaparte's victory at Austerlitz, the statue of Bonaparte in Roman dress being made from melted-down canons captured at the battle. The form is based on a Roman monument, Trajan's column. It had been destroyed once before, in 1814, and was rebuilt after the revolution of 1830 by Louis-Philippe (the citizen-king, so-called) with a new statue. Napoleon III restored it a second time in 1862, substituting a replica of the old statue for the new one. In this restoration it took on three layers of representation: the universality of power conveyed by the monument's Roman form, annexing two millennia of history; the glory of France under Bonaparte; and, trading on both, the power of the bourgeoisie under Napoleon III. Each layer was contestable, particularly the last two. Even for those who remembered, or had heard personal accounts, of Bonaparte's victories, these might have been seen beside the end of the Revolution's radical stage with the fall of the Jacobins. The monument became a central element in Napoleon III's public spectacles, used for military parades, and symbol of a regime known for its increasing corruption. Its destruction abolished all its histories at a stroke, and followed attacks on buildings and monuments, and removal of street signs, associated with the Napoleonic past.[5] The destruction of the column, then, is a key symbolic act alongside other equally symbolic but more everyday acts of erasure, changes in the visual face of Paris to show the shift of power from Empire to Republic.

The unbolting (*déboulonné*) of the Column and removal of its parts to l'Hôtel de la Monnaie was first proposed by Gustave Courbet in a letter to the Government of National Defence in 1870, after the defeat at Sedan. The Column, he argued, was a symbol of war and conquest, antipathetic to the spirit of modern civilisation and the union of universal brotherhood.[6] This was reported in the press, with a suggestion that the metal be turned back into guns to use against the advancing Prussian forces. The letter follows Courbet's wider involvement in issues of art's organisation and conservation. A pacifist at the beginning of the Franco-Prussian war, he was appointed to an arts commission the task of which was to oversee the conservation of works, and investigate previous corruption at the Louvre. Courbet wrote that he was pleased to accept: 'I did not know how to serve my country in this emergency, having no inclination to bear arms' (Chu, 1992: 385, quoted in Roos, 1996: 150). Meanwhile Degas and Manet, both republicans, joined the National Guard; Monet spent the period of the war and Commune in England.

During the Commune, Courbet presided over debates on art education – the abolition of the Academy was proposed as a mark of egalitarianism, along with removal of juries for the annual Salons[7] – and the reorganisation of museums. Following his work in the arts commission he became chair of the new Federation of Artists. On April 16th, 1871 he was elected by the sixth *arrondissement* to the Commune's administrative council,[8] and on April 27th again urged the removal of the Column, this time suggesting its replacement by a statue celebrating the Commune. The removal was agreed, and carried out by contractors in the name of the Federation of Artists (which Courbet chaired). There is some uncertainty as to Courbet's immediate involvement in the event, though it seems clear he argued consistently for it.

The Column was destroyed on May 16th. The Commune's decree states:

> Considering that the imperial column at the Place Vendôme is a monument to barbarism, a symbol of brute force and glory, an affirmation of militarism, a negation of international law, a permanent insult to the vanquished by the victors, a perpetual assault on one of the three great principles of the French Republic, Fraternity, it is thereby decreed:
>
> Article One: The column at the Place Vendôme will be abolished . . .
>
> (Ross, 1988: 5, quoted in Cresswell, 1996: 173)

Here another anecdote can be introduced: that Bonaparte's head broke off and rolled away like a pumpkin.[9] The act was denounced by the Versailles government, Marshall MacMahon writing: 'Soldiers! . . . Men who call themselves French have dared to destroy . . . this witness to the victories of your fathers against the coalition of Europe. Do they hope . . . to erase the memory of the military virtues of which this monument was the glorious symbol?' (attributed to Marshall MacMahon, Commander-in-Chief of the national army, press clipping, Bibliothèque Nationale, Paris; Edwards, 1971: 201, quoted in Roos, 1996: 155).

Courbet's political engagement during the Commune followed a return to images of social injustice in the late 1860s, as a reaction against the regime and its corruptions, and against the triumph of the bourgeoisie under it. Although he made few overtly political works after 1855, one of his entries to the Salon of 1868 – *The Beggar's Charity at Ornans* (Musée d'Orsay, Paris) – marks a return to social criticism and the settings around Ornans of earlier works such as *The Stonebreakers* (1849, destroyed) and *The Burial at Ornans* (1849–50, Musée d'Orsay, Paris), made as representations of the democratic sentiments of the 1848 revolution, when universal suffrage was briefly proclaimed (and later withdrawn by Napoleon III). *The Beggar's Charity at Ornans* shows a beggar on crutches giving a coin to a child while a woman suckles a baby in the background. All are ragged. So, the poor are more generous (in spirit as well as material means) than, by implication, the rich. For the radical critic Jules Castagnary, like Courbet a reader of the utopian socialist Pierre-Joseph Proudhon (see Proudhon, 1969), it represented the endurance of human generosity in adversity:

> For twenty years the poor tramp travels the same land, holding out his hand to all . . . And for the first time in twenty years someone does him the honour of asking him for alms . . . It is the encounter of two miseries . . . the local beggar feels an old forgotten tear well up under his eyelid, takes a sou out of his pocket and gives it to the child who sends him a kiss.
>
> (Castagnary, 1892, vol. I, 287–8, quoted in Roos, 1996: 108)

Zola saw it as representing Courbet's 'gently humanitarian philosophy', again in the manner of Proudhon (Zola, 1991: 219, quoted in Roos, 1996: 106).[10]

Despite the work's negative reception, Courbet was otherwise a widely accepted and popular artist. His work was placed in the room of honour at the 1867 and 1869 Salons; in 1869 he was awarded a gold medal by Leopold II of Belgium, and went to Munich to receive the Order of St Michael from Ludwig II of Bavaria. Yet he declined the *Legion d'Honneur*: 'My opinions as a citizen are such that I cannot accept a distinction which belongs essentially to the monarchical order . . . the state has no competence in the field of art. When it takes on itself to confer rewards, it is encroaching on the sphere of public taste.' (de Forges, 1978: 45, source unstated). Courbet was by now an established artist, selling work to the value of 52,000 francs at the time of the 1870 Salon.[11] At the time of the Commune, then, Courbet was a major figure in French art both for the bourgeoisie who frequented the Salons, and for Parisian artists in their associations. It is not surprising that, given his return to politics and commitment to democracy, he played a key role in the Commune's cultural organisation. The destruction of the Column, however reticent Courbet was about it at his trial, could be seen as the culmination of a development of radical cultural representation and, in the end, action.

The Commune fell on May 28th, 1871. Soldiers of the Versailles government combed the streets rounding up Communards, or anyone suspected, and shot them. Up to 30,000 citizens may have been killed by summary execution.[12] Among them was Eugène Varlin, a 32-year-old bookbinder and socialist, arrested, paraded and humiliated, then shot at Montmartre on May 28th. Harvey records: 'They had to shoot twice to kill him. In between fusillades he cried, evidently unrepentant, "Vive la Commune!". His biographer called it "the Calvary of Eugène Varlin"' (Harvey, 1989: 215). The Basilica of Sacré-Coeur – as penitence for the ills of the preceding years (as seen by the religious right) – was erected on Montmartre, its foundation stone laid in 1875. It was a deliberate erasure of the site of the Commune's first and last days – monumental architecture in service of the suppression of public memories.

Courbet was arrested on June 7th for his part in the destruction of the Vendôme Column, and tried in August. He maintained in questioning that he had simply wanted the column removed on aesthetic grounds, not destroyed.[13] Several critics and established artists testified for him. Only a minor charge was upheld, and he was sentenced to six months' imprisonment and a fine of 500 francs, rashly saying he would pay for the Column to be re-erected if his guilt for its destruction were ever proved. Of the 16 Communards tried with Courbet one was deported, two sentenced to hard labour for life, seven sent to penal colonies, and two executed. In prison he painted a bowl of apples, which was rejected at the 1872 Salon, though he sold several works at an exhibition at the Durand-Ruel gallery that year. Several of his paintings also went missing from his lodgings in Passage de Saumon before his release.

Then disaster struck – in 1873, with a swing to the political right, MacMahon was elected President. Courbet, who fled to Switzerland, was charged in June 1874 with the cost of the Column's re-erection, initially estimated at 250,000 francs but finally assessed at 323,091 francs, 68 centimes, to be paid at the rate of 10,000 francs a year. Works and property were now confiscated, and his hopes of being rehabilitated, and accepted again at the Salon, dissolved when, in 1876 MacMahon dismissed the progressive premier Jules Simon. Courbet's last work was a view of the Alps between Vevey and Montreux. He died of dropsy in 1877, impoverished and with no hope of a return to France.

Two questions arise. Why did the Commune place such emphasis on cultural organisation? And what was left of the avant-garde after its defeat? To approach the first: given the Commune's short life (73 days), most of its projects remained aspirations. There is no major artwork produced in the Commune, no equivalent of the competition for an image of the Republic of 1848,[14] though Courbet had proposed such a monument to replace the Vendôme Column. Manet produced two lithographs in 1871, *The barricade* and *Civil War*, but not until ex-Communard's Jules Dalou's monument to the Republic (1889–99) is there a return, and here in muted form, to radicalism in the arts.[15] The Commune's impact was more in removal of signs of the old regime than in new art, but it devoted much effort to the organisation of journalism, festivals and the theatre, to conservation and to education in the arts. But why all this, when there were barricades to build and defend? The Commune's engagement with culture can be understood in two ways: as extending from a philosophical tradition from Proudhon and Rousseau, in which art is a means of public education, previously employed by David for the Jacobins; and as reflection of the high profile of cultural activities in Parisian life before the Commune, with high attendances at the Salons and a widespread coverage of the arts in the press.

Perhaps to dedicate time to art in the Commune did not seem extraordinary after all, though the example is mirrored 46 years later in the extensive monuments, parades, banners and street decoration of the October Revolution.[16] Just as in Paris in 1871, it seemed necessary in Moscow and Petrograd in 1917 to give material and publicly visible expression to the moment of transformation. A. V. Lunacharski, speaking at the opening of the Free Art Educational Studios in Petrograd in October, 1918, asserted: 'The need has arisen to change the external appearances of our towns as rapidly as possible, in order to express our new experiences in an artistic form as well as to get rid of all that is offensive to the feelings of the people' (Tolstoy, Bibikova and Cooke, 1990: 15). Similarly, in the years leading up to 1968, members of the Situationist International called for the removal of monuments which were, as they put it, irretrievably ugly.[17]

But if the Commune's attention to public spectacle makes it part of the pre-history of 1917, its place in political history is ambivalent. Marx was initially enthusiastic, seeing it as an *enactment* of radical democracy, not merely a regime elected by the working class but the working class *as* the regime: 'The communal constitution would have rendered up to the body social all the powers which have hitherto been devoured by the parasitic excrescence of the "State", which battens on society and inhibits its free movement . . . it would have brought about the regeneration of France' (Marx, 'Address to the General Council of the International on the Civil War in France', quoted in Buber, 1996: 86–7). He may have exaggerated his support to assist the Commune, revising it later.[18] Henri Lefebvre sees the Commune in a different way, more integral to everyday life, representing a reclamation of the inner city by the working class after their peripheralisation by Haussmann:

> Baron Haussmann, man of this Bonapartist State which erects itself over society to treat it cynically as the booty . . . replaces winding but lively streets by long avenues, sordid but animated 'quartiers' by bourgeois ones . . . to 'comb Paris with machine guns'. The famous Baron makes no secret of it.
>
> (Lefebvre, 1996: 76)[19]

Lefebvre also sees the Commune as a *moment* in history, using the term to denote a glimpse of authentic liberation manifest in carnivalesque celebration. The Situationists' incorporation of the Commune into their alternative geography of Paris follows their link to Lefebvre, who set out elements of their discussions of a festive revolution in *Proclamation de la Commune*.[20]

There is, then, a legacy. But is there an avant-garde after 1871? The question of what constituted the avant-garde, and its theoretical content, are discussed below; but I end this first section of the chapter by saying that although the example of Courbet's death in exile – as penalty for his avant-gardism – could mark the end of the avant-garde which began with Realism, the situation is in fact more complex. Some of the Impressionists and Neo-Impressionists – Pissarro, Seurat and Signac, for example – held radical sympathies through the 1870s and '80s. In ways that could be overlooked, their paintings are a reflective if not overtly critical record of the years after 1871. Wood writes of Impressionist street scenes:

> Those streets tell a story of the bourgeoisification of Paris. There is no question of that, but they also contain a memory of the price of that bourgeoisification. There is not a seamless transition between the Second Empire and the Third Republic. Instead there is something like a collective nightmare for the French bourgeoisie. And early Impressionist scenes of urban leisure draw a veil of light across a chasm in French history.
>
> (Wood, 1999: 121)

The physical signs of the Commune's defeat were visible in Paris for several years, and while the province of Alsace was occupied by Prussia the statue of Strasbourg in Place de la Concorde was draped in black, becoming a site of pilgrimage.[21] Degas depicts this by not depicting it in *Place de la Concorde* (1875, Hermitage Museum), concealing the statue by the black hat of Baron Lepic. Manet's *Rue Mosnier with flags* (1878, Getty Museum, Los Angeles), too, is a covert image of defeat – in the guise of a festival.[22] There were also images of reconstruction, such as Monet's *The Railway Bridge at Argenteuil* (1874, Philadelphia Museum of Art), a bridge rebuilt after its destruction by the Prussians, and symbol (for Monet) of the most modern industry.

II A FIRST AVANT-GARDE: FROM THE PAINTER'S STUDIO TO THE BANKS OF THE SEINE

I want now to look back, taking Realism in France in the mid-nineteenth century as a first avant-garde. I differentiate this from a second avant-garde in Modernism (discussed in Chapter 2), which attacks, not bourgeois social values, but art's institutions. In Modernist architecture (discussed in Chapter 3) there is a residual utopianism attached to an aim to engineer a new society. These are broad-brush statements and there are many exceptions. It could be argued, too, that an attack on art's institutions is an attack on bourgeois society. Courbet's proposal to remove juries from the Salons is such a case. But in Modernism, particularly from the 1940s, the process seems more akin to the internal deliberations of a specialist profession, aiming no longer for social justice but to redefine the means of representation. In Clement Greenberg's words, it is an effort to keep art moving.[23] In the end it becomes self-referential, so that today art has a public outside its own circles mainly as an adjunct of the entertainment industry.[24]

This is not to say that the first avant-garde is a model to resurrect. The concept is inherently flawed; but I argue it is worth re-visiting a history of art that sought to act on the conditions of society from within, to contest them and to change them. But what was this first avant-garde?

The military term 'avant garde' denotes a small force ahead of an army. It gained currency in the Napoleonic period as Bonaparte evolved dynamic and innovative military tactics. Its use in art denotes a small group of artists ahead of the mass of society, who foresee society's future development and, instrumentally, lead society towards it. The avant-garde occupies a location paradoxically both inside and outside the wider society: it seeks to represent the condition of society as it is, devising an appropriate visual language for the purpose; and it sees ahead, as if having a vantage point on high ground or looking to the future (and at the present) from a belvedere.

There is a second aspect, also transposed from the term's military origin, of risk. The avant-garde spies out the terrain and may encounter enemy forces before the main army arrives. As a small, intrepid force it is vulnerable but gains (or is graced by) special knowledge. In cultural terms, the idea of a risk-taking avant-garde informs Romantic culture's refusal of the certainties of classicism. Anita Brookner argues that for Stendhal and David risk is found in the act of innovation: 'There are no precedents to fall back on, and this is what distinguishes Stendhal's definition of Romanticism from all those writers and painters who are simply trying to replace the classical tradition with an alternative mythology' (Brookner, 1971: 48). The avant-garde, then, leads the way, and has a celebratory and informative function. It gives form to the moment of change (as in David's festivals during the Jacobin period), and it instils new ideas in a programmatic way.

Claude-Henri de Rouvroy, Comte de Saint-Simon, may have been the first to use the term 'avant-garde', in a dialogue involving an artist, a savant and a scientist. The artist says:

> We, the artists, will serve as the avant-garde; for amongst all the arms at our disposal, the power of the Arts is the swiftest and most expeditious. When we wish to spread new ideas amongst men [*sic*], we use, in turn, the lyre, the ode or song, story or novel; we inscribe those ideas on marble or

> canvas, and we popularize them in poetry and in song. We also make use of the stage, and it is there above all that our influence is most electric and triumphant. We aim for the heart and imagination, and hence our effect is the most vivid and the most decisive. If today our role seems limited or of secondary importance, it is for a simple reason: the Arts at present lack those elements most essential to their success – a common impulse and a general scheme.
>
> (St-Simon, 1825: 332–44, quoted in Harrison and Wood, 1998: 38–9)[25]

Nochlin cites also a passage from the Fourierist critic Charles Laverdant, written 20 years later:

> Art, the expression of society, manifests, in its highest soaring, the most advanced social tendencies; it is the forerunner and the revealer. Therefore to know whether art worthily fulfils its proper mission as initiator, whether the artist is truly of the avant-garde, one must know where Humanity is going.
>
> (Laverdant, *De la mission de l'art et du rôle des artistes*, 1845, quoted in Nochlin, 1991: 2)[26]

The avant-garde, then, must know where society is going. But how?

The development of the concept in art is a critical formulation and not an artists' movement. It derives its vision of a future from French utopian philosophy, and its educative aspect follows Jean-Jacques Rousseau's view that, while modern society decays in luxuries, art can be rescued from this condition by giving form to ideas of virtue – images of virtue are conducive to virtuous behaviour. Saint-Simon says much the same. David's paintings are examples of this, in which the moral fibre of the actors in the drama depicted, as if on a stage, is to be imitated by the spectator. But that is where the difficulty begins: the bourgeois public for David's art knew the histories of the Roman Republic which he uses as coded political statements.

Pierre-Joseph Proudhon, in 1865, asks rhetorically what art can do for the crowd, who by implication are uneducated. He answers that it educates them:

> It could do something most interesting, the most glorious thing of all. Its task is to improve us, help us and save us. In order to improve us it must first of all know us . . . as we are and not in some fantastic, reflected image which is no longer us . . . Man will become his own mirror.
>
> (Proudhon, [1865], 1970: 215)

This both relieves and compounds the difficulty. It relieves it in extracting the representation of ideas and replacing it with that of people as they are, so that they know themselves. It compounds it because to do that, too, is an interpretation of sorts. Do the people not already know themselves? Perhaps not in the drudgery of daily needs, yet to be shown it is not to sense it for themselves. I will return to this later. Here I want to focus a little more on the divergence of art which depicts ideas and that which depicts things.

The issues are clear in Delacroix's *28th July: Liberty Leading the People* (1831, Louvre), his depiction of the July days of 1830.[27] The painting could be taken as the first explicitly avant-garde work. Departing from neo-classicism and using the dynamic compositional devices of Romanticism, Delacroix combines the high sentiment of Liberty with the democratic sentiment of the crowd, and uses two kinds of visual code to stand for these two kinds of subject-matter: the adapted classicism, slightly ruffled and eroticised (both breasts bared instead of the usual one) of Liberty;[28] and the realism of the crowd. The figure of Liberty wearing a red bonnet may or not be credible as 'a lower-class woman purposefully striding barefoot over the rubble . . . *and* the symbol of an abstract idea' as Wood says (Wood, 1999: 37). For me the figure looks like a statue from a museum, a reading not modified by the formal integration of the painting's composition and paint surface. Perhaps the contrary: Liberty forms the apex of the triangle around which the picture's architecture is built, holding its disparate elements in dynamic equilibrium; but the blaze of yellow behind her, the emblem of the tricolour she holds, and her raised position, separate Liberty from the crowd in their murky region, as a military commander might be painted leading the common troops from an exalted position. This may be deliberate, Liberty illuminating the mass consciousness, but underlines the difficulty that Liberty is privileged as representation of a noble, abstract idea. Yet Liberty is not the only invention in Delacroix's painting: the crowd, too, is a carefully selected set of types, a tableau, a staged performance of what might have happened.[29]

The difficulty, then, is that images of abstract ideas, or imagined futures, will tend to draw on past conventions of representation which are not without conceptual baggage. In neo-classicism, the narratives are accessible to those who already know them, the educated classes. For others the pictures must be interpreted, but interpretation – even within an ethos of liberal reform – states power in the knowledge of the interpreter.[30] It seems inescapable that abstract ideas are associated with a socio-cultural elite who, traditionally, have the leisure to discuss and study them as philosophy. This difficulty is compounded by the histories carried within concepts themselves. Liberty, for instance, is a concept of eighteenth-century bourgeois revolutionaries on both sides of the Atlantic, denoting the rights to representation of (mainly) male property owners. Freedom is different, has more radically democratic connotations. Its absence, unfreedom, is not incompatible with Liberty. But can abstract concepts be made into communicable images when allegories of continents, cities, and industries in neo-classical statuary show the difficulties?[31]

Realism can be seen as an attempt – successful or not is beside the point here – to escape the difficulty of representing ideas. To put an example of a painting by Courbet next to Delacroix's Liberty: *The Beggar's Charity at Ornans* (cited above) depicts an act of charity, but is not a painting of Charity (or *Caritas*). It is no less stage-managed than Delacroix's picture, but the intention seems to have been, placing the scene in the landscape around Ornans which Courbet knew well, to show the ravages of poverty and premature ageing which were only too visible in the persons of the very poor – as they were. There is an implicit critique of bourgeois social values, but in a depiction of the conditions those values produce. While, then, the depiction of things has its own art history in genre painting and still life – lesser forms than history in the Academy – what is specific to Courbet's depiction of things (and people) is that they tell stories. This leads to an interesting nuance, between stories and narratives, which is like that between conditions and ideas. To illustrate this I want now to look at two paintings by Courbet, *The Burial at Ornans* (1849, Musée d'Orsay, Paris) and *The Studio* (1855, Musée d'Orsay, Paris).

Beginning work on *The Burial at Ornans* in 1849, in a long and narrow loft where it was difficult to stand back from the painting, Courbet drew his fellow townspeople one at a time, 60 or so in all, from all social classes.[32] The painting is set in a new cemetery, where the first burial took place in 1848. The interment depicted may have been Courbet's grandfather's. Hélène Toussaint writes that Courbet was at pains to make the work as realistic as possible: 'the topography is accurate, the individual figures are identifiable and the forms of the ceremony are carefully reproduced. This grandiose work depicts an event of everyday life' (Toussaint, 1978: 209). She then asserts that he has turned this moment of everyday life into a historical event – no longer a story, it is a narrative. It is a familiar argument that Realism democratises art by painting the incidental on a monumental scale.[33] But Courbet emphasises in a letter to young artists in 1861[34] that he depicts actualities – things, not ideas. If this is an appropriate retrospective reading of this painting, then the superstructure, as it were, of history (making the event into history) seems to miss the point. It is an everyday event, a strand of the texture of everyday life as it is lived. The work does not require an idea to be valid, is not a narrative. The non-hierarchic horizontal arrangement of the figures confirms this. Further, the members of the peasant class whom Courbet depicts in *The Burial at Ornans* were able to see the work when it was exhibited in Besançon, and perhaps to recognise their own lives in it, which reflects Proudhon's idea that art should enable people to see themselves.

I want now to look briefly at this painting, and Courbet's *The Studio* (1855), and then return to the problematics of the avant-garde. In looking at *The Painter's Studio: A real allegory summing up seven years of my artistic life*, to use the full title, my purpose is not to give an account of the derivation of the picture from Courbet's previous work, or identify the figures in it. Both are done well elsewhere.[35] I will simply summarise. Courbet sits in the centre of a large room, painting a landscape. Behind him stands a model in semi-undress, while in front of him are a boy and a cat (I do not understand the cat's role). On the left, as the spectator sees it (the right for Courbet), are artisans, peasants, a gamekeeper and the poor; on the right (or left), are Courbet's patrons and members of the intellectual class, including Baudelaire quietly reading a book. It is clearly not a moment of life as lived, or art as made. The room could be at the Salon, given the number of people present. But it is not. It is a no-place (*utopia*). The painting requires a key to be understood, which is found in Charles Fourier's utopian idea of Universal Harmony, in which work and leisure, as well as the classes and genders of society, are reconciled and work becomes a pleasurable activity and location of human co-operative affection. Linda Nochlin makes a convincing case for this reading in an essay first published in 1968:

> Courbet's painting is 'avant-garde' . . . in terms of its etymological derivation, as implying a union of the socially and artistically progressive. Far from being an abstract treatise on the latest social ideas, it is a concrete emblem of what the making of art and the nature of society are to the Realist artist. It is through Courbet, the specific artist, the Harmonian demiurge, that all the figures partake of the life of this pictorial world, and all are related to this direct experience; they are not traditional, juiceless abstractions like Truth or Immortality, nor are they generalised platitudes like the Spirit of Electricity or the Nike of the Telegraph; it is, on the contrary, their concreteness which gives them credibility and conviction as tropes in a 'real allegory', as Courbet subtitled the work.
>
> (Nochlin, 1968: 17–18; 1991: 12)[36]

I agree. The setting, then, is not the studio, nor the Salon or the Louvre where students copied Art, but the phalanstery. Fourier modelled the architecture of his ideal community in part on the form of a rural estate, but also on the glass- and iron-roofed arcades of Paris which thrived as new spaces of consumption and sociation from the 1830s to the 1850s.[37] He describes a building with a central area of quiet contemplation, and two wings, one for workshops and children's activities, the other for ballrooms, meeting rooms, and rooms in which to receive outsiders. He writes: 'This precaution of isolating outsiders and concentrating their meetings in one of the wings will be most important in the trial Phalanx. For the Phalanx will attract thousands of curiosity-seekers whose entry fees will provide a profit that I cannot estimate at less than twenty million' (Fourier [1851, 1966–8], quoted in Beecher and Bienvenu, 1971: 241). To charge admission might seem odd today, but Courbet did this at a modest level in the provincial exhibitions he organised in places such as Besançon, and the issue is coloured for us by debates on free access to museums and education. Taking the statement on its own terms and putting it beside Courbet's painting, the studio (as he calls it) may be precisely the space in which Harmony is demonstrated to interested parties, or through the monument of a large painting to society as a whole, in its complexities and differences. But here a divergence emerges between depictions of conditions which tell the stories of those conditions, and narratives, such as the depiction of an imagined future. Courbet, in these different paintings, does both. Proudhon, in the passage above, cites him as using humour and directness to portray people as they are;[38] but Courbet adds a note of improvement which hints at the function of narrative as both moral and political education.

Proudhon is, like Saint-Simon and Fourier, a utopian. There is a link to Marx, who attended a Fourierist group in Paris in 1843;[39] but Marx turned against utopianism while revising his narrative of the Commune. He has little to say on art, most of it deriving from his studies in art history in Germany in the late 1830s.[40] His anti-utopian stance, however, is reflected in Socialist Realism. Semyon Chuikov's *A Daughter of Soviet Kirgizia* (1950), for instance, depicts a girl in a blue tunic with a book under her arm, in a landscape of cornfields and distant mountains. The conditions are those of the transformed Soviet Union, epitomised by the book denoting literacy in the peasant class.[41] The work does not represent Communism, it simply shows one aspect of it in action. For some observers, of course, it may also be an idealisation of those conditions.

Returning to *The Studio*, the studio space it shows is not a place but a time: tomorrow, as foreseen in Fourier's utopian text. This brings the argument back to the avant-garde's role in spreading new ideas and thereby bringing nearer the realisation of the hopes those ideas carry. But if art represents the future as well as the conditions of the present in which, through appropriate intervention, the future will be made, how does it predict what that future will be, or ought to be? Who says? How do they know? Théophile Thoré, admirer of Courbet and Millet, participant in the 1848 revolution, writes: 'Art is metamorphosed only by the strongest convictions, convictions strong enough also to transform societies' (Thoré, [1857] 1868, VII, quoted in Harrison and Wood, 1998: 384).[42] This implies a climate of change in which new ideas take social and aesthetic forms at the same time. Thoré makes several arguments, among them that a feature of modern society will be its universalism, as frontiers are opened, laws humanised, notions enlightened and energy lavished everywhere; that technical innovation in the arts is exhausted, so that progress will be found in thought (we could say the manipulation or extension of concepts), not in dexterity; and that a form of universal communication is possible: 'Then the fine arts and letters would cease to be a distraction of the erudite and refined . . . to become a common currency for the transmission and exchange of feelings, an everyday language within reach of everyone' (Thoré, [1857] 1868, VIII, quoted in Harrison and Wood, 1998: 386). This reads like a Realist manifesto, proclaiming an art for every citizen in a language open to their access. It also sounds close to Fourier's idea of libidinous sociation in the Phalanx. Thoré adds: 'There can be no danger of an idea being locked into its hieroglyph when everyone has the keys and can set it free' (ibid.);[43] and concludes 'The transmutation of art cannot therefore take place unless the universal mind changes too. Is it changing? Will it change?' (Thoré, [1857] 1868, IX, quoted in Harrison and Wood, 1998: 387). Thoré, in his exile in Brussels, looks forward to a world in which art communicates universally and the advance of knowledge is a common wealth. He leaves explicitly unresolved how to move from present injustices to that future of light. Is the future, in the end, another abstraction?

There is almost (or it may be my projection) a note of despair in Thoré's final question. Like Laverdant he sees progress in art linked to an underpinning development of thought. But if an avant-garde is privileged to know such a future, to whom will it communicate it in the forms of art? The cognoscenti, or the mass public? And if for the mass public, who will interpret the picture which interprets the future?

This is the flaw in the concept of an avant-garde which undermines it: that avant-garde art and the utopian philosophy which informs it tends to involve an act of interpretation *for* others – a going ahead of the mass – rather than facilitating acts of interpretation by others *for themselves*. It is not a difficulty restricted to the nineteenth century: John Roberts, reviewing the exhibition 'Protest & Survive' at the Whitechapel Art Gallery, London, writes:

> Political art (as understood on the social-democratic model) assumes that those whom the art work is destined for (the fantasised working class) need art as much as they need Ideas in order to understand capitalism and class society. There is never a moment's recognition that people are already engaged in practices in the world which are critical and transformative . . . the category Political Art reinstates the inequality in bourgeois culture between those who supposedly know and those who supposedly don't know.
>
> (Roberts, 2001: 6)

Perhaps in his work between 1848 and 1855 Courbet did recognise those critical and transformative moments, which, later, Lefebvre sees as glimpses of liberation within the routines of ordinary life (discussed in Chapter 4). In *The Studio* the emphasis moves towards interpretation, though the public for the work may, in Courbet's intention, have been a circle of radical thinkers who had the key to its understanding. The problem then is less in the art than the philosophy. Yet I would not want to argue against the imagination of possible futures, or the creation of form for hopes which, if formless, remain distant.

Neither do I want to leave this account of an avant-garde in mid-air, pondering an aporia from which there is no exit. The avant-garde did not end with Courbet's flight to Switzerland, nor with his death in 1877. The example of the Commune's defeat and Courbet's treatment by the MacMahon regime were crushing, yet social criticism continues to be made in more covert ways in some areas of Impressionism. From this point, two tendencies begin to diverge: the alienation evident in the work of, say, Manet,[44] and later in French Symbolism's retreat to a world in which the artist's psyche become art's subject matter; and the renewal of utopian aspirations in Neo-Impressionism. A refusal of everyday life in Symbolism and Decadence is, in its way, a refusal of bourgeois society, though at times given to a regressive aspect, harking back to medievalism and aristocracy. But it is in Neo-Impressionism that a new, forward-looking vision is encountered, as in Seurat's *Bathers at Asnières* (1883–4, National Gallery, London). I see this as a utopian image, and see this reading of it as compatible with Nochlin's reading of Seurat's *Sunday on the Island of La Grande Jatte* (1884–6, Art Institute, Chicago) as anti-utopian. As I will explain, the two seem to go together very well as two halves of a story.

In *Bathers at Asnières*, a large painting of people from the artisan class resting by the banks of the Seine, Seurat begins to give avant-garde art a new language. Wood argues that the techniques of the Neo-Impressionists communicate radical intentions. He cites Paul Signac, the leading Neo-Impressionist after Seurat's death in 1891, under the pseudonym *camarade impressionist* in the anarchist journal, *La Révolte* in 1891: 'It would be an error – an error into which the best informed revolutionaries, such as Proudhon, have too often fallen – systematically to require a precise socialist tendency in works of art' (Signac, 1891, quoted in Wood, 1999: 129).[45] Signac sees the depiction of working-class subjects *and* the decadence of bourgeois society as appropriate to radical art, but, as Wood points out, also sees radical witness to social development in the form of a new artistic language. This is one aspect of Seurat's and Signac's work. But equally significant is that Signac is writing in an *anarchist* journal.[46] And this is where I bring the chapter full circle, to the anecdote of the Communards shooting the public clocks. It does not matter whether the anecdote is true or not, it serves to illustrate a glimpse of utopia, a society in which the day is no longer ruled by the regulation of toil. I speculate that Seurat's painting of bathers – though most of the figures are not in the water but reclining on the banks – is a depiction of a utopia of ease.

The painting is set in a dormitory suburb of Paris, near the industrial district of Clichy. The factory chimneys in the background are those of Clichy, and represent the mass production of goods, which, potentially, will end the economic problem of scarcity. There will be enough for all according to their needs; and leisure for all when modern technology replaces the grind of labour – a vision advanced by anarchists such as Peter Kropotkin[47] in the late nineteenth century, and in another way by Herbert Marcuse in 1968.[48] Today such ideas seem fanciful, and the precondition for social harmony may be a radical revision of wants in terms of needs, an end of consumerism, rather than a simple equality of distribution of the goods produced. But harmony, with its Fourierist associations as well as those to anarchism and syndicalism, is the content which permeates the painting. Nochlin sees the other painting, *La Grande Jatte*, as anti-utopian. Taking Ernst Bloch's critique of it as depicting utter boredom,[49] she comments that the work 'should not be seen as only passively reflecting the new urban realities of the 1880s or the most advanced stages of the alienation associated with capitalism's radical revision of urban spatial divisions' (Nochlin, 1991: 171). I agree, again; here is alienation and anomie in figures who express no relation to each other but stare ahead, in contrast to the informal poses of the bathers. The *Bathers at Asnières* was rejected by the official Salon, and exhibited at the Salon des Indépendants. Were the two paintings ever to be shown together, *Bathers* on the spectator's left (where Courbet put the artisans), *La Grande Jatte* on the right (where Courbet put the patrons), the two groups would look at each other across the Seine – on one bank ease, on the other alienation. That ease might, fancifully, have been produced had the Commune succeeded. As idea, it is a light which articulates the darkness of the real history of the Commune's failure and suppression.

NOTES

1 Illustrated in Michalski, 1998: 12–14, fig. 2. In 1871 it was surrounded by canons.

2 'If the mechanical clock did not appear until the cities of the thirteenth century demanded an orderly routine, the habit of order itself and the earnest regulation of time-sequences had become almost second nature in the monasteries' and 'The gain in mechanical efficiency through co-ordination and through the closer articulation of the day's events cannot be overestimated: while this increase cannot be measured in mere horsepower, one has only to imagine its absence today to foresee the speedy disruption and eventual collapse of our entire society. The modern industrial régime could do without coal and iron and steam more easily than it could do without the clock' (Mumford, 1956: 4, 9).

3 Facing political opposition (45 per cent of the votes in the 1869 national elections were for candidates opposed to his regime), Napoleon III engineered a war with Prussia; but between its outbreak and the final surrender at Sedan, the French armies suffered a series of defeats due to inept command. On September 4th, 1870 the Third Republic was declared by a Government of National Defence at Versailles. On September 19th, the Prussian army besieged Paris, engaging in street battles until the armistice in January 1871. During the siege many people lived in basements and improvised shelters, food supplies dwindled, and cats and dogs were eaten. Harvey notes that the zoo's elephant Pollux was butchered – the meat fetched 40 francs a pound, while the price of rats increased from 60 centimes to 4 francs. Most of the bourgeoisie left the city, but for those remaining flour was adulterated with bonemeal made from bodies in the catacombs, though champagne remained available: 'While the common people were thus consuming their ancestors without knowing it, the luxuries of café life were kept going, supplied by hoarding merchants at exorbitant prices' (Harvey, 1989: 210). The terms of armistice included a payment to Prussia of 5 billion francs, which the Versailles regime borrowed. The bankers then pressed for the disarming of Paris, and soldiers from the Versailles government began to collect canons on the hill of Montmartre. On March 18th, 1871 a crowd climbed the hill to reclaim them, at which General Lecomte ordered his troops to fire. They refused and he was taken prisoner. He and General Thomas (remembered for his role against the 1848 revolution) were shot at rue des Rosiers 6 (Harvey, 1989: 211).

4 Tim Cresswell cites the term 'prank' from anti-Communard poet Catulle Mendès. He refutes this: 'Instead the demolition of the monument was just one – very visible – act to demolish the hierarchy of social space' (Cresswell, 1996: 173). Mulvey notes that, in Eisenstein's *October*, an attack on a statue of the Tsar marks the beginning of the February Revolution. Eisenstein reverses the film to show it being rebuilt as the uprising fails (Mulvey, 1999: 220).

5 Roos notes that (shortly before the Commune) Napoleonic eagles were removed, a relief of Napoleon was plastered over and another covered with a shroud, an imperial

eagle on the Palais de l'Industrie was re-carved as a winged globe, and a statue of the Emperor converted to a figure of Minerva (Roos, 1996: 152–3, fig. 118).

6 The text of the letter is given in French by Roos: 1996: 260, n. 10.

7 Wood, citing Rifkin (1979), writes: 'The governing principle was independence: independence from juries, censorship, and what was seen as the interference of the Academy' (Wood, 1999: 117). Wood notes that the terms 'Intransigent', 'Impressionist' and 'Independent' were all used after 1871 to describe radical artists, and quotes an extract from the founding document of the Artists' Federation [which I edit further]: 'The artists of Paris who support the principles of the Communal Republic will form themselves into a federation . . . based on the following ideas: The free development of art without government protection or special privileges. Equal rights for all members . . . The realm of the arts will be controlled by the artists' (Wood, 1999: 114, 117). Their purposes were to conserve heritage, facilitate creation, and stimulate future art through education.

8 'The people of Paris have plunged me into political affairs up to my neck. President of the federation of artists, member of the Commune, delegate to the mayor's office . . . delegate for public education . . . I get up, I eat breakfast, I attend and preside at meetings twelve hours a day . . . I am in heaven. *Paris is a true paradise*; no police, no nonsense, no oppression of any kind, no disputes' (Courbet, Bibliothèque Nationale, Paris, in Chu, 1992: 416–18, quoted in Roos, 1996: 154).

9 Wood quotes an eye-witness account, which expresses anxiety that the falling column will damage the sewers, and remarks that no mention was made of Austerlitz, the battle it commemorated. It goes on: 'The column lies on the ground, split open . . . Caesar is lying prostrate and headless. The laurel wreathed head has rolled like a pumpkin into the gutter' (from Edwards, 1973: 147–8, quoted in Wood, 1999: 119). This may be fanciful – Roth (1997: 14, fig. 8) includes a photograph of the fallen statute in which it is intact (albumen print, Bruno Braquehais, 1871, Getty Research Institute for the History of Art and the Humanities, acc. 95R.102; see also Roos, 1996: 155, fig. 120).

10 The conception of the picture dates to 1854, when Courbet mentions an image of a gypsy and her children in a letter to his patron Bruyas. He writes of it in 1868: 'My picture will make a great impact at the Salon' (quoted in Toussaint, 1978: 182). The link to Proudhon is through the latter's view of art's educational role: 'To paint men [*sic*] in the sincerity of their nature and their civic and domestic functions, with their actual physiognomy . . . to surprise them, so to speak, in the nakedness of their mentalities . . . with goal of general education . . . this appears to me to be the true point of departure of modern art' (from Rubin, 1980: 92, quoted in Roos, 1996: 108; cf. Proudhon, 1969: 214–17). Proudhon also writes: 'The budget of the banker . . . is raised by taxation on labour. The money spent on luxury is likewise raised by taxation on necessities . . . the happiest of men [*sic*] are those who best know how to be poor' (Proudhon, 1969: 259). The painting was attacked by critics and caricatured in the press as a snub to the Salon. Roos cites the following from *Le Petit Figaro*, June 7th, 1868: 'M. Courbet wanted to prove that a great artist can easily do without form, colour and style when he is sustained by a great and generous idea. This old beggar is deprived of everything, even of the most necessary drawing . . . The woman in a bundle of dirty laundry is a masterpieces' (Roos, 1996: 106).

11 Courbet continued to organise regional exhibitions of his work, as at Dijon in 1870. See de Forges, 1978: 45.

12 Leslie cites three estimates: Maxime du Camp's 6,000; Lissagaray's 17,000; and Louise Michel's 30,000 (Leslie, 2000: 180). See Toussaint, 1978: 232–3 for Courbet's sketchbook drawings during the last days and defeat of the Commune, possibly made after the events.

13 'The initiative did not come directly from me . . . The column seemed to me badly placed; there ere even some who found it hazardous; however, I only considered the thing from an artistic point of view' and 'This column as a feeble replica of the Column of Trajan, badly put together in its proportions. There is no sense of perspective' (*Gazette des Tribuneaux*, August 14–15th, 1871, quoted in Roos, 1996: 156). Roos records that Courbet's lawyer denied any political awareness on the part of his client.

14 See Clark, 1973.

15 Dalou escaped to England in 1871, was sentenced in his absence to hard labour for life in 1874, and returned to Paris following the amnesty of 1879, entering a competition organised by the Paris Municipality for a monument marking the ninetieth anniversary of the Revolution. The brief defined the period to be commemorated as 1789–92, eliminating the Jacobin years. The competition was won by Leopold and Charles Morice with a neo-classical *Monument to the French Republic* – a female figure holding an olive branch, in a pose like that of Bartholdi's *Liberty*, supported by Liberty, Equality and Fraternity. A lion guards a bronze ballot-box. Dalou's design, *Triumph of the Republic*, came a close second. Marianne wears a Phrygian cap (a sign of the left), marching on a globe with arm outstretched over a chariot pulled by two lions; one is ridden by Freedom holding a torch, the other by Labour as a worker with hammer. Warner notes the bare shoulder (the slipped chiton; cf. Delacroix's *Liberty Leading the People* discussed below): 'Her undress coheres with her headdress to express her state, poised between the reality of her identity as the French Everywoman . . . and her lack of personal identity as an emanation of the idea' (Warner, 1987: 267–8). The City Council decided in 1880 to commission Dalou's monument as well as the Morices', siting it in Place de la Nation. It was unveiled as a plaster cast in 1889, the day before elections in which the right was again defeated, and in bronze in 1899 (see Michalski, 1998: 17–26).

16 Lenin was persuaded of the importance of public monuments as a means of education by Lunacharski, whose source is Campanella's *City of the Sun* (Bown and Taylor, 1993: 16–33). Vladimir Tolstoy makes a connection to the Commune: 'The link between Lenin's monumental propaganda plan and the general enthusiasm of the Revolutionary period for festivals is very significant. It is also important that the roots of Lenin's ideas date back to the humanistic traditions of the Renaissance and the experience of previous revolutions, in particular the Paris Commune and the French Revolution. The idea coincides with that of Robespierre . . . that the motherland ought to educate its citizens and use popular festivals as an important means of performing such civic education' (Tolstoy, Bibikova and Cooke, 1990: 13). As an indication of the extent of such activity, the paper *Northern Commune* (October 23rd, 1918) lists 70 sites in Petrograd to be decorated for the first anniversary of the Revolution (Tolstoy, Bibikova and Cooke, 1990: 69–70). This took place in an ambience of shortages, conflict and uncertainty: Kuzma Petrov-Vodkin writes of his painting *Stenka Rasin and Vasilisa the Wise* (1918, destroyed), which he calls an important work, that 'according to a resolution of the Art Workers' Trade Union, was to have been preserved, but it somehow found its way into the backyard of some local Soviet and was late used for foot-bindings, because the canvas was relatively good . . . You must remember that at this time nothing was available and we had to resort to such measures as highjacking horses and cabs and driving round the city confiscating whatever we could' (Tolstoy, Bibikova and Cooke, 1990: 70).

17 Sadler cites the *Lettrist International*: 'Monuments whose ugliness is irretrievable in any part (the Petit and Grand Palais genre) will have to make way for other constructions' (Anon., 'Projet d'embellissements rationals de la ville de Paris', *Potlatch*, 23, October 1955, in Sadler, 1998: 99). But the Vendôme Column had figured, too, in Surrealism. In André Breton's *Nadja*, a set of texts and images recounting the poet's imaginary encounters with a women of that name he meets by chance in Paris, the sites of meetings and wanderings are a geography of repressed struggle, including the Place Vendôme: '*Nadja* is a tour and detour of the non-monumental history of repressed popular struggles, struggles that can be seen as the eruption of everydayness in the everyday' (Highmore, 2002: 54).

18 Marx continues, in Buber's quotation, to argue that the Commune introduced the co-operative ownership of the means of production, land and capital, establishing a possibility for Communism in face of widespread doubt as to its viability (Buber, 1996: 87–9). Geoghegan comments that a significant difference occurs between Marx's first and second drafts of *The Civil War in France*: in the first he sees early utopian groups as aspiring to aims such as the supersession of the wages system, while the organisation of labour (in the Commune) found a means to realise them; comparing the aims of the Commune and the International, he says 'Only the means are

different and the real conditions of the movement are no longer clouded in utopian fables' (Marx and Engels, 1980: 166, quoted in Geoghegan, 1987: 31). In the revised version he says: 'The working class . . . have no ready-made utopias . . . no ideals to realise, but to set free the elements of the new society with which old collapsing society itself is pregnant' (Marx and Engels, 1980: 76, quoted in Geoghegan, 1987: 31; see also Lasky, 1976: 36–43). Benjamin includes a text by Engels in his section on the Commune in the *Arcades Project* in which Engels admits that Marx 'upgraded the unconscious tendencies of the Commune into more or less conscious projects' in a report to the General Council of the International and differentiates factions in the Commune, one following Auguste Blanqui – 'nationalistic revolutionaries who placed their hopes on immediate political action and the authoritarian dictatorship of a few resolute individuals' – and another influenced by Proudhon who 'could not be described as social revolutionaries, let alone Marxists' (Mayer, [1936] 1969: 220, quoted in Benjamin, 1999: 793). Franz Mehring, one of the founders (later) of the German Communist Party, writing in 1896, sees the Commune's failure as resulting from a continuation of bourgeois attitudes and lack of a 'solid organization of the proletariat as a class and the principled clarity about its world-historical role' (quoted in Leslie, 2000: 214).

19 Lefebvre continues that 'One strong aspect of the Commune (1871) is the strength of the return towards the urban centre of workers pushed out towards the outskirts and peripheries, their reconquest of the city.' (Lefebvre, 1996: 76; also cited in Highmore, 2002: 139). Harvey, like Lefebvre, sees the remodelling of Paris as instrumental in the Commune – in the economic boom of the Second Empire, contrasts between affluence and poverty 'were increasingly expressed in terms of a geographical segregation' while signs of social breakdown were widespread in the economically less stable 1860s. He adds: 'To top it all, Haussmann, at the Emperor's urging, had set out to "embellish Paris" with spacious boulevards, parks, and gardens, monumental architecture of all sorts' (Harvey, 1989: 206). Harvey, again like Lefebvre (and Marx), sees the Commune as a working-class movement. Tajbakhsh (2001: 74–8) takes issue with him on this, seeing it as more diverse. Tajbakhsh cites Gould to the effect that the Commune was 'more a revolt of city dwellers against the French state than of workers against capitalism' (Gould, 1995: 4, quoted in Tajbakhsh, 2001: 76).

20 See Plant, 1992: 63–4; Kofman and Lebas, 1996: 11–18; Shields, 1999: 91; Highmore, 2002: 113–44. Shields records that the publication caused a rift between Lefebvre and the Situationists, who were annoyed that their deliberations had been reported (as they saw it, despite their willingness to plagiarise in other contexts). He summarises: 'the study of the Commune allowed Lefebvre's idea of an ecstatic moment in which totality was experienced in a manner that was fully authentic to be linked firmly to the idea of revolutionary fervour. Thus the notion of the 'revolutionary festival': if presence could be experienced during the disorder of carnivalesque festivals and Mardi Gras, why not also during parades, demonstrations, riots and mass occupations? The stage was set for the student occupations of May 1968' (Shields, 1999: 103). Highmore states: 'For Lefebvre, carnival is a moment when everyday life is reconfigured, but this different order of things is present in everyday life itself' (Highmore, 2003: 123) adding that Lefebvre's interest in carnival is in context of that also of Bakhtin and Bataille.

21 Warner notes that its model was the actress Juliette Drouet, Victor Hugo's mistress (Warner, 1987: 32).

22 The painting depicts the celebrations of June 30th, 1878, a date picked as less inflammatory than either May 1st or July 14th (Bastille Day). Wood writes: 'The flags are there but they are pushed to the edge . . . it is a large and empty space, a blinding slice of light rather than a fluttering atmosphere; and we can see the roadworks, the reconstruction in progress, being done of course implicitly by workers, But most of all we can see the crippled veteran in the blue blouse, typically worn by workers' (Wood, 1999: 128). See also Roos, 1996: 204–20.

23 In his essay 'Avant-Garde and Kitsch', first published in 1939, Greenberg writes of the nineteenth-century avant-garde as reacting against bourgeois society by constructing their own bohemian milieu, in the 1850s and 1860s immersed in revolutionary ideas. From this separation of the bohemian from the bourgeois, which is also a separation of art production from the art market which

replaced aristocratic patronage, comes eventually a separation of art from politics: 'The revolution was left inside society, a part of that welter of ideological struggle which art and poetry find so unpropitious.' From this Greenberg argues that avant-garde art's function is 'to find a path along which it would be possible to keep culture *moving* in the midst of ideological confusion and violence' (Greenberg, [1939], 1988: 7–8).

24 'The mass culture machine and its engines of celebrity have long redefined the other structures of cultural meaning, so that patterns of behaviour and estimations of worth in the art world are more and more similar to those in the entertainment industry' (Rosler, 1994: 57).

25 Nochlin uses a compressed version of the text from D. D. Egbert, 'The Idea of an "Avant-Garde" in Art and Politics', *The American Historical Review*, vol. 73, no. 2, December 1967, p. 343.

26 Nochlin cites the passage from Poggioli, 1968: 9.

27 The Bourbon monarch Charles X was deposed in July 1830, when workers, students, artisans and bohemians fought for three days in the streets of Paris; Louis-Phillipe, Duc d'Orleans, was invited to become head of state, known as the citizen-king. Delacroix's painting, depicting fighting at the barricades, was produced in the autumn of 1830, shown at the 1831 Salon, and taken out of storage and exhibited during the 1848 revolution. See Miles, 1997: 70, a few lines of which have been revised for this book. See also Warner, 1987: 271; and Wood, 1999: 35–8.

28 'in the classical costume of a goddess of victory, and her lemony chiton has slipped off both shoulders. Her breasts, struck by the light from the left, are small, firm, and conical, very much the admired shape of a Greek Aphrodite' (Warner, 1987: 271). See note 15.

29 Clark observes that barricades were not used in 1789 but specific to nineteenth-century revolutions: 'The barricade was quickly represented. The makers of popular prints added a few stones and spars to the old format of the battle scene, placed a mass of men on top, and the barricade was done. It was the barricade as stage rather than barrier; not something which blocked roads' (Clark, 1973: 16). He cites Manet's *The Barricade* (1871) as an exception, and illustrates a popular print (anon., 1830; Clark, 1973: fig. 3) in which a figure of a dead soldier at the bottom left has a pose identical to that of a semi-clothed figure in Delacroix's painting.

30 The new public museums established in the nineteenth century, such as the Tate at Millbank on the site of a penitentiary, have an educational function. In a reformist tradition, they bring culture to people of all social classes and, as Taylor argues, define culture (Taylor, 1994).

31 See Warner, 1987: 32 on Hittorf's redesign of Place de la Concorde and the statues of cities placed in it under Louis-Philippe; and 63–88 on gendered representation of industrial subjects such as mechanics and the telephone.

32 For identification of the figures, see Toussaint, 1978: 208.

33 Toussaint maintains that the work is not anti-clerical but profoundly religious in sentiment, a position she derives from Proudhon's defence of the work (Toussaint, 1978: 209).

34 'For painting especially, art can be nothing other than the representation of objects visible and tangible to each artist' and repeats the thought later in the letter: 'I hold . . . that painting is a quite concrete art, and can consist of nothing but the representation of real, tangible things. It is a physical language, whose words are visible objects. No abstract, invisible, intangible object can ever be material for a painting' (Courbet, 1861, in Harrison and Wood, 1998: 403–4). The painting, however, also derives its composition from a current of popular imagery in woodcuts and broadsheets (*l'imagerie d'Epinal*), as do later Realist works – see Nochlin, 1991: 21. Toussaint draws attention to the idea that all humanity is reconciled in Christ's resurrection, the promise of which is spoken at Christian funerals (Toussaint, 1978: 212).

35 See Toussaint, 1978: 251–79; Nochlin, 1991: 1–18.

36 Nochlin notes Courbet's link to François Sabatier, a Fourierist who retreated

from Paris to his estate near Montpellier, and associate of Courbet's patron Alfred Bruyas. Sabatier commissioned Dominique Papety to make a work celebrating the abolition of slavery, under a sketch for which are notes for a Fourierist programme. Sabatier drew up plans for phalansteries – Fourier's term for the unit of society to replace the city – on his estates (Nochlin, 1991: 8). Courbet states in a fragmentary autobiography that he is a Fourierist (Nochlin, 1991: 9), and had depicted the Fourierist missionary Jean Journet; there is a Fourierist aspect to *Bonjour Monsieur Courbet* (1854, Musée Farbre, Montpellier), which shows the artist meeting Bruyas (see Toussaint, 1978: 111–2). Nochlin sees *The Studio* as a Fourierist association of capital, work and talent (Nochlin, 1991: 10), and of the four affective passions of friendship, love, ambition and family feeling, plus the four ages of life to which these correspond, childhood, adolescence, maturity and old age. A key element is the fifth stage, the pivotal years between 35 and 45, of virility. Courbet was 36 in 1855 (Nochlin, 1991: 11).

37 'The edifice occupied by the Phalanx bears no resemblance to our urban or rural buildings; and in the establishment of a full Harmony of 1600 people none of our buildings could be put to use, not even a great palace like Versailles . . . The street-galleries [which] are a mode of internal communication . . . would alone be sufficient to inspire disdain for the palaces and great cities of civilization' (Fourier, [1851, 1966–8] 1971: 240–3). A treatise on Fourierist architecture was produced by Victor Considérant in 1834, laid out like a vast neo-classical palace with street-galleries and arcades (illustrated, Sadler, 1999: 119, fig. 3.6; cited in Markus, 1993: 296–7; see also Kruft, 1994: 286–7). Sadler sees Constant Niewenhuys' *New Babylon* as 'a global phalanstery for the twentieth century' – it was only a model though his photographs of it lend it a sense of reality, as Sadler notes (1999: 140–1, fig. 3.25).

38 'It is to Courbet's credit that he is the first painter who, by imitating Molière's genius in the theatre, has seriously tried to warn us, chasten us and to improve us through portraying us as we really are; who, instead of amusing us with fables or flattering us by adding a lot of bright colours, has had the courage to depict us not as nature intended us to be, but as our passions and our vices have made us' (Proudhon, [1865] 1970: 215). Proudhon also proposed plans for the education of workers as a means to change their conditions: 'By this method the industrial worker, the man [*sic*] of action and the intellectual will all be rolled into one' (Proudhon, [1858] 1970: 80).

39 Marx writes: 'you would have to attend one of the meetings of the French workers to appreciate the pure freshness, the nobility which burst forth from these toil-worn men' (cited in Geoghegan, 1987: 25, 143, n. 20).

40 See Rose, 1984. Among the examples of art seen by Marx in his student years were works by the German Nazarenes, a group whose dedication to Christian morality was shared, slightly later, by the English Pre-Raphaelites. Marx saw their work as antipathetic to Enlightenment philosophy, while identifying the Greek roots of Enlightenment culture with the rationality of the French Revolution. Although briefly interested in an avant-garde function for art, Marx moves in the 1840s from a critique of religious art to investigation of economic production and exchange, his social avant-garde being one not of artists but of worker-producers. As Rose notes, Marx saw art as one among many forms of alienating labour: 'This was of course to bind again his Saint-Simonist argument for an avant-garde, reforming role for art to both a critique of alienated production and to the proposition that art, together with other forms of production, would always be the victim of exploitation under industrial capitalism' (Rose, 1984: 95).

41 Illustrated, Bown and Taylor, 1993, plate II.

42 Thoré was writing on art in the Paris Universal Exposition of 1855, but from exile in Brussels. His essay was first published there in 1857, but in France not until 1868.

43 J. F. Champollion's *Précis*, a grammar of hieroglyphs, was published in 1824, following the excavation and appropriation of antiquities during Napoleon's campaigns in Egypt. The *Description de l'Egypte*, describing Napoleon's expedition in 24 volumes and produced by the team of scientists who accompanied him, was also published in the 1820s (Said, 1994: 37–9). Edward Said sees the reconstruction of Egypt in the European mind

as a precondition of archaeology: 'Egypt had to be reconstructed in models or drawings, whose scale, projective grandeur . . . and exotic distance were truly unprecedented . . . First the temples and palaces were reproduced in an orientation and perspective that staged the actuality of ancient Egypt as reflected through the imperial eye; then . . . they had to be made to speak, and hence the efficacy of Champollion's decipherment; then, finally, they could be dislodged from their context and transported to Europe for use there' (Said, 1994: 142).

44 Nochlin initially locates the beginning of the avant-garde in Realism but towards the end of her essay writes 'Yet if we take "avant-garde" out of its quotation marks, we must come to the conclusion that what is generally implied by the term begins with Manet rather than Courbet. For implicit . . . to our understanding of avant-gardism is the concept of alienation . . . While Courbet may have begun his career as a rebel and ended it as an exile, he was never an alienated man, that is, in conflict with himself internally or distanced from his true social situation externally, as were such near-contemporaries as Flaubert, Baudelaire, and Manet. For them, their very existence as members of the bourgeoisie was problematic' (Nochlin, 1991: 12–13).

45 Signac emphasises the social commentary of Impressionism: 'By their pictures of working-class housing . . . by reproducing the broad and strangely vivid gestures of a navvy working by a pile of sand, of a blacksmith in the incandescent light of the forge – or better still by synthetically representing the pleasures of decadence . . . as did the painter Seurat who had such a strong sense of the great social debasement of our epoch of transition – they have contributed their witness to the great social process which pits the worker against Capital' (Signac, 1891, 'Impressionists and Revolutionaries', quoted in Wood, 1999: 129; see also Harrison and Wood, 1998: 797).

46 Harrison and Wood note the 'major influence' of anarchism on Neo-Impressionism, as on Pissarro (Harrison, Wood and Gaiger, 1998: 876).

47 See Buber, 1996: 38–45. Buber sees Kropotkin as simplifying Proudhon's anarchism (see Proudhon, 1971: 88–102). One of his key themes is the antipathy of the state towards human capacity for self-organisation (equated with order); examples of that capacity in effect include the communes and guilds of medieval Europe. Another is that mutual aid (the title of his best known work) is the foundation of human survival, rather than competition or social atomism. In this he follows Proudhon, who argued for a co-operative, or syndicalist, organisation of labour (Buber, 1996: 31).

48 See Marcuse, 1969: 17–30. Marcuse agues that the productive capacity of industrial economies has been diverted into consumerism's production of ever-expanding demand, and dissipation of the demand for freedom in consumption: 'For freedom indeed depends largely on technical progress . . . But this fact easily obscures the essential precondition: in order to become vehicles of freedom, science and technology would have to change their present direction and goals; they would have to be reconstructed in accord with a new sensibility – the demands of the life instincts' (p. 28).

49 From Bloch's *The Principle of Hope*. Bloch dislikes Seurat's work, and sees this work as depicting joyless leisure: 'The result is endless boredom, the little man's [*sic*] hellish utopia of skirting the Sabbath and holding onto it too' (Bloch, [1959] 1986: 814, quoted in Nochlin, 1991: 170). But for Bloch the Sunday has significance, reflecting the bounty of the Land of Cockaigne: 'As an eternal Sunday, which is one because there is no sign of any treadmill, and nothing beyond what can be drunk, eaten boiled or roasted is to be found' (p. 813). The tranquillity and repose of the residual Sunday is found, for Bloch, in Cézanne's pictures of fruit 'in which happy ripeness has settled' (p. 815).

2

1912

RED FLAGS AND REVOLUTIONARY ANTHEMS

•

My aim in this chapter is to examine some of the contradictions that characterise the Modernist avant-garde in Europe. The chapter begins in 1912, a date that becomes its fulcrum, and looks from there back to Symbolism in Paris in the 1880s and '90s, to ask whether the avant-garde derives its claim to autonomy from Symbolist aestheticism. The concept of an avant-garde within Modernism is then re-examined. The second part of the chapter reconsiders aspects of Cubism in Paris and Expressionism in Munich, drawing attention to ambivalent political and social attitudes, and to overlooked continuities with the avant-garde of Realism, which is discussed in Chapter 1. A clear political alignment is seen in Italian Futurism, which is discussed in Chapter 3 together with Le Corbusier and the Modernist architectural avant-garde.

I STOCKHOLM: STRINDBERG'S BIRTHDAY

> In January 1912 a torchlight procession, headed by members of the Stockholm Workers' Commune, celebrated the sixty-third birthday of August Strindberg. Red Flags were carried and revolutionary anthems were sung.
>
> (Williams, 1989: 49)

This description of the celebration of what, as it happens, was Strindberg's last birthday opens Raymond Williams' essay 'The Politics of the Avant-Garde'. The juxtaposition – Strindberg, the playwright of bourgeois anxiety, and red flags emblematic of class struggle – seems strange. But as Williams says, 'No moment better illustrates the contradictory character of the politics of what is now variously . . . called the "Modernist movement or the avant-garde"' (Williams, 1989: 49). Williams does not, as he elaborates his argument, conflate Modernism and the avant-garde, but sees the avant-garde as a special case, an advanced tendency, within Modernism.

My use above, from Williams, of the term 'within' makes Modernism a more encompassing term than avant-garde. In general, Modernism refers to those elements of the arts from the mid or late nineteenth century – from Baudelaire or Post-Impressionism – which are consciously of their period.[1] Avant-garde includes being of the time but also implies a purpose beyond it. The question then is whether the purpose is to revolt against bourgeois society or, in a more selective way, to refuse its cultural institutions; and whether the second possibility is a development of or a departure from the first. To put it simply, are the transgressions of the avant-garde aesthetic or social, and do aesthetic transgressions merely stand for aspirations for social change, or actively destabilise bourgeois society by exposing the contradictions of its values?

I take up the question below in the context of more recent discussions of the avant-garde, in Chapter 4 in relation to Herbert Marcuse's concept of an aesthetic dimension, and in Chapter 6 in terms of issues of power and gender through writing by Rosalind Krauss and others. But the seeds of a dichotomy between art and society appear in the late nineteenth century, and perhaps in the origins of Modernism as the culture of late modernity. Perhaps, after all, Modernism is not merely a period but, just as the avant-garde is contained in Modernism, a specific attitude within a more encompassing modernity defined as the experience of living in modern times. But what are modern times? And is a set of dates all that is involved?

I suggest that when we look at different chronologies we find different attitudes to being alive and conscious in the world, which imply different intellectual projects. For instance: Wolfgang Welsch notes that in English, as for Charlie Chaplin, modern times are the 1920s, while in French they begin in the seventeenth century in 'a programme typical of the modern age, one of a new, universal science' (Welsch, 1997: 104). Other points of departure include Humanism, and a new economic relation of city and countryside in thirteenth-century Tuscany, given form by Giotto and Masaccio;[2] the colonisation of the Americas in the sixteenth century, allied to Francis Bacon's concept of knowledge as dominion;[3] René Descartes' image of regular places drawn on a blank ground, the reduction of the world to a system of signs which allows the invention of new worlds in the act of their inscription;[4] and modernisation as the reorganisation of society for its industrialisation.[5] Perhaps modernity is an attitude to the self as subject in a mutating relation to objects from which it is religiously, intellectually and economically estranged. Baudelaire's Paris is modern, then, because it is, as Walter Benjamin reiterates, a site of the phantasmagoria of commodity production,[6] a site in which art and literature, also, are commodities. There seem to be many modernities. Take your pick, take the money or open the box. For Welsch, the box is empty: 'Modernity *per se* . . . does not exist . . . [only] varying concepts of modernity' (Welsch, 1997: 104) the relation of which is continuous and reactive, smooth and fracturing at once.

Once we see modernities as conceptual formations rather than delineations of a period, we can see a play of continuities in the place of boundaries. For instance, the concept of knowledge as power mutates, long after its first rehearsal, as the privileged knowledge of a future enjoyed by an avant-garde thus able to lead society towards it. Hence the model of power-over tends to be replicated even when the aim is to overthrow it. Similarly, a Cartesian objectification of the world, its reduction to a system of signs, of representations, leads to the privileging of visuality as the sense that gives, as Doreen Massey has put it, most mastery,[7] yet is inseparable from the concept of a critical distance in Modernist theory. Continuities appear, then, but contain ambivalences, which if seen in isolation from a thematic rather than periodised history are confusing. Part of the present work is to understand this.

There is one factor, however, that unites diverse modernities: they were all conceived in cities, just as Modernism and its avant-gardes belonged, more specifically, to metropolitan cities such as Paris, Berlin and Munich, which underwent modernisation in this period.[8] Nowhere else was there the critical mass of artists, intellectuals, collectors and publics to make new milieux and markets, nor the new technologies of movement and energy, or ubiquity of graphic forms thrown up by consumption to feed into collage and montage. Nowhere else could the gatherings in cafés and apartments, such as Mallarmé's Tuesday evenings, have taken place. And nowhere else were there so many strangers to interrupt self-perception.[9] But this is not to say that all Modernisms or avant-gardes respond in the same way to their metropolitan surroundings, or that there is any agreement in the positions adopted even by participants in a specific grouping. David Cottington writes of the Parisian avant-garde of the 1910s that it was 'composed of many groupings centred on a bewildering variety of aesthetic practices and positions' (Cottington, 1998b: 11).[10] To those present we could add those who are encapsulated in history but who continue outside of themselves to exert an influence through exhibitions, texts and visual, verbal or personal memories. Strindberg writes in a letter to Gauguin: 'When, in 1883, I returned to Paris a second time, Manet was dead, but his spirit lived in a whole school that struggled for hegemony with Bastien-Lepage', and adds that on his next visit in 1885 he saw a Manet exhibition (Strindberg [1895] quoted in Harrison and Wood, 1998: 1035).[11]

To return to Strindberg in Stockholm in 1912: the event Williams describes is not entirely strange. Williams notes that in his early work Strindberg opposed the ruling class, and returned to radicalism after 1909 to again attack 'the rich, militarism, and the conservative [Swedish] literary establishment' (Williams, 1989: 49). Perhaps the red flags were appropriate. If Strindberg was an establishment figure, Secretary of the Stockholm Library on his first visit to Paris in 1876, he also enjoyed an international reputation and notoriety. Yet that international reputation, and his periods of residence in Paris and Berlin in the 1880s and 1890s, are evidence of his alignment, not with political agitation or emerging political philosophies such as anarchism and syndicalism,[12] but with

Symbolism and art for art's sake. As within any grouping, Symbolism contained a range of persuasions, but is marked by a general withdrawal from political and social life linked to an appeal to a mythicised past and mysticism.[13] Since Strindberg is part of that history, his plays having been performed in Paris in the 1880s, we do 'have to think again of the torches and red flags . . . to look beyond these singular men to the turbulent succession of artistic movements and cultural formations.' (Williams, 1989: 50), noting as well the rise of workers' movements.[14] A syndicalist congress took place in Paris, in November 1912, to discuss opposition to the threat of war. A general strike was called for on December 16th but failed, followed by the arrest of activists after which Parisian anti-war feeling tended to subside, eventually subsumed in the patriotism of the war itself. Political and cultural histories thus intertwine, their threads crossing in variegated patterns that render conditions complex.

Analysis of conditions is not, however, the only critical means. There is also the evidence of the work that embodies and enacts those conditions and in an engaged view of art intervenes in them. This is the nub of Ernst Bloch's criticism of Georg Lukács in their contention over Expressionism: Bloch says Lukács has not looked at the paintings: 'there is no mention of a single Expressionist painter . . . even the literary works have not received the attention they merit' (Bloch, 1980: 18).

The radical content of Strindberg's plays is not obvious now. A century after they were performed in Paris they were described as of interest only to academics and critics,[15] yet Strindberg's refusal of bourgeois dramatic conventions to reveal the anxieties of bourgeois domesticity appealed to radical audiences in his time.[16] His audiences in Paris were, at least, made to watch the plays rather than continue their conversations.[17] What they saw was a distorting mirror of their world. While this is a literary retreat from engagement with the causes of dysfunctionality in social and economic organisation, from another viewpoint it is a realism that reveals the contradictions of nineteenth-century bourgeois society.[18] Williams sees in *Lady Julie* a naturalism adapted to convey a sense of fragmented, multiple character,[19] and Strindberg could be compared with Beckett in his refusal of Ibsen's emancipatory content for something nearer the edge.[20] Strindberg's is a realism, too, of a repression of drives – the unmediated content of Freud's *Es* – which threaten social accord.[21] Bloch writes in *The Spirit of Utopia* of Strindberg's 'sphere of a pure soul-reality', which foresees a bourgeois apocalypse (Bloch, 2000: 117). In a more elusive way, Strindberg's interiority can be compared with that found in the paintings of Edvard Munch, of whose work Polish playwright Stanislaw Przybyszewski wrote in 1894:

> I found myself confronted with the naked revelation of an individuality, with the creative products of a somnambular and transcendental consciousness, what is commonly called 'the unconscious' . . . I call it 'individuality'. . . It is like a great wave which grows everlastingly, a germ which

> perpetuates itself to infinity in constantly changing metamorphoses . . . a pangenesis in the sense understood by Darwin: every original cell carries the whole of human kind and all its characteristics within itself.
>
> (Przybyszewski [1894], quoted in Harrison and Wood, 1998: 1045)

Munch's paintings were exhibited at the Association of Berlin Artists (*Verein Berliner Künstler*) in 1892 where their reception caused a split in the membership.[22] Strindberg was in Berlin at the time, and Munch joined the circle around him and Przybyszewski (whose wife was the object of Munch's and Strindberg's desires). Strindberg's impact was felt, too, a decade later in Vienna.[23] But, though Strindberg and Munch, with French Symbolist writers and artists, contribute to an understanding of an unconscious realm of the psyche before Freud's formulation of psychoanalysis, the somnambular unconscious of which Przybyszewski writes is not Freud's.

If, that is, psychoanalysis is a liberating method (as I take it to be), it depends on a specificity of interpretation in analysis so that the image, for instance, of a dream is interpreted in the context of the dream and the dreamer, not as a generalised symbol. Przybyszewski's interpretation of Munch deals instead with universals, as comes to the fore a few paragraphs later: 'This is the mysterious source of that most intimate of all feelings, our love for our native soil, for our fatherland – and this is also the mysterious source of that specific kind of feeling which animates the great and truly powerful artist' (Przybyszewski [1894], quoted in Harrison and Wood, 1998: 1045).

This is an appeal less to a sense of place than to a notion of authenticity dependent on power, echoed later in Nolde's primitivism.[24] In Germany in the 1930s, not dissimilar notions of a mythicised unmediated realm projected onto a twilit past were received by the petit bourgeois class.[25] But it is not easy to differentiate traces of this *ersatz* intensity from those of a potentially liberating recognition of a discomforting profundity,[26] nor an exoticism that trades on otherness from a wild use of colour that frees art from a mimetic tradition.[27] Strindberg says 'I can get quite wild sometimes, thinking about the insanity of the world'; and his response to Gauguin's paintings in 1895 renders wildness an exotic route to new visionary worlds, but worlds of the titan, the repressed and repressive masculine, which pervade the writing of Friedrich Nietzsche.[28] His interiority has, too, a savage aspect, which is not exotic but deeply threatening. In *The Father*,[29] the Captain utters to his daughter Bertha:

> You see, I'm a cannibal and I want to devour you. Your mother wanted to eat me but I prevented her. I am Saturn, who ate his own children because it had been foretold that otherwise they would eat him. To eat or be eaten. That is the question. Unless I eat you, you will eat me.
>
> (*The Father*, adapted by J. Osborne, 1989: 45)

Replacing the Shakespearean question 'to be or not to be' with that of a primitive destiny in which the subject's autonomy is erased (despite the expressiveness of the statement), Strindberg sees life as turmoil. Society is universalised as a quasi-Darwinian world of survival of the most aggressive, though for Strindberg aggression pertains to a war not of species, races or nations but of sexes.

If Strindberg's plays rupture a conventional representation of bourgeois society in a depiction of instability in the mental lives of the inhabitants of bourgeois interiors, the difficulty is that his revolution was not against the dominant socio-economic order that produced the reaction, but against himself: 'I am engaged in such a revolution against myself, and the scales are falling from my eyes' (from O. Lagercrantz, *August Strindberg*, 1984, quoted in Williams, 1989: 49–50). Here Strindberg and Nietzsche coincide: Nietzsche wrote to Strindberg in 1888, citing *The Father*: 'It has astounded me beyond measure to find a work in which my own conception of love – with war as its means and the deathly hatred of the sexes as its fundamental law – is so magnificently expressed' (from M. Meyer, *August Strindberg*, 1985, quoted in Williams, 1989: 50). Strindberg responded that Nietzsche was 'the modern spirit who dares to preach the right of the strong and the wise against the foolish, the small (the democrats)' (ibid.). This elitism of the bourgeois, male outsider brings to mind passages in *Thus Spoke Zarathustra* (1883–5):

> It is time for man to fix his goal. It is time for man to plant the seed of his highest hope.
>
> His soil is still rich enough for it. But this soil will one day be poor and weak; no longer will a high tree be able to grow from it.
>
> (Nietzsche, 1969: 46)

In this preliminary section, Nietzsche contrasts the Superman who embodies a modern spirit (after the death of God, taking on the Christian image of Son of Man) with the Ultimate Man standing for the masses:

> 'What is love? What is creation? What is longing? What is a star? thus asks the Ultimate Man and blinks.
>
> The earth has thus become small, and upon hops the Ultimate Man, who makes everything small. His race is as inexterminable as the flea; the Ultimate Man lives longest.
>
> (Nietzsche, 1969: 46)[30]

Nietzsche adds, in a caricature of mass society, that 'Everyone wants the same thing, everyone is the same: whoever thinks otherwise goes voluntarily into the madhouse' (ibid.). After his breakdown in 1909, Nietzsche was, in fact, taken into the care of female family members; the point remains that although Nietzsche's philosophy contains strands of a rationalist modernity, it is also a version of the Romantic rejection of Reason after the French Revolution:

> Uncanny is human existence and still without meaning: a buffoon can be fatal to it.
>
> I want to teach men the meaning of their existence: which is the Superman, the lightning from the dark cloud man.
>
> (Nietzsche, 1969: 49)[31]

The lightning that strikes out of the mass of the cloud may be the voice of the radical philosopher following in the path of the prophet whose visions of doom foreshadow violent, apocalyptic upheaval. Or, it may be avant-garde art. But this avant-garde, however much it seeks a new language for the new reality of an industrialised, urbanised society, does not coherently align itself with a programme of political action. But then coherence is not necessarily the point. But is Strindberg's version of cultural Darwinism[32] an element in the avant-garde, most often defined as oppositional, or of Modernism?

In 'Language and the Avant-Garde' (1986), Williams separates a Modernism beginning with Baudelaire from an avant-garde of the 1910s in Futurism. Linking the latter to new technologies of communication such as radio, photography and film, he sees only new forms, not a new language or new messages;[33] but in 'The Politics of the Avant-Garde' goes further:

> We can distinguish three main phases which had been developing rapidly during the late nineteenth century. Initially, there were innovative groups which sought to protect their practices within the growing dominance of the art market and against indifference of the formal academies. These developed into alternative, more radically innovative groupings, seeking to provide their own facilities of production, distribution and publicity; and finally into fully oppositional formations, determined not only to promote their own work but to attack its enemies in the cultural establishments and, beyond these, the whole social order in which these enemies had gained and now exercised their power.
>
> (Williams, 1989: 50–1)

The first phase corresponds to the informal cultural milieux of Paris and other large European cities from the 1850s onwards; the second to the organisation by artists of their own exhibitions, such as the Salon des Indépendants in Paris, and the Secessions of Munich, Vienna and Berlin in the 1890s; and the third is constituted by movements, perhaps stating their oppositionality through manifestos, making work which cannot be assimilated by advanced taste, and thereby seeking to destabilise bourgeois society's institutions. Williams continues: 'Thus the defence of a particular kind of art became first the self-management of a new kind of art and then, crucially, an attack in the name of this art on a whole social and cultural order' (Williams, 1989: 51).

Williams makes a coherent case,[34] but there are complexities: not all avant-gardists share a given outlook and some are selective in their allegiances. While

Secessionists tend to internationalism in the 1900s, some Cubists, if more avant-garde in their art, were nationalists.[35] There are differences, too, in relation to tradition and the decorative arts. Williams sees Modernism as following a Romantic 'victorious definition of the arts as outriders, heralds, and witnesses of social change' but later abandoning its anti-bourgeois stance for 'comfortable integration into the new international capitalism' (Williams, 1989: 32, 35). As Cottington writes:

> The borders of the avant-garde's territory were in fact poorly marked, and there was considerable traffic in both directions across them. Montmartre . . . was composed not only of villains, prostitutes and bohemians but also of young men of the middle class – artists and poets as well as critics and collectors – in temporary flight from bourgeois morality. Elsewhere in the city the traffic was in the other direction.
>
> (Cottington, 1998b: 10)

Cottington notes that writers either began with small reviews and moved to established journals, or worked between both; and that the state used the arts to support the economy, freeing the Salon from its control in 1881 to assist a growth of private-sector galleries. The self-organisation of artists' groups and exhibitions during the period is, then, a claim for economic as well as aesthetic autonomy in face of a growing art market. Modernism thus has an economic base which allows the diversity necessary for the emergence of an avant-garde. The appearance of loft-culture in New York's SoHo in the 1970s is, I suggest, a recurrence of this economic Modernism in the efforts of artists to construct a means to continue their practices under their own control.[36]

But what of the avant-garde within this scenario? I look now to accounts of the twentieth-century avant-garde by Renato Poggioli and Peter Bürger. Poggioli writes of two avant-gardes, artistic and political, which diverge from Realism. He dates the split to the Commune and growth of small reviews in the 1880s (citing *La Revue indépendante*, founded in 1880, as the last that unites cultural and political progressives): 'Only then did it begin to designate separately the cultural-artistic avant-garde while still designating . . . the sociopolitical avant-garde' (Poggioli, 1968: 10). But for Roger Shattuck avant-gardism retains an urgency:

> a comment on the character of French civilisation since the Revolution . . . Like the anarchists, the artists of the avant-garde took liberties with the structure of life itself, defied convention and lethargy in order to assert a new order of things. This tendency to violent dissent is a prime attribute of the celebrated French critical spirit.
>
> (Shattuck, 1969: 41–2)

Shattuck's remark is sweeping, yet conceals a characteristic noted in a different way by Poggioli:

> If the essence of activism lies in acting for the sake of acting; of antagonism, acting by negative reaction; then the essence of nihilism lies in attaining nonaction by acting, lies in destructive, not constructive, labor. No avant-garde movement fails to display . . . this tendency.
>
> (Poggioli, 1967: 61–2)

He means Futurism and Dada, citing Tristan Tzara to the effect that Dada means nothing, a destructive task of sweeping out and cleaning up. But what is swept out and cleaned up in this iconoclasm? The remains of the nineteenth century. Yet this nihilism can be seen in more conventional works, too, such as Picasso's *Demoiselles d'Avignon* (1907, Museum of Modern Art, New York), as a strident primitivism.[37] This brings me to a wider ambivalence: if Modernism begins with Baudelaire or the independent art movements of the 1880s, its avant-garde extends to the Communism of Mayakovsky and the Fascism of Marinetti, and was always fractured.[38] As Williams writes, 'the commitment to a violent break with the past, most evident in Futurism, was to lead to early political ambiguities' while 'The renewed rhetoric of violent rejection and disintegration' of Germany in the 1920s 'produced associations within Expressionism . . . which . . . led different writers to positions on the extreme poles of politics: to both Fascism and Communism' (Williams, 1989: 58). In view of this, Poggioli's argument that 'In the case of the avant-garde, it is an argument of self-assertion or self-defense used by a society in the strict sense against society in the larger sense' (Poggioli, 1968: 4) fails for two reasons: first, although the formulation is neat, avant-garde groups are not always tightly bound, like societies with membership, which attack the wider (bourgeois) society around them, but are more fluid formations overlapping with other formations; and second, there is no more *case of the avant-garde* than a single modernity, only several contenders of varied, even variegated, persuasions.

Are there other characteristics, apart from the condition of metropolitanism, that help us see the aims and significance of European avant-gardes in the 1910s? A possibility is that avant-garde groups attack art's institutions, following a pattern set by the Secessions and withdrawal of later groups from them in a succession of departures. This entails not simply a reform of language – collage replacing painting as a means to introduce the city's materiality into art – but an offensive against taste and the structures, such as museums, criticism and the market, which legitimate Art by rendering its history coherent. Like Poggioli and Williams, Peter Bürger, in *Theory of the Avant-Garde* (1984),[39] sees Dada as epitomising this antagonism. Like Futurism, Dada attacks not the art of previous periods as such, but Art as a category of bourgeois aestheticism: 'The European avant-garde movements can be defined as an attack on the status of art in bourgeois society' (Bürger, 1984: 49). Among his few illustrations is Duchamp's *Fountain* (1917, Museum of Modern Art), a urinal signed R. Mutt, now a museum exhibit used to legitimate later avant-garde art in a linear art history. Although he does not make the same differentiation as Williams between an oppositional avant-garde and an independent Modernism,[40] Bürger nonetheless affirms the radicalism of the avant-garde in its moment of self-realisation: 'In bourgeois society, it is only with aestheticism that the full unfolding of the phenomenon of art became a fact, and it is to aestheticism that the historical avant-garde movements respond' (Bürger, 1984: 17). In a note, he adds that 'In their most extreme manifestations, their primary target is art as an institution such as it has developed in bourgeois society' (Bürger, 1984: 109, n. 4). For Bürger, then, the avant-garde rejects past art in its entirety, not piecemeal in dislikes of individual artists, sweeping the lot out of the studio. But it is still the studio which is the theatre of this drama of rejection.

The difficulty is that this formulation of an avant-garde is contradictory: new art must reject old art because it represents bourgeois values and institutions, but at the same time – *as art* – requires validation by those institutions. To an extent, Bürger answers this difficulty in course of a discussion of Habermas, arguing that 'it is necessary to distinguish between the institutional status of art in bourgeois society (apartness of the work of art from the praxis of life) and the contents realized in works of art' (Bürger, 1984: 25). This allows recognition of the conditions in and point at which art becomes self-critical and, like critical theory, able to interrogate its own assumptions. But there is no guarantee. As Bloch writes of art that gives form to wishful dreams: 'is there anything more .. than a game of appearance? . . . In aesthetic ringing or even jingling is there any hard cash, any statement which can be signed?' (Bloch, 1986: 210). Yet Bürger offers an insight worth citing at length:

> For reasons connected with the development of the bourgeoisie after its seizure of power, the tension between the institutional frame and the content of individual works tends to disappear in the second half of the nineteenth century. The apartness from the praxis of life that had always constituted the institutional status of art in bourgeois society now becomes the content of works. Institutional frame and content coincide. The realistic novel . . . still serves the self-understanding of the bourgeois. Fiction is the medium of a reflection about the relationship between individual and society. In Aestheticism, this thematics is overshadowed by the ever-increasing concentration the makers of art bring to the medium itself . . . At the moment it has shed all that is alien to it, art necessarily becomes problematic for itself . . . social ineffectuality stands revealed as the essence of art in bourgeois society, and thus provokes the self-criticism of art. It is to the credit of the historical avant-garde movements that they supplied this self-criticism.
>
> (Bürger, 1984: 27)

However, the refusals and interruptions become an evolution of formal solutions leading, via the reductionist criticism of Clement Greenberg, to the self-referential art of late Modernism. This does not revolve around criticality but, as Greenberg puts it, keeping art moving as a means to avoid what he regards, in the 1940s, as the kitsch of Socialist Realism.[41] For Greenberg the avant-garde becomes progressively withdrawn until revolt is set aside in the society from which the avant-garde has set itself aside: 'Hence it developed that the true and most important function of the avant-garde was not to "experiment", but to find a path along which it would be possible to keep culture *moving*' (Greenberg, 1986: 8). It is not my purpose to interrogate Greenberg. At this point I suggest simply that Greenberg's art criticism is a specialist practice suited to an art market requiring, as with any commodity, ever-new variants of form. As each new wave comes onto the scene others are discarded (built-in obsolescence) or promoted to senior status (history, which is more expensive but not as expensive as silence). I do not wish to be cynical.

II CUBISM IN PARIS, EXPRESSIONISM IN MUNICH

To sum up: the Modernist avant-garde is produced in conditions in which art has a growing economy outside state institutions; artists form affiliations or groups, in some cases movements, through which to promote themselves and claim control over the reception of their work; and the avant-garde is not a single entity but a multiplicity of formations divided in its approach to internationalism and the possibilities for intervention in social change. Overt criticism of bourgeois society has for the most part shifted into oblique criticism through the construction of a new aesthetic. This aesthetic is radically different from that which precedes it, may reject art entirely, yet remains viable as a critique only because it is validated by a discourse of art, and hence by art's institutional structures and the assumptions which permeate them.

The case of Cubism in Paris in the early 1910s – in particular the group around Albert Gleizes, Jean Metzinger, Roger La Fresnaye, Sonia and Robert Delaunay and Fernand Léger in Montparnasse, who formed a gallery-based (rather than salon-based) grouping within what is not a movement but is more than a style – shows some of the ambivalences of the new situation. Cubism has no manifesto, but it has the essay by Gleizes and Metzinger – *Du Cubisme* – written in preparation for a major exhibition, *Section d'Or*, and a showing of Cubist works at the autumn Salon, published in 1912.[42] Its writers, themselves painters, sought to establish Cubism as a movement which could be seen as accessible, hence legitimate, while remaining advanced and requiring an educated, perhaps intellectually advanced, spectator.

I now look at this essay as a means to open some questions around Cubism in context of the above discussion of an avant-garde, in preparation for a slightly longer discussion of Expressionism.

The essay begins with a reference to Courbet, which sets up an expectation of radicalism which is at the same time respectable, given Courbet's status.

Yet unlike Courbet, who showed at the Salon or rented his own spaces for public exhibitions when rejected by it, the artists of *Section d'Or* operated more (not exclusively) through private galleries and a small network of dealers and collectors – itself a characteristic of Modernism. Gleizes and Metzinger soon rubbish Courbet anyway: 'he remained a slave to the worst visual conventions . . . accepted without the slightest intellectual control everything his retina communicated' (Gleizes and Metzinger, 1912, quoted in Harrison and Wood, 1992: 188). Manet gets off lighter because he dares more, transgressing 'the decayed rules of composition' (ibid.); then, it is Cézanne who is claimed as the foundational source of Cubism.[43]

The true engagement, it seems as the essay progresses, is not with social formations but with the invention of a new pictorial language. This is no longer linked to the imitation of natural appearances in art, nor to Euclidian space, but is now implicit in form itself. The task of painting is therefore logically to reinvent the language of representation.

There are echoes of both Bergson's concept of duration[44] and Nietzsche's sense of being ahead of the masses. For instance:

> pictorial space is defined: a sensitive passage between two subjective spaces.
>
> The forms which are situated within this space spring from a dynamism which we profess to dominate. In order that our intelligence may possess it, let us first exercise our sensitivity. There are only *nuances*.
>
> [. . .]
>
> To compose, to construct, to design, reduces itself to this: to determine by our own activity the dynamism of form.
>
> (Gleizes and Metzinger, 1912, quoted in Harrison and Wood, 1992: 191)

This is post- rather than anti-Cartesian – the design of regular places on a blank ground, the drawing of lines which did not exist – but what has changed radically is that the world no longer consists of objective realities which can be represented, only of the representations, as it were, themselves, the signs set free from their referents:

> It therefore amazes us that well-meaning critics explain the remarkable difference between the forms attributed to nature and those of modern painting by a desire to represent things not as they appear, but as they are. And how are they? . . . An object has not one absolute form, it has several; it has as many as there are planes in the domain of meaning.
>
> (Gleizes and Metzinger, 1912, quoted in Harrison and Wood, 1992: 194)

These artists also argue that although their art is for a mass public this is only an ultimate end, to be achieved through a new language of the avant-garde artist, inevitably misunderstood but of special status.[45] As Wood says, 'An avant-garde is intended to lead the way' (Wood, 1999: 195).

The background to the essay is complex. A dispute with Germany over Morocco in 1911 provoked renewed nationalism, in a nation for whom the defeat at Sedan in 1870 and loss of Alsace and Loraine, and the war indemnity paid to Germany in 1871, was not forgotten. Cubist work was attacked as anti-patriotic, a threat to society.[46] It clearly diverged from the statues of Jean d'Arc, which remained, for both left and right, icons of a nation whose glory was now necessarily projected back to a suitably distant past, after Sedan.[47] Today the allusion to saboteurs seems fanciful, when the avant-garde has become a self-referential force within late Modernism; but in 1911 the relation between art and society was more open.[48] Art exhibitions, too, tended to internationalism, and some of the strongest cross-currents were Franco-German. Meanwhile, the far-right *Action Française* had been intimidating suspect publics, particularly the rapidly increasing number of students who inhabited the Latin Quarter of Paris, since 1908.

Within art criticism in Paris in 1911–12, there were aestheticist and philosophical positions, including accusations that Cubism was merely decorative, a superficial mirror to everyday life. But the more interesting difference is between Gleizes and Metzinger and those who, like the dealer Kahnweiler, adopt a Kantian position in which Cubism reveals an underlying essence of reality, a true form hidden by everyday appearances.[49] This could be mapped back onto Cézanne, in my view no more appropriately, but is confounded by both the formal invention of the multiple viewpoint, and the insistence on multiple moments of perception of equal value which are perhaps Cézanne's legacy to Cubism. Alongside this was, too, Apollinaire's near hagiography for Picasso in a new magazine, *Les Soirées de Paris*. No doubt Gleizes and Metzinger wanted to even the balance of critical recognition, to set Montparnasse beside Picasso's Montmartre. But, if Picasso develops primitivism into an almost classical analysis of form which then incorporates a patriotic statement (see note 46), while retaining some connection to the anarchist and syndicalist politics of his background in Barcelona in the 1890s, Gleizes can be better understood – in contrast – as having spent a period in 1906–7 living in an artists' and writers' commune, the *Abbaye de Créteil*, in a Parisian suburb. There, at a time when Mallarmé was frequently discussed, 'they hoped to escape from both politics and the city into an aesthetic arcadia' (Cottington, 1998b: 52). One of Gleizes' companions was Henri-Martin Barzun, founding editor of *Poème et drame* in 1912, whose editorials condemned any art of social engagement or the confusion of aesthetics with class struggle.[50] For Cottington, and I agree, Barzun's expression of autonomy is a product of free-market liberalism; but he notes that some of the commune's members were previously associated with left extra-parliamentary campaigns. The commune was precarious but much reported and visited. It seems a halfway house between Symbolism, with its withdrawal, mysticism and disdain for everyday life, and the metropolitanism, excitement in new engineering and machinery, and Cubism's depiction or direct incorporation into collage of everyday objects such as playing cards and bits of newspaper, though both tendencies are essentially metropolitan.

Very briefly, before turning to Expressionism, I want also to make a connection between Cubism and the Nietzschean current of Modernism's avant-garde. Gleizes and Metzinger write that 'the artist' tries to 'enclose the quality of this form (the unmeasurable sum of the affinities perceived between the visible manifestation and the tendency of his mind) in a symbol likely to affect others' (in Harrison and Wood, 1998: 190). They continue that, to paraphrase, the artist makes the crowd adopt the same relation to nature, though while the artist moves on to new symbolic images the crowd continues to see the world through the first. Later, they argue for forms in art removed from both natural appearances and popular imagination. This is in 1912, the year of Kandinsky's *Über das Geistige in der Kunst* (actually published in late 1911 but dated 1912).

Kandinsky's extended essay on the spiritual (or it could mean the intellectual) in art derives part of its case from Wilhelm Worringer's doctoral thesis, *Abstraction and Empathy* (1908), in which distortion (abstraction) is a representation of anxiety to be taken as seriously as the empathy with forms of a classical sense of beauty. Although the two are seldom compared, Kandinsky's text shares with Gleizes and Metzinger's essay an intellectual avant-gardism – a going ahead of the mass society in the invention of a visual language which opens a possibility for a new consciousness. For Gleizes and Metzinger this aim remains implicit, is perhaps rhetorical in context of a search for recognition; for Kandinsky it is central, and links his intentions to those of the first avant-garde, if from a reactionary political stance.

Kandinsky writes of the desolation of materialism, as if finding himself in an alien condition, and describes a movement upwards towards a new era of the spirit. Artists, as revealers of new forms, are its conceivers:

> The life of the spirit may be fairly represented in diagram as a large acute-angled triangle divided horizontally into unequal parts with its narrowest segment uppermost . . .
>
> The whole triangle is moving slowly, almost invisibly forwards and upwards . . . what today can be understood only by the apex and to the rest of the triangle is an incomprehensible gibberish, forms tomorrow the true thought and feeling of the second segment.
>
> At the apex . . . stands often one man, and only one. His joyful vision cloaks a vast sorrow. Even those who are nearest to him in sympathy do not understand him. Angrily they abuse him as a charlatan or madman . . .
>
> In every segment of the triangle are artists. Each one of them who can see beyond the limits of his segment is a prophet to those about him.
>
> (Kandinsky, 1977: 6–7)

He has a special word for socialists; a base segment of the triangle has reached the point – in this Hegelian pre-ordained trajectory – at which its occupiers sing the materialist creed. Although Christians and Jews, they are really atheists, and in economics they are socialists about to hew off the head of the hydra of capitalism, as he puts it. But they 'have never solved any problem independently, but are dragged as it were in a cart' (Kandinsky, 1977: 10).

Much of the second, longer section of the book, titled 'About Painting', is a technical investigation as to how to construct the new language of forms and colours, a language of potentially universal communication transcending nation, class and period. The cover of *Über das Geistige in der Kunst* shows an abstracted image of a city on a hill, one of its towers falling before a sun. A white shape against a solid background within the city's form resembles a horse and rider. Kandinsky, a theosophist, explains that his book, like his and Franz Marc's almanac for *Der Blaue Reiter* (1912), was designed to awake a capacity 'absolutely necessary in the future' for experiences of the spiritual – which I take liberally as a realm of thought as well as mystery (Kandinsky, 'Rückblick' *Kandinsky, 1909–1913* [1913], quoted in Washton-Long, 1975: 221). The point of comparison with Cubism is not in a vocabulary of forms – Kandinsky discards the everyday objects of still life (except in his study the everyday referents are woodcuts, glass paintings and icons of apocalyptic content) – but in the role of the artist and purpose of art. That is, in the artist's use of veiled images to deepen consciousness and thereby usher in a more advanced world. Again, Kandinsky sees that world in a quite different, non-metropolitan way; but in both cases the special status of the artist is assured. So, art changes the world – a recurrence of the aim of the first avant-garde of 1848 – but for a different direction which can now be encased in an aesthetic of pure form and colour: abstraction.

1912 is also the date of Emil Nolde's notes for a book never published, *The Artistic Expression of Primitive People*, following visits to the Berlin Ethnographic Museum.[51] His notes and Kandinsky's text lean in opposing directions: the former to an abstraction of cosmic significance, a revelation, though part of his early training was in ethnography; the latter to a populist celebration of religious themes and natural wonders, eventually sympathetic to an aesthetic of blood and soil. Both, nonetheless, draw on vernacular imageries and imaginaries rather than the incidents of urban life, seeking an authenticity escaped from a life of fluidity and mobility, the qualities Georg Simmel sees, in his essay on 'The Metropolis and Mental Life' (1903) as the frenetic condition of cities such as Berlin. Kandinsky, a Russian, worked not in Berlin but in Munich, drawn there because it was by the 1900s a centre of international visual and musical culture. The Secession gallery was a landmark of *Jugendstil*, and the new suburb of Schwabing housed a milieu of artists, intellectuals and academics.[52] There was, too, a blossoming of new drama enabled in part by a more liberal censorship than in Berlin – a performance of an expurgated version of Wedekind's *The Spring Awakening* took place in 1908. But Munich was not a progressive city politically. Neither Fischer nor the Secession were concerned with the conditions of the poor, or the 60,000 immigrants to the city in the 1890s.[53]

Kandinsky fits well in this situation. When he turns away from academic and Secessionist art to seek new forms it is not to the moden city but to folk traditions he encountered in the village of Murnau and in his ethnographic studies in Russia. For Kandinsky, as Rose-Carol Washton Long summarises:

> [The] search for ways to reconcile his need to communicate his messianic visions with his need to spiritualize or abstract the content of the message led to the use of these motifs which to Kandinsky seemed powerful enough to suggest a cosmological resonance even when partially hidden. The theme of the apocalypse had universal connotations, the folk depictions offered a simplified but vital treatment of these eschatological motifs, and the veiling of them provided Kandinsky with a means of involving the spectator.
>
> (Washton Long, 1975: 227)

Though Nolde, too, looked to ethnographic sources, he and Kandinsky differ in the publics they sought: for Nolde a folk reception; for Kandinsky an intelligentsia. Jill Lloyd writes of Nolde's *Life of Christ* polyptych (1911), that it represents a 'spiritual, inner direction . . . coinciding in terms of subject if not style with aspects of conservative, *volkish* ideology' (Lloyd, 1991: 97), noting his defence of rural life, opposing the introduction of customs-houses, pumping stations and dykes in North Schleswig, but equally that much of his formative time was spent in metropolitan Berlin.

The Nazi recategorisation of Expressionism as degenerate in the 1937 *entartete Kunst* exhibition in Munich retrospectively introduces a new critical configuration. Kandinsky may have been a 'politically reactionary religious mystic' (Wood, 1999: 202) and Nolde a Nazi supporter, but Bloch defends Expressionism against Lukács' attack. The issues are brought out in essays published in left journals – *internationale literatur* and *Das Wort* – and I want, finally in this chapter, to summarise the argument as a way to re-present the contradictions of the European avant-garde. In passing, we should recall that Expressionism was not exactly the contemporary art of the 1930s: its key time was the 1910s, that by the 1920s artists such as Otto Dix, George Grosz and John Heartfield had begun to create a politicised avant-garde using new visual techniques such as montage.[54]

For Bloch, however, who left Germany in 1933 to avoid arrest, the key artists are Marc, Kandinsky, Nolde, Heckel, Kirchner, Pechstein, Beckman, Kokoschka, Schmidt-Rottluff and Klee, all cited in 'Jugglers Fair Beneath the Gallows' (1937), his review of the Munich exhibition. They contribute 'everything which has given a new lustre and name to German art' (Bloch, 1991: 77). In his contention with Lukács, two ideas of what constitutes an avant-garde (as an art of intervention in the conditions and consciousness necessary for radical social change) collide. This is not a dispute over ideology: although Bloch was not a Communist Party member in the 1930s, in 1949 he accepted a chair in philosophy at Leipzig in the German Democratic Republic, writing to Lukács 'So we are now in a sense colleagues' (quoted in Geoghegan, 1996: 21). It is a dispute over *tactics*, and the relation between visual language and revolutionary content.

For Bloch, the Nazi attack on Expressionism renders it above class consciousness; the contrast of expressive authenticity and kitsch – 'a similar proximity of evil and good, of corruption and future, of kitsch museum and picture-gallery has not yet existed in the world' (Bloch, 1991: 76) – is intertwined with claims that authenticity derives from adherence to a subjective vision: 'Klee almost alone, the wondrous dreamer, remained true to himself and to his unrefuted visions' (Bloch, 1991: 234). In his response to Lukács, Bloch accepts that Expressionism may suffer 'too little forming . . . a rawly or wildly confused hurled-out fullness of expression' (Bloch, 1991: 249), but argues that it has the force of an inner voice. In *The Principle of Hope*, he cites Marc's remark that 'Painting is surfacing in another place', adding 'the inner voice is presupposed wherever there is artistic form . . . as soon as it has something to say, [it] always speaks outward expression' (Bloch, 1986: 794–5). Bloch's theory, developed from the 1930s to the 1950s, is that hope is always at least latent, accessed in art even in oppressive times so that it becomes an educated hope bringing freedom nearer by grasping it imaginatively. Unlike other critical theorists, he sees hope in both popular novels and high art. Bloch's effort to lend hope an objectivity equivalent to that of the end of history in scientific Marxism, by casting it as equivalent to a Freudian drive, is problematic.

Yet his idea of art as extension of day-dreaming through which hope is shaped is viable, I think, giving art a role in shaping consciousness. Nazi attacks on Expressionists as 'Miserable wretches, daubers, prehistoric stutterers, art swindlers' (Bloch, 1991: 78), terms reminiscent of those used for Jews, Marxists and émigrés, no doubt strengthened Bloch's resolve. His review of the *entartete Kunst* exhibition in 1937 was also a means to expose the false millenarianism of victory runes and thingsteads (pagan-style amphitheatres for fascist ceremonies), fires in the night and songs of old crowns in the Rhine and returning emperors, which constitute the false but emotive politic of Nazism. This falsity requires a contrasting authenticity, which Bloch finds in one way in earlier forms of social transformation such as Joachim of Fiore's thirteenth-century third kingdom,[55] and in another in Expressionism's representation of inner consciousness.

In 'Expressionism, Seen Now' (1937), Bloch writes, in a passage which cites the work of fascist sympathiser Gottfried Benn:

> Here there is no decay for its own sake, but storm through this world, in order to make room for the images of a more genuine one. Here the will

> towards change is not confined to canvas and paper . . . to artistic material that contents itself with shocking artistically . . . no prevalence of the archaic, brooding, no intentionally lightless and forged diluvial elements as so often in Benn's work, but integration of the No-Longer-Conscious into the Not-Yet-Conscious, of the long past into the definitely not yet appeared, of the archaically encapsulated into a utopian uncovering.
>
> (Bloch, [1937] 1991: 238)

The forgotten content of consciousness is hope; new languages make it visible and reveal the decay of capitalism. Collage and montage depict social fragmentation: 'For as a period of bourgeois decay it is also a period of cracked surface . . . as in painting, so in film, the time of a not only subjectively, but objectively possible montage' (Bloch, 1986: 411).[56] But questions arise. First, does the exposure of a time of crisis in montage, or in Marc's fragmented images, induce a vision of the desired new? Second, does the use of a new language of art to express inner realms imply an elitism that requires an educated spectator and counters a claim for freedom to be achieved – in Marxism – by class struggle? And third, does this dichotomy between purpose and means run through the Modernist avant-garde?

In 'Discussing Expressionism' (1938), Bloch cites an article by Alfred Kurella (published under the name Bernhard Ziegler) in which Expressionism is seen to lead to fascism. Bloch regards this as banal given Hitler's view of Expressionism, though, he accepts the non-reliability of the critique. But it is Lukács' 'Greatness and Decline of Expressionism' (1934) which is Bloch's real target: 'Lukács is . . . more cautious . . . But the conclusion *nevertheless remarks that "the fascists* . . . see in Expressionism a useful inheritance for themselves"' (Bloch, 1991: 242).[57] Bloch criticises Lukács for a failure to demonstrate this through cases. Lukács argues, however, in 'Realism in the Balance' that older forms of popular art cannot be assumed to retain currency in modern times, and that much seemingly popular art is not 'genuinely popular' (Lukács, 1980: 53) – which could be a criticism of Kandinsky's or Nolde's vernacularism. Lukács contrasts the work of Mann to novels for the mass market, and notes that *Buddenbrooks* was printed in millions of copies, that 'when the masterpieces of realism past and present are appreciated as *wholes* . . . their topical, cultural and political value [will] fully emerge' to represent contradictions in bourgeois society (Lukács, 1980: 56). Art, then, can be critical of its own class origins to appeal to a mass audience in forms of realism – like Socialist Realism. The new language of specialist cultural production is not necessary, he asserts, and may not only lack embodiment of the new but be obstructive to its recognition.

In the exchange between Bloch and Lukács, the direction of social change is not in question, but the means to achieve the consciousness which will produce it are. In France and Germany this follows failures of political action, not only the Paris Commune but also the Soviet in Munich in 1918 and revolt in Berlin in 1919.[58] Perhaps in response, Bloch's position is to affirm an art of aesthetic innovation which he sees as reflecting its time, exposing it in a way conducive to new insights. At the same time, he remains interested in popular and vernacular sources which demonstrate hope's ubiquity. A difficulty is that formal innovations are likely to deny a mass audience, a problem the Cubists address through a Nietzschean stance – ahead of their time.[59]

NOTES

1 Frisby, who takes Baudelaire as his starting point, cites the latter's remark in 'The Painter of Modern Life': 'I know of no better word to express the idea I have in mind', adding that he saw modernity as both a quality of modern life and an artistic project (Frisby, 1985: 15).

2 Lefebvre links the development of perspectival representation to the new economy, but this does not mean that 'townspeople and villagers did not continue to experience space in the traditional emotional and religious manner – that is to say, by means of the representation of an interplay between good and evil forces at war throughout the world.' (Lefebvre, 1991: 79).

3 See Adorno and Horkheimer, 1997: 3–5.

4 Citing Descartes' *Discours* (1637), in which he describes an engineer (*ingénieur*) making regular places according to his imagination (*fantaisie*), Lacour says: 'The act of architectural drawing that Descartes describes is the outlining of a form that was not one before. That form would combine reason . . . with imaginative freedom . . . It is not only new to the world, but intervenes in a space where nothing was' (Lacour, 1996: 37).

5 Frisby cites Berman's argument that Marx was the first and greatest modernist, who characterised the new configuration of capitalism as a constant revolution of production and disturbance of social relations. He summarises that, for Marx, this produces a situation in which people can confront the conditions of their lives and relations with others anew, but cites Berman's alternative, that such upheavals do not subvert but strengthen capitalism (Frisby, 1985: 21, citing Berman, 1983: 89).

6 See *Das Passagen-Werk* (*The Arcades Project*) – Benjamin, 1999. Leslie notes Benjamin's reference, in a letter to Horkheimer, to Marx's concept of commodity fetishism: 'Society's repression of production, because of the form of fetishized production, makes its representation of itself fetishistic. This thing that the bourgeoisie calls its culture is phantasmagoric. It is a fantasy, a projection, a fabrication that hopes to deny its fabricated provenance' (Leslie, 2000: 192).

7 Massey equates the visuality of perspective and panorama with masculine power (mastery), seeing this gaze as detached while 'Detached does not mean disinterested' (Massey, 1994: 232).

8 See Williams, 1989: 37–48.

9 'The question of the other, is always the question of the stranger, the outsider, the one who comes from elsewhere and who inevitably bears the message of a movement that threatens to interrupt the stability of the domestic scene' (Chambers, 2001: 163–4).

10 Cottington notes publication of nearly 200 small magazines in Paris between 1900 and 1914: 'These milieux . . . encompassed a sometimes debilitating plurality of positions and allegiances' (Cottington, 1998b: 11).

11 'implicit . . . to our understanding of avant-gardism is the concept of alienation – psychic, social, ontological – utterly foreign to Courbet's approach to art and life' (Nochlin, 1968: 18).

12 Cottington (1998b: 35–6, 80–4) writes of syndicalism as being a development of Proudhon's utopianism based on workers' self-organisation and the tactic of the general

strike, and opposed to aestheticism: 'in 1912 the socialist Jean-Richard Bloch dismissed the idea of art for art's sake as comparable to that of charcuterie for charcuterie's sake; art's importance, he argued, was determined by its utility' (p. 81).

13 In *Axel's Castle*, Wilson cites a scene from Villiers de l'Isle Adam's play *Axel* (1890), in which the protagonist, Count Axel of Auersburg, is reproached by his would-be lover Sara: 'Those who fight for Justice say that to kill oneself is to desert.' He replies: 'The verdict of beggars . . . for whom God is but a way to earn their bread.' Of what Sara calls 'the general good' he says 'The universe devours itself; at that price is the good of all.' They drink a goblet of poison together and perish in a rapture (Wilson, 1961: 210). Another key work of literary Symbolism is Joris-Karl Huysmans' *A Rebours* (1884), the protagonist of which is the aristocratic Des Esseintes: 'He was constantly coming across some new source of offence, wincing at the patriotic or political twaddle served up in the papers' (Huysmans, 1959: 22).

14 'The Futurist call to destroy "tradition" overlaps with the socialist call to destroy the whole existing social order . . . [it is] a world away from the tightly organized parties which would use a scientific socialism to destroy the hitherto powerful and emancipate the hitherto powerless' (Williams, 1989: 52).

15 '. . . adaptations of this kind, often of "unperformed" European masterpieces', were best left to academics . . . Besides . . . these unperformed masterpieces had remained unperformed for one reason. They were . . . 'bloody boring', of no interest to anyone except professional theatre critics' (Osborne, 1989: ix). Osborne's adaptation of *The Father* was performed at the Cottesloe Theatre, London on October 26th, 1988.

16 Jackson notes, in *The Eighteen Nineties* (1913), that in England the public for new drama was one of 'intellectuals . . . [who] belonged very largely to the literary fringe of the Fabian Society and other reform or revolutionary organisations' (Jackson, 1950: 211).

17 Shattuck observes that at the Théâtre Libre, Paris, where Strindberg's plays were performed in the 1880s, innovations included extinction of the house lights during the performance 'so that the attention of the audience would have to be directed at the stage' (Shattuck, 1969: 9).

18 Lukács, in *The Sociology of Modern Drama*, sees Strindberg as constructing a world in which the personality turns inwards due to the alienation of an outer world of forces beyond control, devising a fragmented language for its expression as a personal pathology (Lukács, 1965: 425–30).

19 Williams cites Strindberg: 'I have . . . let people's brains work irregularly, *as they do in actual life*' (in Williams, 1989: 65–6). See also Lukács, 1965: 425–30.

20 'It is obvious that Strindberg repressively inverted Ibsen's bourgeois-emancipatory intentions. On the other hand, his formal innovations, the dissolution of dramatic realism and the reconstruction of dreamlike experience, are objectively critical. They attest to the transition of society toward horror . . . To this extent they are also socially progressive, the dawning of self-consciousness of that catastrophe for which the bourgeois individualistic society is preparing' (Adorno, 1997: 257). While Beckett 'draws the lesson from montage and documentation, from all the attempts to free oneself from the illusion of a subjectivity that bestows meaning. Even where reality finds entry into the narrative, precisely at those points at which reality threatens to suppress what the literary subject once performed, it is evident that there is something uncanny about this reality' (Adorno, 1997: 30–1).

21 Freud writes: 'At the very climax of my psychoanalytic work, in 1912, I had already attempted in *Totem and Taboo* to make use of . . . analysis in order to investigate the origins of religion and morality' (Freud [1935] cited in Chasseguet-Smirgel and Grunberger, 1986: 34). Freud moves beyond this to see human history as a reflection of dynamic conflicts within the psyche, his findings published as *The Future of an Illusion* (1927) and *Civilization and its Discontents* (1930).

22 Rosenthal writes that Munch's importance: 'lies in his creation of a shocking and archetypal image of an alienated northern European society, which has obvious affinities with the work of Strindberg and Ibsen', to

whom he ascribes 'a universal mythic potential' (Rosenthal, 1979: 151).

23 Vergo notes Strindberg's influence on Kokoschka, whose play *Mörder Hoffnung der Frauen* (*Murderer Hope of Women*) 'brings to mind the stage works of Strindberg' and prefigures Expressionist theatre (Vergo, 1975: 192).

24 'Nolde understood native art to result from an organic, unmediated relationship between producer and product, capable of expressing subjective emotions in objective form' (Lloyd, 1991: 100).

25 Bloch writes: 'Here is the Tavern of Nordic Blood, there the castle of the Hitler-duke, there the Church of the German Reich, an earthly church in which even city folk feel themselves to be a fruit of the German soil and worship soil as holy . . .' (Bloch, 1991: 101–2).

26 Frisby notes Baudelaire's sense of a savagery which exists within civilisation, revealed in *Spleen* (Frisby, 1985: 19).

27 Pollock takes Gauguin's *Manao Tupapau* (1892) as a case of 'sadistic voyeurism' re-orientalising Manet's *Olympia* (Pollock, 1994: 68). She argues that Manet painted a white woman with a black maid as a working partnership in the Parisian sex industry, but Gauguin overlays a European primitivism of death and sexuality on an image of his thirteen-year-old Tahitian wife Teha'amanaand. His tropical journey is marked by oppositions: 'here and there, home and abroad, light and dark, safety and danger' (Pollock, 1984: 66).

28 In a letter to Paul Gauguin declining to write a catalogue introduction for his final show at Hôtel des Ventes in 1895, which Gauguin used anyway, Strindberg writes 'He is Gauguin, the savage, who hates a whimpering civilisation, a sort of Titan . . . I, too, am beginning to feel an immense need to become a savage and create a new world' (Strindberg [1895], quoted in Harrison and Wood, 1998: 1036).

29 *The Father* is set in a provincial town. The Captain is in dispute with his wife Laura over the education of their daughter, Bertha, whom he wishes to see trained as a teacher and boarded with a free-thinker in the town. The Captain's progressive attitude is at odds with his sexism: 'This house is filled with women, all intent on raising my daughter . . . *I* should have the final voice, and all I get is opposition . . . It's like living in a cage full of tigers: if I didn't keep a red-hot poker under their noses they'd tear me apart' (Osborne, 1989: 3–4). In the final scene of the Captain's delirium he laments the rupture of old certainties: 'In the old days you got married and you got yourself a wife. Now you go into a business partnership with a career woman . . . As it is, there are only shadows' (p. 48).

30 This passage leads to the frequently cited section in which a tightrope-walker begins his traverse above the marketplace, to be distracted by a buffoon. The square clears as the tightrope-walker falls, but Zarathustra remains to bury him with his own hands (Nietzsche, 1969: 48). For an exceptional critique of Nietzsche, see Luce Irigaray's *Marine Lover of Friedrich Nietzsche* (1991): 'Perched on any mountain peak, hermit, tightrope walker or bird, you never dwell in the great depths. And as companion you never choose a sea creature. Camel, snake, lion, eagle, and doves, monkey and ass, and . . . Yes. But no to anything that moves in the water. Why this persistent wish for legs, or wings? And never gills?' (p. 13).

31 Griffin notes the influence of Nietzsche on the German neo-conservatives of the 1930s – see the extract from Moeller van den Bruck's 'The Eternal German Reich': 'The German nationalist of this age is, as a German being, still a mystic, but as a political being he has become a sceptic . . . It is in this sinking world . . . that the German is attempting to save what is German' (Griffin, 1995: 105–6).

32 Williams sees Strindberg and Nietzsche as informing Futurist manifestos: 'In the same language of cultural Darwinism, war is the necessary activity of the strong, and the means to health of a society' (Williams, 1989: 51). This position was shared by Kandinsky until 1914, writing to Marc that war was bound to purify Europe (Marc and Kandinsky, *Briefwechsel*, 1983: 44, quoted in Werckmeister, 1989: 13).

33 See 'Language and the Avant-Garde' (1986): 'For suppose we say, conventionally, that Modernism begins in Baudelaire, or in the period of Baudelaire, and that the avant-garde begins around 1910, with the manifestos of the

Futurists, we can still not say, of either supposed movement, that what we find in them is some specific and identifiable position about language' (Williams, 1989: 66). See also 'Culture and Technology' (1983): 'The original innovations of Modernism were themselves a response to the complex consequences of a dominant social order, in which forms of imperial-political and corporate-economic power were simultaneously destroying traditional communities and creating new concentrations of real and symbolic power and capital in a few metropolitan centres' (Williams, 1989: 131).

34 Cottington cites Williams as giving a 'cogent and valuable analysis', except that he assumes an identification between the avant-garde and progressive ideas which 'fails to address the specificity of the formation' (Cottington, 1998b: 197). Cottington also cites Duncan (1993) on the sexism of the avant-garde.

35 Wood notes that Julius Meier-Graefe, an internationalist critic and supporter of Munch in 1892, was forced to leave the periodical *Pan* in 1895, spending much of the next ten years in Paris; and that the Director of the National Gallery in Berlin resigned in 1909 after buying a painting by Delacroix (Wood, 1999: 102). In 1911–12, however, after a dispute between France and Germany over Morocco, nationalism strengthened in both countries. Cottington cites Gleizes' attraction to 'the increasingly hegemonic discourse of a broadly traditionalist nationalism' as a basis for a populism of 'time-honoured artisanal skills' (Cottington, 1998b: 162).

36 See Zukin (1989) for discussion of loft living in SoHo.

37 Cottington writes that Picasso's motivation in this work arose from 'an appetite for iconoclasm and a profound sense of his own artistic ability . . . but also from an attitude that was substantially a product of that pre-First World War decade: avant-gardism' (Cottington, 1998a: 12–13).

38 'Modernism thus defined *divides* politically and simply – and not just between specific movements but even *within* them. In remaining anti-bourgeois its representatives either choose the formerly aristocratic valuation of art as a sacred realm above money . . . or the revolutionary doctrines, promulgated since 1848, of art as the liberating vanguard of popular consciousness' (Williams, 1989: 34). Michael Hamburger defines a similarly bifurcating history from Baudelaire to Pound and Brecht: 'Baudelaire was the prototype; not least because he wavered between the aristocratic and the revolutionary positions, sure only about his bitter rejection of the bourgeois and capitalist order that had no place for him' (Hamburger, 1969: 2). And: 'Ever since Baudelaire, poets have felt themselves to be pariahs or aristocrats – if not both at once – in societies dominated by bourgeois values . . . Baudelaire's gibes at "democratization" . . . are typical reactions of an aristocrat-pariah who is excluded from the benefits of capitalist industry as much as from solidarity with the working classes' (Hamburger, 1969: 89).

39 The English edition (1984) is based on the second German edition, (1980, Frankfurt, Suhrkamp Verlag), with 'Theorie der Avantgarde und Theorie der Literatur', and 'Hermeneutik-Ideologiekritik-Functionsanalyse' in *Vermittlung-Rezeption-Funktion* (1979, Frankfurt, Suhrkamp Verlag). Schultz-Sasse, in his Introduction, emphasises the divergence of Bürger's theory from Poggioli's, noting the latter's sweeping criteria and his failure to make a distinction between an attack against tradition and an attack 'meant to alter the institutionalized commerce of art' (in Bürger, 1984: xv). Further reference to Bürger is made in Chapter 7.

40 Pinkney writes in his Introduction to Williams (1989) that Bürger, like Williams, is concerned to go 'beyond internal-formal analysis of avant-garde artifacts' (Williams, 1989: 17). He adds, however, that while Williams is interested in formations of production, Bürger is more concerned with those of reception.

41 Greenberg was a left critic, writing an (unpublished) ode to Trotsky on receiving news of his assassination in 1940. Only during the cold war does he call himself an ex-Marxist (see O'Brian's Introduction to Greenberg, 1986).

42 Cottington emphasises the importance for this group of the weekly meetings which took place at Puteaux, in the home of the Duchamp brothers. He cites Marcel Duchamp: 'the group that spent Sunday afternoons at Puteaux was far from homogeneous' (Cottington, 1998b: 158).

43 This judgement is echoed by Fernand Léger: 'Cézanne will occupy the place that Manet held some years before him' ('The Origins of Painting and its Representational Value', in Harrison and Wood, 1998: 196–7).

44 'No doubt it is possible . . . to conceive the successive moments of time independently of space; but when we add to the present moment those which have preceded it . . . we are not dealing with these moments themselves, since they have vanished forever, but with the lasting traces which they seem to have left in space on their passage through it' (Bergson, [1910] 1971: 79). Bergson's argument on numbers above can be loosely applied to Cézanne's multiple sensations, each traced as a brushmark; but a different framework through which to consider the Modernist reconfiguration of reality as a series of *fragments* (which may or not be like Cubist facets – a discussion for which I lack space here) is found in Simmel, as summarised by Frisby: 'Modernity consists in a particular mode of experiencing the world, one that is reduced not merely to our inner responses to it but also to its incorporation in our inner life . . . The fleeting, fragmentary and contradictory moments of our external life are all incorporated into our inner life' (Frisby, 1985: 62).

45 'in order to move, to dominate, to direct, and not in order to be understood' (Gleizes and Metzinger, in Harrison and Wood, 1998: 195).

46 'The cubists play a role in art today analogous to that sustained so effectively in the political and social arena by the apostles of anti-militarism and organised sabotage . . . the excesses of the anarchists and saboteurs of French painting will contribute to reviving . . . the taste for true art and true beauty' (Gabriel Mourey, review of the 1911 *Salon d'Automne* in *Le Journal*, in Cottington, 1998b: 145). But in contrast to the non-political stance of the Montparnasse Cubists, Picasso's *Notre Avenir est dans l'air* (1912, private collection) includes the red, white and blue cover of a pamphlet of that title promoting military uses of aviation. Krauss sees the *tricolore* as stating Picasso's adopted nationality (Krauss, 1985: 31).

47 Images of Jean d'Arc reflect flamboyance or sobriety for, respectively, a religious, royalist right and a secularist, republican left. At Jean's birthplace of Domremy, André Allar's *Jean d'Arc listening to her voices* (1892) is a flamboyant group outside the basilica, and Antoin Mercié's (1901) sculpture, commissioned by socialist Jules Ferry, shows a meeker Jean with sword in hand, in the embrace of France. For both camps she remains a heroine of a vanquished France. See Warner, 1981: 255–6; plates 33, 35 and 36; Michalski, 1998: 14–16.

48 Henri Guilbeaux, for instance, writing in *Les Hommes du jour* on the 1911 *Salon des Indépendants*, sees the work of Léger, Metzinger and others as grotesque, possibly funny or an insult to the bourgeoisie (Cottington, 1998b: 146).

49 Cottington notes Olivier Houcard's essay (1911) proposing a Kantian view, as philosophical justification of Cubism, in which essence and appearance are differentiated (Cottington, 1998b: 151). Kahnweiler writes that the essay by Gleizes and Metzinger did not reflect either Picasso's or Braque's ideas: 'it [the essay] expresses a completely different point of view' (Kahnwelier [1961] 1971: 43).

50 Other members were the writers Alexandre Mercereau, René Arcos, Charles Vildrac and George Duhamel. A printing press was installed as a means of economic support (not the virtue of labour as in anarchist communes), while the group sought 'an abode too high and vertiginous for the impure, where we shall be free from shallow society' (in Cottington, 1998b: 74).

51 'The products of primitive peoples are created with actual material in their hands . . . The primal vitality, the intensive, often grotesque expression of energy and life in its most elemental form – that perhaps is what makes these native works so enjoyable', contrasting this quality to the mechanical reproduction of art and design in industrial culture (Nolde, *Jahre der Kämpfe*, 1958: 177, translated in Miesel, 1977: 34, quoted in Lloyd, 1991: 100); see also note 27.

52 Meller writes of Schwabing's planning, as part of the city's extension, by Theodor Fischer from 1893: 'He started from the premise that the people of Munich liked their city and that city extensions should by just that: "city" extensions. New areas were to become "epicentres" of Munich itself' (Meller,

2001: 61–2). Meller notes (p.64) objections to Fischer's proposal for a metropolitan railway by the Artists Association.

53 The main contenders for power were the Bavarian Centre Party and the liberals. After repeal of anti-socialist legislation in 1890, two Social Democratic Party members for Munich were elected to the Reichstag; but their influence was less at city level: 'the artistic and literary scene of Munich was heavily focused on the lives of the bourgeoisie' (Meller, 2001: 59).

54 Bloch argues that the work of Grosz and Dix was received with more comprehension than that of the Expressionists, who did not attain the communicability they sought. He cites Heartfield as producing collages 'so close to the folk [art] that many educated people do not want to have anything to do with montage' (Bloch 1991: 250).

55 See Bloch, 1986: 509–15. Joachim's commune was characterised by the abolition of property and office, a free dwelling in the Spirit, an immediate and pervading (immanent not imminent) transformation.

56 Bürger notes Bloch's differentiation of montage in late capitalism and in socialism: 'Even though the concrete determinations . . . are occasionally imprecise, the insight that procedures are not semantically reducible to variant meanings must be held onto' (Bürger, 1984: 79).

57 See also Adorno *et al.*, 1980: 16–27.

58 The republic in Munich began as a socialist state in 1918, power passing in April to a Soviet. This was repressed by troops and Freikorps irregulars called in by the moderate socialists, leading to the killing or imprisonment of its leaders. In Berlin, Rosa Luxemburg and Karl Liebknecht, founders of the German Communist Party, were murdered (Willett, 1982: 46–7).

59 In Bloch's writing a dichotomy lingers which he masks by eclecticism. It is that art reveals, but does so obliquely. Perhaps this is tenable because he maintains an orthodox Marxism in which the end of history is objectively given, the path already inscribed in a trace to be read back, like redemption, from the future.

3

1938

CAP-MARTIN

•

This chapter begins with an account of *Graffite à Cap-Martin*, a mural by Le Corbusier in a villa by Eileen Gray. The mural is incidental to Le Corbusier's work as an architect and urban planner, yet I argue that an analysis of its content reveals his attachment to an orientalism that informs his plan for the redevelopment of Algiers. By putting his urbanism in this context and, in the second part of the chapter, investigating its links to political currents in Paris in the 1920s (the period of his key texts), I find ambivalences in Le Corbusier's approach to the city. This leads me to reconsider his legacy to urbanism in the post-war period. Finally, I contrast Le Corbusier's technocratic approach with an alternative Modernism in the work of Hassan Fathy in Egypt in the 1940s. More recent departures from Modernist art and architecture are considered in Chapter 7, and alternative constructions of urbanism for a post-colonial and post-industrial society in Chapter 8.

I E.1027: GRAFFITI BY CHARLES-EDOUARD JEANNERET (LE CORBUSIER)

> Le Corbusier explained to his friend that 'Badou' was depicted on the right, his friend Eileen Gray on the left; the outline of the head and the hairpiece of the sitting figure in the middle, he claimed, was 'the desired child, which was never born'.
>
> (Colomina, 1994: 84–8)

This reading of the mural *Graffite à Cap-Martin* is from Beatriz Colomina's *Privacy and Publicity: Modern Architecture as Mass Media* (1994), in which she cites Stanislaus van Moos, who quotes the new owner of E.1027, a villa designed by Eileen Gray for Jean Badovici at Cap-Martin in 1927–9. Gray had vacated the villa by the time Le Corbusier inscribed eight murals on its walls in 1938, though Colomina states they were made without Gray's permission and that she saw them as an act of vandalism. Colomina evaluates *Graffite à Cap-Martin* as 'a defacement of Gray's architecture and perhaps even an effacement of her sexuality' (Colomina, 1994: 88).[1] She adds that Le Corbusier claimed the villa as his own.[2] But Colomina is less concerned with rights to the ownership and integrity of a design than with Le Corbusier's invasion of Gray's space. Apart from a few reservations I note below, I accept Colomina's case; but a reading of the mural in conjunction with Le Corbusier's plan for Algiers – suggested by one of the sources for the mural, as I indicate below – leads me to put both in another, broader context: that of orientalism, for which I draw on Zeynep Çelik's 'Le Corbusier, Orientalism, Colonialism'.[3] While not discounting Le Corbusier's obsession with E.1027 and the lives of its occupants,[4] I suggest that orientalism offers a context in which his work as a whole can be viewed as conditioned by a masculine gaze that is entirely compatible with, has certain characteristics in common with, the voyeurism seen by Colomina in the mural. But orientalism, of course, is inseparable from colonialism, another terrain of the objectification of others.

The colonial frame introduces a more politicised approach. This is necessary because if, as Peter Hall says, the legacy of Le Corbusier in post-war planning is between the questionable and the catastrophic,[5] the difficulty is beyond personal obsession. It may be that, as Colomina argues, there is a fetishism in Le Corbusier's preparation for a figure composition he never makes and for which the mural is an interim statement,[6] and in his invasion of Gray's space, but perhaps Le Corbusier's urbanism, as in *The City of Tomorrow and its Planning*,[7] exhibits an equivalent gaze. Evidence of this is found in his feminisation of the landscape of Algiers. Perhaps in such cultural and political cross-currents it is possible to arrive at an explanation as to why Le Corbusier's seemingly utopian vision could have such a catastrophic impact if taken uncritically. It may be necessary, that is, to excavate strands which do not add up from Le Corbusier's writing and planning as well as his mural, to discern the utopian and the authoritarian (which are not always the same). This adds to Colomina's reading rather than competing with it, while avoiding speculation on the intimacies of Le Corbusier's mental life.[8]

To begin, then, with *Graffite à Cap-Martin*: the description cited above suggests two main figures, with a child between them. My reading differs, but I start with the two figures. That on the right is square-shouldered, almost male except for pendulous breasts; it is inscribed with a strange, rectilinear geometry hinting at a swastika (I do not know why), and has a prominent right thumb. The figure on the left, Gray in the above account, is fragmentary, like Picasso's female nudes of 1932–3. Its head is thrown back, the right arm bent above it in the pose of an odalisque – a pose often used by Matisse,[9] and by Ingres in *Odalisque with Slave* (1858, Louvre). This figure lies back with breasts upward and legs spread, the knees bent back in a sexually inviting position. It fits Carol Duncan's category of images attesting male virility in early Modernism:

> these paintings forcefully assert the virile, vigorous and uninhibited sexual appetite of the artist . . . [They] often portray women as powerless, sexually subjugated beings. By portraying them thus, the artist makes visible his own claim as a sexually dominating presence, even if he himself does not appear in the picture.
>
> (Duncan, 1993: 81)

Or perhaps the artist appears by proxy: if the right-hand figure which seems to switch gender is 'Badou' (I take as Badovici), does he stand in for Charles-Edouard Jeanneret? Leaving aside what can only be speculation, what of the third figure? A reading as the unborn child depends on the diminutive head and seems fanciful, whatever its source. On a closer view the head belongs to a woman seen in back-view, superimposed on the left-hand figure. This third figure has full, round hips and rests an arm on her ample thigh.

Colomina quotes van Moos citing the owner of E.1027: seems like a game of Chinese whispers which re-invents the message. But the discrepancy throws attention onto the image itself and its various sources, to emphasise, not a fantasised family life, but something between that and a fantasy of a harem.

The mural can also be seen beside early Modernist figure compositions in which the (usually female) figures are de-contextualised, set in spaces which are purely pictorial, autonomous domains without reference to location – such as Picasso's *Demoiselles d'Avignon* (1907, Museum of Modern Art). Picasso's figures are, as the title says, the inmates of a brothel in rue d'Avignon. But in their non-places such figures are universalised as nudes, as they were in Cézanne's bathers and Gauguin's Tahiti. In nineteenth-century orientalism, it is places from Morocco to the Ottoman Empire that meet the requirement for a non-place of the imagination, a utopia of sorts lent credibility by incorporation of visual clues derived as much from literary tales as from colonial histories. A case of this fusion of the real and the imagined is Delacroix's *Les Femmes d'Alger* (1833, Louvre).

Le Corbusier made repeated drawings and tracings from a postcard of Delacroix's painting, which he combined with original drawings made in Algiers, and postcards of women bought in the kasbah.[10] Given the mix of sources it is not surprising that comparing the mural to Delacroix's painting does not produce a correlation. Le Corbusier's figures are nudes while Delacroix's are clothed, and his right-hand figure borrows the profile head of the equivalent figure in Delacroix's painting but gives it a new body. While Delacroix separates his figures, Le Corbusier superimposes them, though the gesture of the hand on thigh noted above is found in Delacroix's left foreground figure. Delacroix's painting includes a fourth figure – a servant whose negritude emphasises the quasi-whiteness of the Arab women – which Le Corbusier omits. Colomina sees the mural as a fetishistic sign of Le Corbusier's 'abuse of Eileen Gray', while 'the endless drawing and redrawing is the scene of a violent fetishistic substitution that in Le Corbusier would seem to require the house, domestic space, as prop' (Colomina, 1994: 88).[11] But why does Le Corbusier, in 1938, use exotic sources, when Picasso, for one, drew on contemporary urban life? I think it follows from his visits to Istanbul in 1911 and Algiers in 1931. Colomina, citing Peter Adam, reads the mural as an image of the conquest of Gray: 'It was a rape. A fellow architect, a man she admired, had without her consent defaced her design' (Adam, 1987: 311 in Colomina, 1994: 355, n. 12). But there is enough evidence of a link to orientalism to suggest, too, that the image is multi-valent, and that one dimension of its meaning is in the orientalist tradition, where male sexual conquest goes in hand with colonial conquest, the women of the conquered nation being seen as sexually available in a way prevented by social conventions (within the bourgeois class) at home. If an orientalist reading is viable, it is a context in which Le Corbusier's plans for north African cities – Algiers and Nemours – can also be considered. This approach is supported by Çelik's reading of Delacroix's painting as 'a symbol of the conquest of Algeria' (Çelik, 2000: 327), which took place in 1833.

The story of orientalism, in which Delacroix's painting is an icon, could begin with Napoleon's expedition to Egypt in 1798 but includes earlier expeditions and topographic descriptions, and portrayal of individuals – such as Jean-Etienne Liotard's *A Turkish Woman and her Slave* (1742–3, Musée d'Art et de l'Histoire, Geneva) – which do not romanticise their subjects. It includes, too, Mozart's *The Magic Flute* (1793) from Tobias Gebler's novel *Thamos, King of Egypt* (1773). But it is in the nineteenth-century period of colonial expansion that description gives way to the creation of an Orient as counter to an Occident by a selective re-presentation of material according to the view of colonising soldiers and administrators, travellers and the artists and writers who manufactured the myth.[12]

For Napoleon, Egypt's attraction was that its monuments could, unlike the objects collected in ethnographic museums, be put beside those of Greece and Rome to bolster aspirations to imperial status. The monuments, and knowledge of them, are accoutrements of power's sublime cultural expression. This continues to inform French engagement in Egypt after Napoleon's defeat at Aboukir in 1801. An obelisk was sited at Place de la Concorde in 1832, for instance. There are travellers' stories, such as Flaubert's travel notes of 1850, and representations in art;[13] and an economic interest evident in the Suez Canal, a French engineering project that opened in 1875. The cultural appropriation of Egypt and the wider Orient was, however, highly coded.[14] Edward Said argues that the French appropriation of Egyptian antiquities followed production of texts and drawings which produced a context into which the antiquities themselves could be imported and be legible.[15] This leads me to see Le Corbusier's absorption of orientalism as the coding that, re-expressed in Modernist terms, makes his mural and plan for Algiers legible. But while Egyptian antiquities were made to speak the aspirations of the state, orientalism also offered a space of projection for more intimate fantasies, as in its imagination of the interior of the harem – a space to which European men were not admitted and which they could only imagine.[16] Its occupants are depicted in varying degrees of decorum in Ingres' *Odalisque* (1814, Louvre) and later *Odalisque with Slave* (1858, Louvre), Delacroix's *Les Femmes d'Alger* (1833, Louvre), Renoir's *Odalisque* (1870, National Gallery of Art, Washington), and Jean-Jules-Antoine Lecomte du Noüy's *The White Slave* (1888, Musée des Beaux-Arts, Nantes),[17] to give but a few cases. In the 1880s and 1890s the paintings are overtly erotic, in parallel to a renewed and compensatory French engagement in north Africa after the defeat at Sedan.[18] And it does not end there; in the early twentieth century, while the Cubists and Expressionists made sorties into ethnographic museums, Matisse, following visits to Algeria in 1906, the 1910 Munich exhibition of Islamic art, and Tangier in the winters of 1911–12 and 1912–13, extended the tradition of orientalism in a Modernist language of autonomous colour and form.[19] A photograph of him in his studio in 1928 shows him in dark suit, tie and white shirt with cuff-links, drawing a model dressed in baggy (Turkish) trousers reclining on a divan, the scene draped with patterned fabrics.[20]

Orientalism, then, was current during Le Corbusier's formative years and those of his key texts such as *Urbanisme* (1925). His visit to Istanbul in 1911 links him directly to this tradition, and its influence is seen in the Ottoman elements of his early work, notably Villa Schwob (1916); scenes of Istanbul appear, too, in *Urbanisme*. Islamic architecture continues to influence his designs into the 1950s.[21] The attraction of the Orient is, however, more than formal. For Le Corbusier, the orient offers both new forms and new, transgressive experiences:

> Le Corbusier was immersed in the discourse that attributed a lascivious sexuality to Islamic culture. This was one of the attractions that had drawn him to Istanbul in his youth. Re-enacting the scenes he had read of in books and had seen in paintings and repeating another favourite association between prison and palace, he fantasized about life in the seraglio, which would be filled with 'divine, thrilling odalisques . . . [wearing] around their naked ankles and arms . . . solid gold rings . . . like serpents . . .'
>
> (Çelik, 2000: 326)

This corresponds to the exoticism of Flaubert's travel notes from Egypt[22] or Gautier's *Le Roman de la Mômie*. Le Corbusier's guide in the kasbah, then aged eighteen and working for a French planner, later Curator of the Fine Arts Museum in Algiers, Jean de Maisonseul, recalls their visit:

> Our wanderings through the side streets led us at the end of the day to the rue Kataroudji where he was fascinated by the beauty of two young girls, one Spanish and the other Algerian. They brought us up a narrow stairway

> to their room; there he sketched some nudes on – to my amazement – some schoolbook graph paper with coloured pencils; the sketches of the Spanish girl lying both alone on the bed and beautifully grouped together with the Algerian turned out accurate and realistic; but he said they were very bad and refused to show them.
>
> (quoted in Colomina, 1996: 83)

These were the drawings that Le Corbusier merged with transcriptions from Delacroix. Çelik notes that prostitution was rife in the kasbah, encouraged by the French authorities while families not so engaged put notices to that effect on their houses, or dressed their daughters in European clothes.[23] Le Corbusier bought postcards of Algerian women, too, de Maisonseul expressing surprise that he wanted such vulgar images, though the vulgarity may have been in the colour reproduction as much as the subject-matter.[24] The experience of the kasbah seems to fit the construction of an orient offering, as Said says, 'Sensuality, promise, terror, sublimity, idyllic pleasure, intense energy' (Said, 1991: 118), in which Le Corbusier could play the *flâneur* abroad, finding there perhaps furtively an eroticism not on offer at home.

In the 1920s and 1930s, French north Africa was a site of economic expansion. Colonial plans combined city extensions in European style with the selective preservation of Arab quarters. The latter were valued not as habitats but for the economic benefits their picturesque qualities could generate through tourism.[25] Algiers presented its face to arriving travellers as a European port, a terrace of four-storey arcaded buildings like those of a European city standing above the waterfront. At one level, Le Corbusier uncritically extends this colonial policy. At another, the Orient becomes not just a site in which to consume exotic others (or their images) but also on which to project his vision of a new urban world. As the caption to an aerial view of his project for a new town at Nemours (1934, also in north Africa) states: 'The scheme controls the entire development of the new town for a population of 38,000 European inhabitants and ensures rational and healthy future extension' (Martin, Nicholson and Gabo, 1971: section III, plate 15). Looking at the plan, a European quarter of huge white blocks connected by sweeping boulevards contrasts with an indigenous area confined to a few houses next to the port. Next to the native quarter are gas, electricity and water plants, and between them and the European quarter an industrial zone flanked by a broad communal zone of empty space. On the other side of the port, adjacent to the European quarter, are recreational areas and a sports stadium (Martin, Nicholson and Gabo, 1971: section III, plate 16). The segregation is obvious and the position of the Arab population is marginal.

Le Corbusier, like Marshall Lyautey (Governor of Morocco) saw urbanisation as shaping the lives of dwellers. For Lyautey it was even an agency which could replace military force;[26] both believed in a segregation of European from Arab populations,[27] as in Le Corbusier's Nemours. According to Lyautey:

> Large cities, boulevards, tall façades . . . upset the indigenous city completely, making the customary way of life impossible. You know how jealous the Muslim is of the integrity of his private life . . . the narrow streets, the façades without opening behind which hides the whole of life, the terraces upon which the life of the family spreads out . . . sheltered from indiscreet looks.
>
> (Abu-Lughod, 1980, quoted in Çelik, 2000: 323)

This was the ethos applied in Algiers, and discussed in 1931 at an International Congress on Urbanism in the Colonies. At times it sounds liberal, seeking to preserve indigenous quarters, but this was for pragmatic reasons.[28]

In the plan for Algiers, on which Le Corbusier worked with his son Pierre Jeanneret, the city's redevelopment as the French capital of north Africa was to be the culmination of a century of occupation.[29] A central boulevard, under which were to be homes for 180,000 (white) people in new blocks, cuts through a landscape Le Corbusier viewed as 'a magnificent body, supple-hipped and full-breasted . . . A body which could be revealed in all its magnificence . . . through the judicious influence of form and the bold use of mathematics to harmonize natural topography and human geometry' (Çelik, 2000: 326).[30] The boulevard runs between massive, curved white apartment blocks (also for white people) on the hillside, an organic departure from previous tower forms, and links them to a massive office block at the waterfront. Dealing with the overcrowding of the kasbah that resulted from inward migration from rural areas, and seeking to sanitise what in the view of the colonial administration was a site of undifferentiated otherness,[31] Le Corbusier's plan preserves the upper section of the kasbah with reduced habitation density by selectively converting buildings from residential to cultural uses – a museum without walls. The rest of the kasbah was to be cleared, apart from a few mansions which were to become museums, the nameless poor presumably fleeing to peripheries as usual. The resulting open spaces act as a cordon between European and residual Arab zones.

In *Aircraft*, published in 1935, a caption to a photograph of a gas station in the desert asserts the following without irony: 'The white race goes its conquering way. The filling station is a symbol of white civilization' (Le Corbusier, 1987b: plate 107). Is the plan for Algiers (1931) a colonial escapade, or a re-application of the *Voisin* plan (1925) for Paris?[32]

Is his pursuit of a conquering way the flaw in his legacy to European Modernism? To approach this I look in the second part of this chapter at the viewpoint from which Le Corbusier sees the city, and at his political associations; I seek ambivalences in his utopianism, and finally contrast it to an alternative Modernism in Hassan Fathy's work in Egypt in the 1940s.

II A MODERN UTOPIA, ANOTHER MODERNITY

In *Aircraft*, an aerial view of Le Corbusier's plan for Algiers appears below the lines 'Sweep away the refuse with which life is soiled, clogged, encumbered.

Let us undertake the great tasks of the new machine civilization' (Le Corbusier, 1987b: plate 110). On the previous page, above a photograph of demolition on Boulevard Haussmann,[33] is the heading 'Cities must be extricated from their misery, come what may. Whole quarters of them must be destroyed and new cities built' (plate 108), which echoes a passage in *Urbanisme*, section VII, 'The Great City':

> Therefore my settled opinion, which is quite a dispassionate one, is that the centres of our great cities must be pulled down and rebuilt, and that the wretched existing belts of suburbs must be abolished and carried further out; on their sites we must constitute . . . a protected and open zone, which when the day comes will give us absolute liberty of action.
>
> (Le Corbusier, 1987a: 96)

Plans and aerial views of cities illustrate a history in which 'a preconceived and predetermined plan embodying the then known principles of science' is contrasted with 'the pack-donkey's way' of piecemeal growth in Paris (Le Corbusier, 1987a: 91–2). There is the *tabula rasa* on which Descartes' engineer imagined regular places;[34] and a futuristic vision enabled, as Le Corbusier's use of illustrations of cars, aircraft and industrial machines attests, by new technologies. To dispassionately sweep away the past is a Cartesian gesture,[35] but here it ceases to relate to cognition and becomes a cleansing, a sweeping away that, far from being dispassionate, is as passionate as the rhetoric of Italian Futurism – a parallel suggested, too, by a passage in *Urbanisme* in which Le Corbusier recalls being in the Champs Elysées surrounded by cars. His moment on the road to Damascus comes when he sees that machine power is the answer as well as the problem:

> I was overwhelmed, an enthusiastic rapture filled me. Not the rapture of the shining coachwork under the gleaming lights, but the rapture of power. The simple and ingenious pleasure of being in the centre of so much power, so much speed. We are part of it. We are part of that race whose dawn is just awakening.
>
> (Le Corbusier, 1987a: xxiii)

In his first manifesto, published in *Le Figaro* in 1909, Marinetti relates his experience with a group of friends, racing their cars at dawn when they hear the tramcars, his car overturning in a ditch but running again at the touch of his caress. Then: 'We affirm that the world's magnificence has been enriched by a new beauty: the beauty of speed' and a racing car is more beautiful than the *Victory of Samothrace* (quoted in Harrison and Wood, 1991: 147). Marinetti wrote this sixteen years before publication of Le Corbusier's *Urbanisme*, yet there is the same enthusiasm for the dynamic and radically new, and a complete rejection of the past. It is a dangerous enthusiasm which, after the 1914–18 war in which architect Sant' Elia-Mario Chiattone was killed, turns into alignment with fascism. Marinetti even wrote a eulogy for Mussolini.[36]

In France an emergent fascism, which was founded on November 11th, 1925 at the tomb of the unknown soldier, also reacted to the war. It was a catastrophic victory in terms of the death-toll,[37] but the Versailles Treaty afforded reparation for the defeat at Sedan and war indemnity imposed by Germany in 1871. As in Italy, there was a sense of national resurgence. The movement – *Le Faisceau* – grew to 60,000 members but did not become a platform for mass action, and was dissolved by its founder, George Valois, in 1927.[38] Valois saw modernisation as a way to re-order society on the model of Taylorism in North America.[39] Adopting a technocratic solution for the problem of scarcity, which had preoccupied nineteenth-century anarcho-syndicalists,[40] Valois described Le Corbusier's plans as 'an expression of our profoundest thoughts' (Antliff, 1997: 137),[41] featured him as an *animateur* of the new France on the front page of *Nouveau siècle* (January 9th, 1927), and used part of the Voisin plan in *Nouveau siècle* (May 1st, 1927).

I refrain from the easy option of labelling Le Corbusier a fascist. There are factors that separate him from it;[42] as James Donald writes, Le Corbusier's taste for polemic 'led him to say some fairly daft things . . . to flirt with odious political regimes if he thought it would help him get his buildings built' (Donald, 1999: 54–5). Mies van der Rohe, too, seems to have been as content to design a pavilion for the 1929 Barcelona International Exhibition as to make proposals for a building for the Nazis. It is more interesting to ask to what extent an attraction to technocracy on the part of authoritarian regimes, and of their supporters such as Le Corbusier and Valois, results from an inherent characteristic of the technocratic ethos. Is there something decidedly not value-free in the expertise with which technocrats justify their authority? Is there some logic in the way that authoritarian regimes use a value-free, scientific approach to mask value judgements?

The legacy of Le Corbusier to post-war planning may, then, be evaluated in the context of his attraction to Taylorism as well as to a centralised state, the two being linked, and the model of which is the colonial regime of military (for which in Taylorism read economic) imperative – the *only* state able to enact urban change on the scale he envisaged. This leads me to ask whether there are common if coincidental assumptions between a fascist appropriation of technology in the 1920s and 1930s and the rational planning model of 1950s North American urban planning.[43] The planners, Robert Moses in New York, for example, were of course liberals. But while they believed in a beneficial technology, they saw its management as being reserved to a technocratic elite who were the only actors in the situation without self- or vested interest and who were able to solve a city's problems.[44] There is a parallel here to Donald's critique of Le Corbusier's concept of the city *as a problem to be solved* rather than a set of conditions in which dwellers and others negotiate possibilities. Le Corbusier, despite the one-dimensionality of the technocracy model, still sees things from several contradictory viewpoints.[45] Renata Salecl draws attention to a confusion of external necessity with personal vision:

> For Le Corbusier, his project was thus a fulfilment of the demands coming from some greater order, the principles of industrial society. These principles were . . . the big Other for whom his fantasy was staged; or, more precisely, Le Corbusier had posited this principle in the place of his Ego Ideal from where he then observed himself in the way he wanted to be seen, as a dutiful creator who would make reality accord with the Ideal.
>
> (Salecl, 1999: 107–8)

Which to me seems an exact and informative criticism, linking Le Corbusier to an authoritarian mentality delineated by Adorno in his efforts to understand how it was possible for Nazism to come about, to operate with mass support and the complicity of millions.[46] Salecl sees two comparable cases: Ceauşescu's new Bucharest and Disney.[47]

The latter, however, are extreme cases, each terrorising in its way. Le Corbusier's participation in the Congrès Internationaux d'Achitecture Moderne (CIAM) situates him in a rational forum uniting architecture and planning. But it is difficult to extricate Le Corbusier's influence from that of other Modernists in CIAM, founded in 1928, such as Mies van der Rohe and Gropius, and from the wider context of urbanism in the inter-war period. What can be done is to discern in that context a functionalism that continues in post-war developments. New urban environments were seen as a means to a new and better world, but on the assumption that those who inhabit the new environments produced by advanced planners and architects – an avant-garde in full possession of a vision – are incapable of self-organisation. This is the lesson of the Thamesmead Estate in south-east London, for instance, built with adequate resources as a demonstration of innovative thinking, yet a sterile environment in which the functions of everyday life are allocated separate spaces, and the spaces planned for public mixing are voids because there is nothing to mix for. Edward Robbins contrasts this with the uses of the inner city street as a multi-layered extension of living space, and sees the scheme as embodying an ideology which is on examination disempowering and anti-urban.[48] This attitude is encountered in the nineteenth-century construction of model villages such as Port Sunlight and Bourneville, and in the Amsterdam School's liberal-progressive housing schemes in the south of the city during the 1920s. New flats for low- and middle-income families provide clean, well-designed spaces, with tree-lined streets and shared courtyards, in a variety of styles. But the interiors determine a certain way of living; cooking space is separated from eating and sitting spaces, with no balconies for hanging out washing.[49] Other such schemes of the inter-war period include Karl Ehn's 1930 Karl Marxhof in Vienna, and the Weissenhof housing settlement in Stuttgart (1927), directed by Mies van der Rohe.[50] There were, too, other mega-plans apart from Le Corbusier's, such as Ludwig Hilbersheimer's (1927) proposals for Berlin. Le Corbusier's legacy, then, must be seen within this varied terrain.

CIAM's fourth meeting took place on a boat on the Mediterranean in 1933 and produced a set of principles that informed urbanism in the post-war period. The CIAM ethos included provision of high-density urban housing, demarcation of discrete urban neighbourhoods with separate zones for living, working and recreation, and development of highways for rapid transport. The eighth congress took place at Hoddesden, England, in 1951, by which time the cleansing operation of Modernism was over.[51] Yet other assumptions were maintained. Although there was, as Barry Curtis examines, much talk of spaces for democracy, these were to be like the formal (bourgeois) public open spaces characteristic of nineteenth-century cities. Mass communications, which today open new forms of public space as resistance becomes globalised through the internet, were seen in CIAM as a threat, a cause of public acquiesence.[52] So, too, were speculators.[53] There were counter-arguments in CIAM,[54] but the dominant view was of a universal modern male citizen inhabiting those public spaces. In the early 1960s, the more citizen-centred proposals of Jane Jacobs and Kevin Lynch might have seemed refreshing departures from this universal accord, this *Pax Romana* of the universal (male, white, bourgeois) citizen given form in technically advanced solutions.

The universal perspective has a long history. I will try next to draw out four interconnecting but different aspects of it: a naturalising technocracy; a narrative of cleansing; a privileging of visuality; and the gendering of such viewpoints. To bring them together in a perhaps simplistic way, for illustration, the purpose of technological progress is to order the world, to cleans it of dirt (things out of place); the project can be undertaken because a view is afforded of the whole, reduced not only to a set of signs, a representation, but also to a unity, like a city skyline; this viewpoint from which all, including tomorrow, can be seen, is visual and distancing, an overview which is also a viewpoint of masculinity and power – the latter the term which runs through all this. Yet when I say all this, as if I can see it all, I fall into my own trap.

A naturalising explanation of the conditions of a city is given by E. W. Burgess in 'The Growth of the City: An Introduction to a Research Project' (1925).[55] Burgess introduces the concentric ring diagram for which he is known, and sees types of natural metabolism as a metaphor for the transitional states of zones following waves of migration into a city. The biological metaphor puts the process outside history just as the universalism of the diagram puts city form outside intervention, and replaces a sense of cities with that of *the* city.

The Chicago School to which Burgess, Robert Park and Louis Wirth were the major contributors pioneered this approach which evolved into the rational comprehensive planning model[56] of the 1950s, in which decisions are made by professionals on the basis of data to which they have privileged access. Le Corbusier, however, is a technocrat with 'attitude', and tends to see the efficient management of space as having more than a technical, formally soluble dimension. It has a moral aspect: 'We struggle against chance, against disorder, against a policy of drift and against the idleness which brings death; we strive for order, which can be achieved only by appealing to what is the fundamental basis on which our minds can work: geometry' (Le Corbusier, 1987a: 93). For Descartes geometry represented hope of certainty in face of doubt. Here it is a prescription. Reading Le Corbusier's texts I see authoritarianism masked as a celebration of technology:

> There is a degree of error that cannot be exceeded. It is the moment when the conditions which have plunged persons and society into apathy, misery, and misfortune, must be revolutionized. The brief and rapid history of aviation . . . explains to us the hostile elements surrounding us, and provides us with the certainty that soon the very laws of life will justify us.
>
> (Le Corbusier, 1987b: 11)

The laws of life seem messengers of an apocalypse, and there is a note of Calvinism in the fear of excess and aversion to present conditions; but I think this also reflects a specifically European history of urban exclusions and confinements. It is evident as follows: in the institution of the *Hôpital Général* (1656) in which the non-productive vagrant and insane were confined and excluded from visibility in the street; in the removal of graves from churchyards to peripheral cemeteries in the eighteenth century; in the fear of another miasma in the odours of the poor in the nineteenth; in the fear of otherness in white North American suburbia; and in the continuing treatment of travellers and gypsies.[57] If strategies for public order are reactions to fear, there is an equivalent personal reaction in a clinging to the known, to order as public safety. This is where authoritarian solutions appeal. Richard Sennett writes: 'The work of authority has a goal: to convert power into images of strength. In doing this work, people often search for images that are clear and simple. The search for clear and distinct images of authority, however reasonable, is dangerous' (Sennett, 1980: 165). Sennett quotes Mussolini, but I want to link this notion of authority, too, to a specific point of vision. For Le Corbusier, imagining himself in the air, the vantage point above legitimates the desire for purification. In *Aircraft*, he writes: 'We desire to change something in the present world. For the bird's eye view has enabled us to see our cities . . . and the sight is not good' (Le Corbusier, 1987b: 11). The cities are tanneries that make people serfs. And in the introductory text for *Aircraft*: 'The airplane is an indictment . . . By means of the airplane, we now have proof, recorded on the photographic plate, of the rightness of our desire' (Le Corbusier, 1987b: 11). Aircraft appear, too, in *Towards a New Architecture*,[58] with cars such as the Voisin sports torpedo (made by the sponsors of the Voisin plan). The proof confirms a preconstructed knowledge. As orientalism provided the code through which to read an Orient, so the aerial photograph reproduces the gaze established in cartography.

An analogy can be made with tourist photography: John Urry sees images in brochures and television programmes as producing a preconception of a scene that is re-encountered, far-away but familiar, when tourists seek out the views they know; the photograph is the proof of the correspondence as well as of the visit.[59] The aerial photograph similarly corresponds to a view of the city known before flight in the conventional city plan, which adopts a viewpoint in the sky in which, later, aircraft fly. It is the viewpoint of power from which a new city can be drawn on paper and then inscribed on the surface of the land, or re-inscribed on an extant city. A systematisation of this view occurs in Alberti's use of a device to measure a city's streets from a vantage point on the surrounding circuit of walls, and its sense of power, gendering and purification permeate Alberti's text on architecture.[60] The overview is only enhanced by aerial photography, more so when montage and the moving image are introduced, but is not created by it.[61] Incisive critiques of this distancing perspective are found in feminist geographies and cultural criticism. Doreen Massey writes of visuality as an authoritative, masculine way of seeing, which gives mastery, diminishes our awareness of other senses, and states detachment: 'Such detachment, of course, can have its advantages, but it is also necessarily a "detached" view from a particular point of view. Detached does not here mean disinterested' (Massey, 1994: 232).[62] And Marsha Meskimmon writes that western knowledge systems traditionally privilege sight 'as the most perfect and truthful sense and the one best suited to rational knowledge claims'; she continues that paradigms of sight underpin the concepts of universal knowledge and the rational subject who knows (sees) objectively: 'If the whole . . . can be seen completely and with total objectivity, its truth will be revealed' (Meskimmon, 1997: 17–18).[63]

In the film he made with Pierre Chenal, Le Corbusier drives to the entrance of the Villa Stein à Garches. Wearing a dark suit and bow tie, his hair oiled, smoking a cigarette, he enters the house; in the roof garden, a boy plays with a toy car and women sit; Le Corbusier reappears on the far side of the terrace and climbs the stairs to the lookout point.[64] He does not speak.

But what else? If the Modernist intention to create a better world in accordance with rational principles were to be delivered through what are, to my mind, denials of reason in uncritical assumptions affirmed by a perspective of male desire and masculine power, are there alternatives or is the concept of building a better world inherently reactionary? I wonder if the question hangs on the word building, whether the separation of concept from process, design from making, in a division of labour between designer and builder, between planner and dweller, is part of the difficulty. Perhaps this is where the flaw in the concept of an avant-garde is most clearly evident, in the desire to lead society upward, onward, in a way which retains power (and expertise) in the hands of those leading, does not transfer it to those being led who become only nominal beneficiaries of change, their state remaining one of disempowerment.

Seeking an alternative, I look now to Hassan Fathy's experimental village of New Gourna, in the agricultural area of the west bank of the Nile opposite

Luxor.[65] Fathy was employed by the Egyptian Department of Antiquities from 1945 to 1949 to supervise the construction of New Gourna, to which it was intended to move *en masse* the inhabitants of the old village, whose houses stood over tombs in the Theban necropolis. Their incomes came from selling things found in tombs, and employment in archaeological sites. By the 1940s, the level of thefts (or finds from another viewpoint) had reached proportions unacceptable to the government, and it was decided to resite the villagers. This undermines the project at the outset, and the Gournis in any case refused to go; most still live in the old village, in a structure of extended families.

Fathy was, in any case, the professional from the city, arriving at Luxor station with his gramophone, a product of the Beaux-Arts tradition, though he harbours sweet memories of childhood visits to rural Egypt.[66] His role is in part to civilise. He writes of the peasant house as holding 'a large variety of bulky stores and the owner's cattle as well . . . [with] hens running in and out among the dust and babies' (Fathy, 1973: 92).[67] Although he went to some lengths to persuade the local population to work with him, seeking their knowledge of social structure and employing them as labour, he saw villagers as unable to think conceptually: 'The Gournis could scarcely discuss the buildings with us. They were not able to put into words even their material requirements in housing' (Fathy, 1973: 40). At the same time he used his status to cary out the scheme despite opposition from government offices (where the import of materials and technologies provided income), seeing in the use of mud brick and traditional skills, which could be found or developed within a local population, a 'no-cost solution to Egypt's housing problem' (Fathy, 1984: 16).[68] Today, after the building of the Aswan High Dam has stopped the annual inundation, there is no renewal of the supply of mud, but the proposal for a vernacular which radically reduces building costs and transfers most of the work to local masons, and actively invites dwellers' participation, seems a model which could be widely adapted for other circumstances, as in self-build housing schemes.[69] The credibility of the material for a site such as Gourna is enhanced by the survival of mud vaults in the granaries of the nearby Ramasseum (XVIIIth dynasty) – the methods used by the masons Fathy brought to Gourna from Aswan do not differ in any major way from those of pharaonic times – but mud is only one more or less free material. There are many others, from straw to old car tires and rubble. The architectural style of New Gourna is derived from mosques and tombs in the Aswan area (not exactly from Luxor), and the use of courtyards may have been inspired by medieval houses in Cairo.[70] But Fathy's role was more to draw up a plan for the village and act as intermediary with higher and remote authorities than to design in detail – a matter for the masons.[71]

The plan reflected Fathy's research in the old village, where each family group (*badana*) had its own cafe, barber and grocer, and shared the use of baking ovens as well as participating together in feasts for marriages and male circumcisions. The new village consists of several groups of houses (a quarter of those planned) sited irregularly in quarters reflecting the old family groupings, around small squares linked by streets which have blind corners to discourage strangers. There is a mosque, a theatre, a market and – the only concrete building – a recent government school.[72] The mosque also has a small schoolroom for Koranic studies. Thresholds of public and private space follow a gradation from the large square before the mosque to streets, shared courtyards, houses, benches outside front rooms where men drink tea and rooms where women spend their time. Fathy does not interrupt the traditional gendering of spaces in the village, and perhaps romanticises it in a feminisation of the space of the courtyard, in its way equivalent to Le Corbusier's of Algiers:

> the courtyard is more than just an architectural device for obtaining privacy . . . It is, like the dome, part of a microcosm that parallels the order of the universe itself . . . the four sides of the courtyard represent the four columns that carry the dome of the sky. The sky itself roofs the courtyard and is reflected in the customary fountain in the middle.
>
> The inward-looking arab house, open to the calm of the sky, made beautiful by the feminine element of water, self-contained and peaceful . . . is the domain of woman . . .
>
> Now it is of great importance that this enclosed space with the trembling liquid femininity should not be broken. If there is a gap in the enclosing building, this special atmosphere flows out and runs to waste in the desert sands.
>
> (Fathy, 1973: 57)

But – differentiating Fathy from Le Corbusier – this takes place within a colonial situation in which, allied to a folk tradition in Egyptian visual culture in the 1940s,[73] Fathy's traditionalism is a reclamation of national culture, a disavowal of all that is European. He tirades against the effects of importing western technologies and styles, and sees tradition not as a binding aesthetic but as something constantly reinvented in everyday life.[74]

New Gourna was left unfinished in 1948, squashed by government delays and obstruction. The first inhabitants were squatters displaced by the High Dam and flooding of villages under what is now Lake Nasr. If Fathy's affirmation of tradition echoes a romantic image of rural life, his social concern seems genuine and practical, and his vernacularism radical.[75] While Fathy counterposes tradition to modernity, however, he remains, for me, a Modernist in as much as his guiding principle of facilitating a no-cost solution to the housing problem of a non-affluent country is utopian. But his Modernism departs from the European and North American project to engineer a new society by design in his relinquishing of the conventional role of the architect, his use of local knowledges and materials, and alignment with a traditionalism based in material culture. It is, foremost, an architecture *by* as well as for the poor, in that respect a utopian *possibility*.

NOTES

1 Colomina cites van Moos, S. (1980) 'Le Corbusier as Painter', *Oppositions*, 19–20; and Rafi, S. (1968) 'Le Corbusier et les femmes d'Alger', *Revue d'histoire et de civilisation du Maghreb* on the mural; she describes Rykwert, J. (1972) 'Eileen Gray: Pioneer of Design', *Architectural Review* December, pp. 357–61 as the first critical recognition of Gray.

2 Citing Adam (1987), Colomina states that photographs of the mural in Le Corbusier's (1948) *L'Architecture d'aujourd'hui* do not credit Gray; other publications call it Maison Badovici or credit it to Le Corbusier; *Casa Vogue* 119 (1981) credits it to Eileen Gray and Le Corbusier, and Gray's sofa is called a unique piece by Le Corbusier (Colomina, 1994: 355, n. 13).

3 Çelik's essay appeared in *Assemblage*, 17, April 1992. I use the version in Rendell, Penner and Borden, 2000: 321–31, and reference it as Çelik (2000). See also Çelik (1986) *The Remaking of Istanbul: Portrait of an Ottoman City in the Nineteenth Century*, Seattle, University of Washington Press; Çelik (1992) *Displaying the Orient: Architecture of Islam at Nineteenth-Century World's Fairs*, Berkeley, University of California Press; and Kinney and Çelik (1990) 'Ethnography and Exhibitionism at the Expositions Universelles', *Assemblages* 13.

4 After 1945 Le Corbusier built a cabin overlooking E.1027 from the edge of the adjacent property: 'He occupied and controlled the site by overlooking it, the cabin being little more than an observation platform' (Colomina, 1994: 88).

5 'The evil that Le Corbusier did lives after him; the good is perhaps interred with his books, which are seldom read for the simple reason that most are almost unreadable . . . [their] impact on twentieth-century city planning has been almost incalculably great . . . Ideas forged in the Parisian intelligentsia in the 1920s, came to be applied to the planning of working-class housing in Sheffield and St Louis . . . in the 1950s and 1960s; the results were at best questionable, at worst catastrophic' (Hall, 1996: 204). The reference to St Louis is to the 1956 Pruitt-Igoe project, a development of social housing in 33 identical towers designed by M. Yamasaki, which was demolished in 1972 (Hall, 1996: 235–40). If the good is in the books, why is their influence (as the repository of Le Corbusier's urbanism, distinct from that of his villas) catastrophic? Donald is more careful: 'a perplexing and controversial figure . . . too easily caricatured as the wicked wizard of hubristic modern planning' (Donald, 1999: 52), citing Lefebvre: 'a good architect but a catastrophic urbanist' (Lefebvre, 1996: 207), perhaps what Hall meant.

6 Colomina cites van Moos (1980) p. 89 on a projected figure composition: 'the plans for which seem to have preoccupied Le Corbusier during many years, if not his entire life' (Colomina, 1994: 84).

7 I use the 1987 Dover reprint of the 1929 translation by Frederick Etchells, from the 8th edition (1929) of *Urbanisme*, first published in Paris in 1927.

8 My approach is similar to Duncan's: 'It is also relevant to ask whether these artists sought or achieved such relationships in reality, whether their lives contradict or accord with the claims of their art. *But this is not the question I am asking here*. My concern is with the nature and implications of those claims . . . in the art and as they entered the mythology of vanguard culture' (Duncan, 1993: 81, my emphasis).

9 *Blue Nude, Souvenir de Biskra* (1907, Baltimore Museum of Art); *Pink Nude* (1935, Baltimore Museum of Art); and *Odalisque with Tambourine* (1926, New York, Paley collection). A related pose is seen in Ingres' *The Turkish Bath* (1862, Louvre), in which a figure reclines with both arms above her head.

10 'Le Corbusier seems to have produced hundreds and hundreds of sketches on yellow tracing paper by laying it over the original sketches and retracing contours of the figures. He also studied exhaustively Delacroix's *Femmes d'Alger*, producing a series of sketches of the outlines of the figures in this painting, divested of their "exotic clothing" . . . Soon the two projects merged . . . He kept redrawing it. That the drawing and redrawing of these images became a lifetime obsession would have been enough of an indication that something was at stake. This becomes even more obvious when in 1963–64 . . . [he] copies a selection of

these sketches onto transparent paper and . . . burns the original sketches' (Colomina, 1994: 84).

11 Colomina cites Freud's argument that fetishism derives from the absent maternal phallus; and quotes from Burgin's *The End of Art Theory*: 'In fetishism, an object serves in place of the penis with which the child would endow the woman . . . Fetishism thus accomplishes that separation of knowledge from belief characteristic of representation; its motive is the unity of the subject . . . The photograph stands to the subject-viewer as does the fetishized object' (Burgin, 1986: 44, in Colomina, 1994: 91).

12 Gautier's 'Art in 1848', first published in *L'Artiste* series V, vol. 1, Paris, May 15th, 1848, states: 'What an endless field of possibility is today open to the artist! Past is the age of three or four Greek and Roman ideals . . . The mysterious Orient is finally becoming accessible, lifting a corner of its veil; those beautiful, pure faces, so calm and dreamy, pale and fresh from the cool shadow of the harem, or bronzed to coppery gold by the fiery sun, faces which previously blossomed in a forgotten, secret solitude, leaving no silhouette in our memory . . . This unknown world, prevented by iconoclastic Islam from translating its thought into shape and colour, thanks to the travel of our artists, is beginning to become familiar to us' (in Harrison and Wood, 1998: 317).

13 Among French artists who travelled in the Orient are Gros (with Napoleon in Egypt, 1798), Decamps (Smyrna, 1828), Delacroix (Morocco and Algeria, 1832–3), and Gérôme (Turkey, Egypt, Palestine, Syria, north Africa, 1854).

14 Miller, referencing Said (1978) emphasises the construction of an Orient as counter to an Occident: 'As Said makes clear, one result of this is that in so far as the Occident actually deals with the Orient it assumes that reality to be an actual expression of the model it has constructed of it, and treats it accordingly' (Miller, 1991: 58).

15 'Egypt had to be reconstructed in models or drawings, whose scale, projective grandeur . . . and exotic distance were truly unprecedented . . . First the temples and palaces were reproduced in an orientation and perspective that staged the actuality of ancient Egypt as reflected through the imperial eye; then . . . they had to be made to speak, and hence the efficacy of Champollion's decipherment; then, finally, they could be dislodged from their context and transported to Europe to use there' (Said, 1994: 142). For Said, this is a specifically French attitude, from the lack of an overseas empire to compare with Britain's. See also Said, 1991: 42–3 and 80–8. Champollion's *Précis*, a grammar of hieroglyphs, was published in Paris in 1824 and *Description de l'Egypte* (24 volumes) describing Napoleon's expedition was published in the 1820s (Said, 1994: 37–9).

16 Verrier calls such scenes 'a convenient fiction for the portrayal of titillating nudes with the added spice of Eastern eroticism' (1979, not paginated). The reclining female nude or semi-nude, often with one or both arms behind the head, called odalisque is not confined to representations of oriental figures: Delacroix's *Woman in White Stockings* (1832, Louvre) depicts a European model, seen by Duncan (1993: 109–10) as evoking a male fantasy of sexual confrontation in which the position of the model's arms above the head is an attitude of surrender.

17 Richon (1996) sees Lecomte de Noüy's painting *Rhameses in his Harem* (1885), based on Gautier's novel *Le Roman de la Mômie* (1858), as 'an academic painting which represents the most repeated theme of Orientalism: the despot and his harem' (p. 247). Richon emphasises that the Orient, as an entity, exists only for a European gaze; and while the scene represents masculine power, it also represents 'that which cannot be seen, as the harem prohibits any foreign look' (p. 250).

18 Said notes a growth in French geographical societies, and a renewed demand for territorial expansion as a sign of national resurgence as scientific geography gives way to economic interest: 'Much of the expansionist fervour in France during the last third of the nineteenth century was generated out of an explicit wish to compensate for the Prussian victory in 1870–71 and . . . desire to match British imperial achievements' (Said, 1991: 218).

19 Among paintings produced during these visits are *Park in Tangier* (1912, National Museum, Stockholm), *Zorah on the Terrace*

(1912, Pushkin Museum) and *Entrance to the Casbah* (1912, Pushkin Museum). His summation of his north African experiences is *The Moroccans* (1916, Museum of Modern Art).

20 In Jacobus (1973), fig. 37. The model's pose loosely resembles that of the central figure in *The Siesta*, a pencil drawing of 1928 (fig. 90). See also fig. 38 *Odalisques* (1928, private collection); fig. 39 *Moorish Woman* (1922, Barnes Foundation); and fig. 40 *Odalisque with Tambourine* (1926, private collection).

21 In *Urbanisme* (translated as *The City of Tomorrow and its Planning*), Le Corbusier includes sketches of Istanbul, admiring the trees and 'noble examples of architecture' (Le Corbusier, 1987a: 62–4 and 77–9). Çelik cites Le Corbusier's travel notes (published as *Journey to the East* (1987), trans. Zaknic, I., Cambridge, Mass., MIT) and notes similarities between Notre Dame de Ronchamp (1950–5) and the Sidi Ibrahim mosque, el Ateuf, Algeria (Çelik, 2000: 321, 328–9, n. 1 and 8).

22 Flaubert writes of a visit (in 1850) with his travelling companion Maxime du Camp to the dancer Kuchuk Hanem's house (March 6th, 1850): 'She had just come from the bath, her firm breasts had a fresh smell, something like that of sweetened turpentine; she began by perfuming her hands with rose water.' After sex there is a musical entertainment: 'She squeezes her bare breasts together with her jacket. She puts on a girdle fashioned from a brown shawl with gold stripes, with their tassels hanging in ribbons.' They return in the evening: '*Coup* with Safia Zugairah – I stain the divan. She is very corrupt and writhing, extremely voluptuous. But the best is the second copulation with Kuchuk' (Flaubert, 1983: 113–17).

23 See Çelik, 2000: 331, n. 45. Çelik cites Rafi (1968) and Gordon (1968), with Lucienne Favre's *Tout l'inconnu de la casbah d'Alger* (1933, cited by Çelik, publication details not given).

24 Colomina reproduces a card from the series *Scenes et Types* (Colomina, 1994: 92). Çelik (1992) quotes a letter of de Maisonseul's (1968): 'horrible . . . raw colours, pinks and greens, representing *indigènes nues* in an oriental decor' (Çelik, 2000: 331, n. 46). See also Said, 1994: 133–4 and 416, n. 78.

25 Çelik cites Marshal Lyautey (Governor-General of Morocco (1915–25), under whom Rabat, Fez and Casablanca were extended using plans drawn up by architect Henri Prost) on the charm and poetry of the Arab town (Çelik, 2000: 323–4). The tourist potential of Algiers is confirmed by my grandfather's visit there with his family in 1935. A photograph in a family album (taken by my mother, then aged 14) shows the colonial-style buildings of the port and the native quarter on the hill behind.

26 'A construction site is worth a battalion' (attributed to Lyautey, Çelik, 2000: 322). But also: 'Le Corbusier wrote that the effect of this plan on Algiers would be like artillery shells' (Barnett, 1986: 121). Çelik cites Rabinow (1989), Abu-Lughod (1980) and Wright (1991) on Lyautey in Morocco, and notes that in 1931, Lyautey claims that Islam gave him 'a taste for great white walls' and that in this he 'could almost claim to be one of the forerunners of Le Corbusier' (from Vigato, 1986: 28–9, in Çelik, 2000: 330, n. 9).

27 'The colonial world is a world divided into compartments. It is probably unnecessary to recall the existence of native quarters and European quarters . . . Yet if we examine closely this system of compartments, we will at least be able to reveal the lines of forces it implies' (Fanon, 1967: 29).

28 'In Algeria, however inconsistent the policy of French governments since 1830, the inexorable process went on to make Algeria French. First the land was taken from the natives and their buildings were occupied; then French settlers gained control of the cork-oak forests and mineral deposits . . . A dual economy came into being [of capital and bazaar] . . . Algerians were relegated to marginality and poverty' (Said, 1994: 206–7).

29 The French government made a decision to renovate the city to mark the centenary of colonisation, in 1933. Çelik writes that most accounts of Le Corbusier's plans for Algiers erase the colonial context: 'They have been explained as a parable of European modernism, as a poetic response to the machine age, to syndicalism, and so forth, and thus abstracted from the "political geography" of Colonial Algeria' (Çelik, 2000: 322, citing Said (1990)).

30 The quotation is from Le Corbusier's *La Ville radieuse* (p. 260); Çelik cites Lorrain, J. (1899) *Heures d'Afrique*, reprinted in Knibiehler, Y. and Goutalier, R. (1985) *La Femme au temps des colonies*, Paris, Editions Stock, which says that Algeria is a wise and dangerous mistress who exudes a climate of caresses and torpor. See also Sandercock (1998a), referencing Hooper (1995), in which the metaphor of the site of architecture as body is seen from a feminist position: 'She "reads" the plans for "the modern city" of Baron von Haussmann and his contemporaries, and of Le Corbusier, as "poems of male desire", fantasies of control, written against the fears and upheavals of the nineteenth century which the female body comes to represent' (p. 49).

31 'The town belonging to the colonized people, or at least the native town, the Negro village, the medina, the reservation, is a place of ill fame, peopled by men of evil repute. They are born there, it mattes little where or how; they will die there, it mattes not where, nor how. It is a world without spaciousness; men live there on top of each other . . . The native town is a hungry town, starved of bread, of meat, of shoes, of coal, of light. The native town is a crouching village, a ton on its knees, a town wallowing in the mire' (Fanon, 1967: 30).

32 After the German occupation of France in 1941, Le Corbusier attempted to persuade the Vichy government to implement his plan for Algiers (Barnett, 1986: 115).

33 In *The City of Tomorrow and its Planning*, Le Corbusier refers to this scheme: 'In 1925 the demolition had made great progress . . . a large and impressive open space which brought many possibilities to one's mind. At that moment the space was there, not yet built over again, and the fact may well be considered one of great importance in the history of town planning, occurring as it did in the very heart of Paris. This bold piece of surgery must be credited to Haussmann . . . When he began to cut Paris about so mercilessly, his contemporaries said it meant the end of the city. But modern motor traffic in Paris to-day is only possible thanks to Haussmann' (Le Corbusier, 1987a: 257).

34 'there is often less perfection in what has been put together bit by bit, and by different masters, than in the work of a single hand. Thus we see how a building, the construction of which has been undertaken and completed by a single architect, is usually superior in beauty and regularity to those that many have tried to restore. So, too, those old places which, beginning as villages, have developed in the course of time into great towns, are generally so ill-proportioned in comparison with those an engineer can design at will in an orderly fashion' (Descartes [1637] 1960: 44–5). For critical commentaries see Lacour (1996: 32–7) and Melehy (1997: 101–8). Lacour, from another translation, sees the free drawing of a line as a foundational gesture of modernity; Melehy sees it as a metaphor for the construction of the text. Descartes is at pains to deny any prescriptive aspect in his writing.

35 'Descartes himself already spoke of the revolutionary achievements which were to await medicine: it would . . . in the end even invent a remedy against death. In the same way one would be able to structure and institutionalize the oppressively impure sphere of emotion . . . With this method people were to be able to cognize everything . . . you sense in these texts of this early modern age an urge to transform the whole world into one of light and universal transparency' (Welsch, 1997: 106).

36 Williams notes the divergence of Futurist calls for the rejection of the past from socialist calls for an overthrow of the social order; referencing the 1909 manifesto (point 11), he states 'But "great crowds excited by work, by pleasure and by riot", "the multicolored polyphonic tides of revolution": these, while they can appear to overlap, are already . . . a world away from the tightly organized parties which would use a scientific socialism to destroy the hitherto powerful and emancipate the hitherto powerless' (Williams, 1989: 52). Marinetti's appreciation of Mussolini is an appendix to A. Beltramelli's *L'uomo nuovo* (1923). Marinetti ends: 'A Futurist orator, who prunes, cuts down, drills through, ties back the opponent's argument, methodically shearing away all the tangled weeds of objections, cutting through the crowd like a torpedo-boat, like a torpedo' (in Griffin, 1995: 45–6). Gentile (1997) sees the attraction to fascism in Italy in the 1920s as a reaction to the 1914–18 war, a purifying of the nation for a national regeneration to echo the *Risorgimento*: 'Fascism was not antimodern, but rather had its own vision of modernity which opposed the visions of

liberalism, socialism, and communism, and which claimed the right to impose its own form of modernity on the twentieth century' (Gentile, 1997: 41).

37 This is reflected in the muted forms of war memorials after 1918 – see Michalski, 1998: 77–106.

38 Valois' movement was not the only form of French fascism. Jacques Dorriot, for instance, a former Communist mayor, founded a movement leading to the Parti Populaire Française in 1936, synthesising socialism and patriotism to combat liberalism and Bolshevism. Its newspaper was *L'Emancipation nationale*. Valois (who invited Jews to write in his journal *Nouveau Siècle*) died in Bergen-Belsen concentration camp in 1945.

39 Taylorism has a utopian aspect in the aim to produce goods for all through mass production, the old call for an ending of scarcity seen by Taylor as a way to dissolve class struggle. Sorel, in contrast, rejected anarchism (and its link to the neo-Impressionism of Signac, which he saw as bourgeois), and any lessening of class war, seeing conflict as a rejuvenating and re-energising force (Antliff, 1997: 137, 143). Antliff writes: 'Valois . . . claimed that the spirit of battle should also animate industrial production, with the result that fascism was to fuse the soldier and the citizen, the combatant and the producer. . . . the task of fascism was to extend the energy created through war into the postwar period. To do so required a return to the ethics of production, reinforced through class conflict' (Antliff, 1997: 155). Valois himself asserts that liberal democracy cannot manage 'a world which has undergone enormous economic transformations, transformations demanding state organs which those who drew up the constitutions of the nineteenth century had never foreseen' and compares the state to a horse-drawn carriage while what is required is a 40-horsepower car (G. Valois, 'Empty Portfolios', first published in Italian, 1926, in Griffin, 1995: 197).

40 Antliff discusses the influence of anarcho-syndicalist Georges Sorel on Valois: 'Valois's embrace of Le Corbusier's technocratic modernism cannot be understood apart from the moral values and antimaterialist precepts animating his fascist politics' (Antliff, 1997: 139); and notes that Valois had been central to a study group known as *Cercle Proudhon* between 1912 and 1915. From Proudhon, they drew a possibility that an antimaterialist approach might unite workers and the bourgeoisie in a national rejuvenation and cleansing of the corruptions of parliamentary democracy. This parallels Taylorism, a form of social engineering to unite classes in a quest for prosperity, but with a different agenda.

41 Antliff quotes from Valois' 'La nouvelle étape de fascism', *Nouveau siècle*, May 29th, 1927. Among Antliff's key sources on Le Corbusier and French fascism are McLeod (1980, 1983 and 1985) and Sternhell (1986).

42 Antliff points to Le Corbusier's internationalism and his call for foreign capital to be invested in his new Paris. A further difference, drawing Le Corbusier more to *Redressement Française* than *Le Faisceau*, was his view of Taylorism as enactable within the Republic not by total revolution. Antliff notes the omission in the extract of the Voisin plan published in *Nouveau siècle* of Le Corbusier's views on class (Antliff, 1997: 139, 157).

43 The product of a technologically driven urbanism during the inter-war years in North America resembles, too, that of Stalin's Soviet Union. Rowe and Koettler (n.d.) illustrate Le Corbusier's design for a Palace of the Soviets (1931) beside Auguste Perret's project for the same competition (Rowe and Koettler, n.d.: 70–1); Susan Buck-Morss compares the Waldorf Astoria (1931) and Leningrad (1949) Hotels (Buck-Morss, 1997: 97–115).

44 See Berman (1983) who points out that Moses loved his city and saw his work as improving it for public benefit.

45 'it is the way of seeing the city, the metaphors through which his vision of the city was mediated, and so the way of conceptualising the city as a problem to be solved . . . [His] way of thinking . . . appears to have been riven by a number of unstable oppositions: between calculation and aesthetics, between polemical plan and normative model, between empirical and ideal, between industrial methods and architectural values, between engineer and artist' (Donald, 1999: 55).

46 Adorno (1994: 102–27).

47 Le Corbusier's concepts also filter into popular culture, for instance in the US magazine *Amazing Stories*: an illustration on its back cover shows a future city of circular towers topped by electric moons, vast chasmic blocks at separate levels of which are pedestrians and cars, and a caption: 'What will the city of tomorrow be like? Here is the giant plastic, metal, and unbreakable glass city of the 21st century. A city of science, of atomic power, of space travel, and of high culture. See page 240 for complete story' (*Amazing Stories*, April 1942, in Taylor, 1990: 165). In the accompanying text (1990), Isaac Asimov makes his own predictions, including mile-high towers, 170 of which could house all Manhattan (p. 164).

48 'Underlying the design of Thamesmead is the assumption that physical environment is a strong cultural and social influence on behaviour, morality and well-being of people. The shift to a precise, ordered and rationalized spatial form comes out of a deep suspicion of traditional working- and lower-class neighbourhood form and possibly unconscious and unstated distrust of the life it is assumed to produce. The segregation of function we see at Thamesmead represents a deeply felt anti-urbanism and a distrust in the chaos of the streets because they were unsafe, unhealthy and threatening' (Robbins, 1996: 289). See also Miles, 2000: 57–66. Hall makes a different but not unrelated claim in regard to Le Corbusier's plan for Chandigarh: 'The only realized Corbusian city design: here a residential quarter, functionalist boxes for Punjabi functionaries, from the pen of the master' (Hall, 1996: 213, fig. 7.4). Le Corbusier worked on Chandigarh with his son and Maxwell Fry and Jane Drew, and with planner Albert Mayer.

49 I am grateful to Joost Smiers of Utrecht University for a walk through the zone. See Casciato (1996: 122–55) and Barnett (1986: 112–19). The blocks, in brick and of three or four storeys, incorporate decorative and structural features of advanced Dutch design. The plan involved co-operative organisations such as the *Algemene Woningbouwvereniging* (General Housing Association). One block, for the middle class, was called *De Harmoniehof* (Harmony Court, 1919–22).

50 See Searing, H., 'Workers' Housing', in Taylor, 1990: 106–8; and Barnett, 1986: plates 102, 109.

51 Citing debates on monumentality involving Léger, Giedion, Set and others, reported in *Architectural Review* (September, 1948), Curtis sees a questioning of the undemocratic aspects of pre-war Modernist planning, and a recognition of a need to humanise urban settlement: 'there was a widespread awareness that "the spring cleaning" phase of modernism was over' (Curtis, 2000: 55).

52 'a premium was placed on casual encounters and free speech . . . The images of meeting places simultaneously captured the hoped-for freedoms which the planners were seeking an architectural language to endorse' (Curtis, 2000: 57).

53 Sandercock points out the affirmation of regulation in the ethos of CIAM and rational planning, in which laissez-faire speculation and land use are countered by organisation. On CIAM, she states: 'Their urban plans would become blueprints, based on a presumed ability to control the future through action guided by rationality, and protected by the authority of the central state' (Sandercock, 1998a: 23).

54 Curtis notes Wiener's contribution on 'the concept of an era of magical abundance, the 'problem' of leisure, the overthrow of the concept of property, a society where remuneration was no longer linked to effort and the new paradigm of a flow-continuous confluence of atoms, molecules and energy no longer linked to location' (Curtis, 2000: 59, citing Wiener, P. L., 'New trends will affect the core', in Tyrwhitt, J., Set, J. and Rogers, E. N. (1951) *The Heart of the City: Towards the Humanisation of Urban Life*, London, Lund Humphries).

55 Burgess, E. W. [1925] (1972) 'The Growth of a City', in Stewart, M. (1972) *City*, Harmondsworth, Penguin, pp. 117–29.

56 See Sandercock (1998a: 87–9) on the development of the rational planning model in the 1950s and 1960s from the Chicago School model of the 1920s and 1930s: 'Here is planning at its most heroic, confident in its capacity to discern and implement the public interest in specific settings' (p. 88).

57 Foucault (1967); Illich (1986: 50–3); Illich (1986: 54–64); Sennett (1970: 26–45); and Sibley (1995: 27–9).

58 Le Corbusier, 1987c: 111–21. The photographs accompany a section redefining the problem of architecture: 'The airplane shows us that a problem well stated finds its solution' (p. 113).

59 'Photography gives shape to travel. It is the reason for stopping . . . much tourism becomes in effect a search for the photogenic . . . What is sought for . . . is a set of photographic images, as seen in tour company brochures or on TV programmes . . . Photography is thus intimately bound up with the tourist gaze. Photographic images organize our anticipation or daydreaming about the places we might gaze upon' (Urry, 1990: 139–40).

60 Wigley notes Alberti's writing on the family – in which the father is like a spider at the centre of a (domestic) web – and his insistence on a privileged and pure visuality; following a passage dealing with whiteness as cleanliness, he states: 'This architecture of vision was already in place in Alberti's text in which the status of the white wall depends upon "the keenest of the senses" with which the rational mind (which is today the masculine eye) is said to "immediately" comprehend the immaterial order within a material object' (Wigley, 1992: 360).

61 Colomina (1994) writes extensively on Le Corbusier's uses of photographic images, seeing this in relation to psychoanalysis. I have two reservations: first, Colomina writes 'The diffusion of photography coincides with the development of psychoanalysis' (Colomina, 1994: 80); but photography begins in the 1830s and was diffuse before publication of Freud's theories from the 1890s, while her other parallel between photography and railways is more accurate (Colomina, 1994: 47). Second, Colomina cites Freud's use of the photographic process from negative to positive as a metaphor for bringing into consciousness; but his metaphors are convenient to lay explanation, not to be taken literally – in *The Question of Lay-Analysis*, he uses war as a metaphor: 'Think of the difference between "the front" and "behind the lines", as things were during the war . . . many things were permitted behind the lines which had to be forbidden at the front. The determining influence was, of course, the proximity of the enemy; in the case of mental life it is the proximity of the external world' (Freud, 1962: 106), which does not mean mental life is war, only that the free play of the drives in the *Es* compared to regulation in the *Ich* can be understood this way. I am more inclined to think that growing familiarity with photographs led in the 1870s to an ability to see them as much as projections of a mental state as literal transcriptions of reality.

62 Massey cites Irigaray (1978), Owens (1985) and Pollock (1988).

63 Meskimmon sees the ideal of objective knowledge as linking sight, power, and rationality, and having 'a tremendous impact upon traditional conceptions of space of which the modernist city is but one example' (Meskimmon, 1997: 18). Later she remarks that feminist challenges to the model have also produced ruptures in other models such as centre-margin – 'by demonstrating that all knowledges are partial and located' (op. cit.: 19). Cf. note 49.

64 *L'Architecture d'aujourd'hui* (1929), see Colomina, 1994: 289–93. Colomina remarks that in another part of the film there is a woman moving through Villa Savoye behind the bars of the ramp to the roof garden. At the point where she would face the camera, she vanishes: 'Here we are literally following somebody, the point of view is that of a voyeur' (p. 293).

65 My observations are based on a visit in 1997. See Miles (2000) pp. 105–27.

66 Fathy writes of the countryside as a lost paradise, though one 'darkened by clouds of flies . . . bilharzia and dysentery' (Fathy, 1973: 2).

67 Fathy is less sympathetic than English expatriate Winifred Blackman: 'In the better houses there is generally a flight of steps leading to an upper storey, where there may be a sitting room . . . The flat roof is a pleasant place on which to sit and watch the life in the streets below' (Blackman, 1927: 27).

68 See also Fathy, 1973: 37–8, 129–30.

69 For a background in self-empowering urbanism see Turner (1976). For a case of self-build housing in South Africa, see Chinedu (2000); and on Walter Segal's self-build scheme in Lewisham, London, see Hughes (2000).

70 Steele, 1988: 74–5; and Steele, 1997: 60–89.

71 See Rapoport (1980) on vernacular design.

72 Fathy included schools for girls and boys, the latter built but allowed to fall into disrepair (Steele, 1997: 77).

73 Karnouk, 1988: 47.

74 'the work of an architect who designs, say, an apartment house in the poor quarters of Cairo for some stingy speculator, in which he incorporates various features of modern design copied from fashionable European work, will filter down, over a period of years, through the cheap suburbs and into the village, where it will slowly poison the genuine tradition' (Fathy, 1973: 21).

75 Fanon writes that colonised people look to remote pasts for legitimation of freedom: 'the past existence of an Aztec civilization does not change anything very much in the diet of the Mexican peasant of today . . . this passionate search for a national culture . . . finds its legitimate reason in the anxiety shared by native intellectuals to shrink away from that Western culture in which they all risk being swamped' (Fanon, 1967: 168–9). Perhaps Fathy's mud-brick architecture, which looks to living, local roots, not those of pharaonic times, is a post-colonial solution outside the knowledge, and technologies of the colonial power. Perhaps, equally, it is not an absorption of the colonial culture in the way described by Fanon as typical of elites within colonised countries, who mimic the ways of the masters as far as they are allowed, and become a quasi-ruling class under domination of the colonial power.

4

1967

WHY TOMORROW NEVER DAWNS

•

In the previous three chapters I set out a range of avant-gardes in art and architecture from the mid-nineteenth to mid-twentieth centuries. These included the politicised avant-garde of Realism in the mid-nineteenth century, the aesthetic avant-garde of early Modernism in art, and the technocratic avant-garde of Modernist architecture and planning. I turn now to radical theory in the late 1960s, a period of affluence as well as protest, when an optimistic counter-culture engendered for a moment a feeling that social transformation might be imminent.

I look in particular at the work of Herbert Marcuse in the late 1960s. This is because his writing is more accessible than that of some other critical theorists, and because his key concerns are the possibility for social transformation, and the function in that context of an aesthetic dimension in which such transformation is imagined or perhaps brought nearer realisation.

But I ask, following discussion of the avant-garde as retaining an old model of power (in Chapter 1), whether his retention of time as the dimension of change is a flaw in his theory, which in the end prevents his being able to explain how the hoped-for transformation will come about. This produces the dilemma indicated in the quotation with which I open the chapter below. As a way out of it, I turn, in the second section of this chapter, more briefly to Henri Lefebvre's theory of moments of liberation within everyday lives, set in the dimension of space as the co-presence in which an incipient revolutionary consciousness already exists – and does not need to be introduced by either an avant-garde nor an intelligentsia. Here, the possible future is situated within rather than after the existing social reality; it is experienced now rather than displaced to a tomorrow that never dawns except as a repetition of today.

I THE FREE UNIVERSITY, BERLIN

> Q. It seemed to me that the centre of your paper today was the thesis that a transformation of society must be preceded by a transformation of needs. For me this implies that changed needs can only arise if we first abolish the mechanisms that have let the needs come into being as they are. It seems to me that you have shifted the accent toward enlightenment and away from revolution.
>
> M. You have identified what is unfortunately the greatest difficulty in the matter. Your objection is that, for new, revolutionary needs to develop, the mechanisms that reproduce the old needs must be abolished. In order for the mechanisms to be abolished, there must first be a need to abolish them. That is the circle in which we are placed, and I do not know how to get out of it.
>
> (Marcuse, 1970: 80)

The difficulty is that for the new society to be introduced the institutions of the old which perpetuate the conditions of unfreedom must be abolished. This will happen when the need for such abolition is felt. But it seems that the new consciousness, which produces new social relations, is a product of a liberation that arises only once those relations arise. The future thus carries the burden of producing the conditions of its own coming into being. There is no exit from the dilemma and it is unsurprising that Marcuse falls back in 1968 on the role of an intelligentsia, expanded to include radical student movements, as the engine of radical change. I see this as another avant-garde.

The difficulty resides in the gap between the imagination of freedom and its social realisation, and preoccupies Marcuse throughout his philosophical development to lead him in his earliest and last works – from his doctoral thesis in 1922 to *The Aesthetic Dimension* (1978) – to dwell on an aesthetic dimension in which freedom is imagined. In between, his position undergoes several shifts: in the thesis he identifies with artistic withdrawal to suggest that the marginal social position of the artist or writer is liberating, if at a cost; then in his 1937 essay 'The Affirmative Character of Culture', written in exile from Nazi Germany, he attacks the tendency in bourgeois culture to displace hope for a better world to a compensatory aesthetic realm; writing in the 1940s on French literature under the Nazi occupation he sees the novel of intimacy as expressing a freedom beyond the grasp of an authoritarian state. Through the 1950s and 1960s, he develops a revision of Freud through a revision of Marx to draw out the liberating possibilities of both in a theory of social and individual development he articulates in *Eros and Civilization* (1956); this informs Marcuse's interventions in radical politics of the 1960s, allied to a revision of the idea of utopia as no longer visionary but now, through new technologies of production, a real possibility. Marcuse foresees society as a work of art, a state in which work and social relations are libidinised. This is a high point of his optimism before a return to preoccupation with aesthetics after the failure of 1968.

In *The Aesthetic Dimension*, the aesthetic is a bastion of critique in the face of a dire political reality; and perhaps throughout his development Marcuse is drawn to the aesthetic dimension as that which remains viable when the rest of life is difficult.[1] But the question is whether an aesthetic dimension in which a future other than that of the given can be imagined is a catalyst to realisation of such a future, or instead a substitute realm in which utopia again recedes to take symbolic form. In 1967, Marcuse is on the cusp of working through the problem of how transformation comes into being, but – as the above extract from the discussion in Berlin after his lecture 'The End of Utopia' indicates – he does not know. He addresses this in the *Essay on Liberation*, informed by his discussions in London, Berlin, and Paris, by proposing a new (Freudian) drive for liberation as the source of a new (revolutionary) consciousness. But it never happened. Then a decade later in *The Aesthetic Dimension* Marcuse announces a second withdrawal into an imaginative realm separate from daily life, comparable to that of his thesis more than fifty years before – if with considerable extension of the argument.

To begin, then, in Berlin during the German Revolution of the first days of the Weimar Republic of 1918, after a mass strike in January, a naval mutiny, admission of defeat by the military, widespread shortages of essentials items, and eventual armistice in November. As a 19-year-old conscript Marcuse was elected to represent the working-class Berlin district of Reinickendorf on a Soldiers' Council, and in December sent to Alexanderplatz to defend the Republic against snipers. He aligned himself neither with the governing social democrats nor the Spartacus League (later German Communist Party) of Rosa Luxemburg and Karl Liebknecht – which he saw as remote from the actualities of working-class lives – but with the Independent Socialists of poet Kurt Eisner and writers such as Ernst Toller and Erich Mühsam (an associate of Dadaist Hugo Ball). A rising of the left in Berlin in January 1919 was suppressed, and during a second attempted rising in March the irregular *Freikorps* were ordered by the Social Democrats to shoot anyone bearing arms, leaving 1,200 dead. Luxemburg was beaten unconscious, shot and thrown in the canal, and Liebknecht was also killed.[2] Eisner, then Prime Minister of Bavaria, was assassinated in February. A wave of popular sympathy led to a *Räterepublik* (Soviet) in April, but troops were again called in and members of the Soviet killed or jailed.[3] The German revolution was over, and Marcuse returned to Humboldt University, Berlin to read literature. There in a radical literature group he met Walter Benjamin and Georg Lukács, and moved in 1920 to Freiburg.[4] His thesis topic was the *Künstlerroman* (artist-novel), a genre in which an artist or writer makes a journey of self-discovery in adversity.[5] Examples include Mann's *Tonio Kröger* and *Death in Venice* (1911), and Rilke's *Malte Laurids Brigge* (1910). Marcuse contrasts portrayal of the authentic struggles experienced in artistic existence with the illusions of bourgeois life, a contrast denoting a separation between realms of necessity and imagination in the society which provides the conditions of the genre's production. If the social environment is unsympathetic and the artist is 'estranged . . . in ecstasy, anger, or despair' (Katz, 1982: 41) then reconciliation is sought either by accepting bourgeois values or, as in Mann's *Death in Venice*, in a refusal that entails disintegration.[6]

Marcuse does not accept this wholesale. In the pre-Socratic world 'where life was itself art, and mythology life' (cited in Katz, 1982: 42) he finds an integration of beauty and necessity; and in the twilight before modernity sees the troubadours as a rootless, subversive artistic class.[7] A strand which binds this cultural history is an idea of alienation as paradoxically able to preserve imaginative freedom: 'even in a time of universal suffering and oppression, the lost values of a world at one with itself, of the immediate unity of the artistic life and the fully human life, are preserved – if in attenuated form – in the shape of artistic subjectivity' (Katz, 1982: 43). The troubadours who 'vanish into the mists of restless wandering, of dissolute vagrancy' (*Der deutsche Künstlerroman*, p. 13, quoted in Katz, 1982: 42) thus negate the given order for an unstable authenticity, perhaps like that of the Hippies later, or the artistic youths of Romanticism.[8]

Marcuse completed his doctorate in 1922, returning to Freiburg in 1928 to work with Heidegger and looking towards an academic career. He realised by 1931, however, that as a Jew he had no prospect of appointment to a teaching post in the German academic system,[9] turning instead (at this point where the political was racial and personal) to the Frankfurt Institute for Social Research where Max Horkheimer (whom he first met in Freiburg) had begun to build a programme for a critical theory of society.[10] Marcuse moved to the Institute's Geneva office – set up to avoid the persecutions of the Nazi regime – and to New York in 1934 where the Institute was re-housed at Columbia University. His essays 'The Struggle Against Liberalism in the Totalitarian View of the State' (1934), and 'The Affirmative Character of Culture' (1937),[11] and book-length study of Hegel and German Idealism,[12] are among his contributions to the Institute's efforts to understand the rise of fascism in Germany after the failure of revolt, the dual spectre which haunts critical theory.

Marcuse worked for the intelligence services in the war years, then until the death of his first wife Sophie in 1951 in the de-Nazification programme. During this period he produced a study of French literature under the occupation, 'Some Remarks on Aragon: Art and Politics in the Totalitarian Era',[13] which takes the theme of intimacy as a realm beyond the grasp of the regime. In the sensuality of Paul Eluard's poetry,[14] and Louis Aragon's novel *Aurélien*, as in Baudelaire's *Invitation au voyage*, he sees intimacy as a form of authentic existence. This seems like the artistic life described in his thesis, the transcendent moments of love taking the role of authentic existence in art.[15] Love's transcendence, however, is not metaphysical but transgressive:

> It is the essential illegality, its transcendence over the established order of life which makes love a political and at the same time artistic a priori . . . In *Aurélien*, the illegality of love lies in its incompatibility with all normal relationships . . . in its disproportionate character which absorbs all other contents, in its impossibility to adjust itself to the requirements of sanity and reasonableness.
>
> (Marcuse, 1998: 210–11)

The representation of love between individuals is seen as celebratory, while political art is liable to be subsumed in the dominant culture: 'All indictments are easily absorbed by the system which they indict' (Marcuse, 1998: 201). Picasso's *Guernica* (1937) is taken as a case of a radical image turned museum piece. The question then is how art retains or regains a possibility for negation of tyranny, how it sets itself apart from the authoritarian state while indicating the dehumanising effects of that state. Marcuse argues that such a possibility is found in sensuality, and interprets Aragon's novel as reflecting its epoch in the personal (almost casually tragic) history of its protagonists Aurélien and Bérénice.[16] After a critical reading of the novel Marcuse writes that the call for political action negates the call for joy, but that 'the negation reveals at the same time the true relation between the two realities: their final identity . . . Political action is the death of love, but the goal of political action is love's liberation' (Marcuse, 1998: 210–11).

This leads Marcuse to a model of argument also characteristic of Adorno in *Aesthetic Theory* (incomplete on his death in 1969) and which Marcuse reiterates in *Essay on Liberation* and *The Aesthetic Dimension*: tensions between art's social and aesthetic dimensions appear as mutually destructive[17] and a tendency to reconciliation is another bind in which the negations, too, are negated: 'Art may well try to preserve its political function by negating its political content, but art cannot cancel the reconciliatory element involved in this negation' (Marcuse, 1998: 212)[18]. This is not a defeat. The negating power of art may be repressed in an unfree world and art may reproduce the beautiful illusions of bourgeois society – but its vital autonomy stands nonetheless for that of the subject to hint at a world freed from domination, to glimpse *bonheur*.[19] In the conditions of a totalitarian regime when the political reality is expressed outside the content of the work, art is shaped 'in such a manner that it reveals the negative system in its totality and, at the same time, the absolute necessity of liberation' (Marcuse, 1998: 203). What is unavoidable is that in aestheticising its material art separates its content from the historical world: 'In the medium of the artistic form, things are liberated to their own life – without being liberated in reality' and this applies, even especially, to anti-art: 'Art creates a reification of its own. The artistic form, however destructive it may be, stays and brings to rest' (Marcuse, 1998: 213).[20] Art may transpose reality as transfiguration, but is inevitably a source of gratification that undermines resistance to the regime sedimented in it. Marcuse writes of

Guernica that 'Darkness, terror and utter destruction are brought to life by grace of the artistic creation . . . they are therefore incomparable to the fascist reality' (ibid.). Part of the difficulty, notably in a painting such as *Guernica*, which semi-abstracts its pictorial vocabulary, is the generalisation implicit in concepts such as humanity which such art (like political rhetoric) tends to celebrate. What does this leave?

> The incompatibility of the artistic form with the real form of life may be used as a lever for throwing upon the reality the light which the latter cannot absorb, the light which may eventually dissolve this reality (although such dissolution is no longer the function of art). The untruth of art may become the precondition for the artistic contradiction and negation. Art may promote the alienation, the total estrangement of man from his world. And this alienation may provide the artificial basis for the remembrance of freedom in the totality of oppression.
>
> (Marcuse, 1998: 214, partly cited in Katz, 1990: 163–4)

The words 'the remembrance of freedom' in the last sentence above, which is the last in the essay on Aragon, indicate the key direction of Marcuse's work in the 1950s and 1960s: a revision of Freud in terms of dialectical materialism, and of dialectical materialism in terms of Freud, which he sets out in *Eros and Civilisation* (first published in 1955, to which Marcuse added a political preface in 1966).[21]

By then a US citizen and academic (while Adorno had returned to Frankfurt in the west and Bloch to Leipzig in the east), Marcuse proposed that a memory of *bonheur* is found in the repressed material of the unconscious, and that it is from this that a new revolutionary consciousness springs. This departs from the Marxist concept of class struggle to give the revolutionary consciousness a ground that is perpetual but not biological, and mutable but not economic. It departs, too, from psychoanalytic theory's preoccupation with the mental states of individuals, and from a perceived closure of psychoanalytic practice to the impact of social problems. While Freud saw in consciousness a mediation of the urges of the pleasure principle, so that unpleasurable experiences might be repeated or pleasurable experiences delayed, Marcuse saw a retrieval of the repressed memory of gratification as a means to the reinvestment of the pleasure principle as a path to freedom at the level of society as well as subject.

The basis for Marcuse's integration of social and psychological theory is found in Freud's meta-psychology.[22] But, drawing on *Beyond the Pleasure Principle* (1920) and *Civilisation and its Discontents* (1930), Marcuse adds to Freud's model of unconscious drives competing in the *Es* (id) and mediated in the *Ich* (ego)[23] an emphasis on the possibility for new drives, and insists on the *social* production of these drives – in this applying the model of dialectical materialism to psychoanalysis.[24] Marcuse finds in Freud's late work a correlate in the phantasy of mental life for a desire for a free society, just as for Freud repression in mental life correlates with oppression in society. Marcuse's starting point is Freud's model of mental development in which the pleasure principle gives way to the reality principle (repeated performance of the demands of which is called the performance principle) to produce civilisation.[25] But, as immediate gratification is replaced by delay, pleasure by restraint, play by work and receptiveness by productivity, the organisation and technologies of civilisation are bought at the cost of a sublimation of joy.[26] While, then, the consciousness-producing *Ich* orders the wild and competing drives of the *Es* to produce social organisation, that organisation is historical:

> Psychoanalysts have correctly emphasised that Freud's last metapsychology is based on an essentially new concept of instinct: the instincts are defined no longer in terms of their origin and their organic function, but in terms of a determining force which gives the life processes a definite 'direction' (*Richtung*).
>
> (Marcuse, 1956: 27)

A question then arises as to whether that direction can best be described, if metaphorically, as a fall or an ascent. For Freud, in *Civilization and Its Discontents*, the reality principle acts to validate a primal competitiveness moderated only to produce a necessary social organisation which will minimise pain and enable common projects but is not genuinely mutual.[27] Marcuse sees a rational end of history as a state of joy. Since it is not-yet, a potential to achieve it as a renewal of the primary urge to gratification repressed as a no-longer conscious memory, produces a new drive for liberation.

As Marcuse reads Freud, then, a pleasure principle of immediate but uncertain gratification gives way in early development to a reality principle of a predictable tomorrow; then individual aspirations are subsumed in social ordering to create a stable civilisation. But if the operation of the pleasure principle is repressed or displaced to phantasy and myth, it is recoverable.[28] The key to Marcuse's theory of liberation in which the realm of pleasure shapes social organisation through a new drive is, then, in Freud's theory of phantasy (see note 25). Freud writes that a mode of mental activity is split-off from reality-testing at an early stage and retains allegiance to the pleasure principle. Evidence of this is found in the daydreams of children and extends to the wishful constructions of adult life or in aesthetic experience. Marcuse sees in the persistence of the memory of pleasure in phantasy a possible spark which under certain conditions leads to adaptation of the structure of the drives.[29] A utopian imagination can then be seen as derived from memories of an archaic or primal state in which *bonheur* is reality[30] – the Rousseauesque vision which colours Baudelaire's *Invitation au voyage* – just as oppression is the reality of the administered world. This memory of gratification survives in the forms of art,[31] folk tales, and myths of Eden, Arcadia, Utopia and Nirvana, which express anticipatory memories of freedom.[32]

The advantage of a psychoanalytic approach is that it confers a scientific status on social theory, and means that the new drive which presages revolution arises through processes which are unconscious and, if still conditioned by external realities, not products of instrumental agency on the part of a revolutionary class or intelligentsia. But I wonder if this is a replication of the objectively given end of history by other means; and whether there is an aspect of projection when Marcuse sees, for instance, a state of unity in pre-Socratic Greece.[33] This may be, in one way, a projection of a memory of infantile gratification (as a repressed wholeness) onto a remote situation, conflating the infantile and the archaic as primal states just as it conflates individual repression (ontogenesis) and socialised repression (phylogenesis); and in another way, a statement of the lack of whatever is projected in the present. The content of Eden, thus, is the absent but phantasised content of today. I have reservations also on the establishment of liberation (or Bloch's hope) as a drive. This is not a matter of whether or not liberation arises naturally under certain conditions, though it could be that, but of a need to not obscure the responsibility for engagement. Marcuse adopts an unconscious process, which happens in but is abstracted from social reality, and may weaken the case for struggle. In *Eros and Civilization* he rethinks class consciousness to find a way, other than through class agency, in which a transformation will occur. But does he turn critical theory into a psychological theory limited to investigation of the mental development of individual subjects? Or is the developmental model he adapts from Freud, with its liberatory drive, a common myth (of the kind sometimes said to be lacking in late modernity)? I have neither the space nor expertise to go into this properly. There is I suggest in any case a further factor: Marcuse reintroduces into social theory an erotic dimension in the vital pleasure of social relations, which contrasts with civilisation's self-destructiveness.[34]

This brings me to Marcuse's Roundhouse paper and his lectures in Berlin, in July 1967.[35] Two themes run through them: the need to redefine revolution for an affluent society; and the vision of a libidinisation of social relations, a return to Eros which constitutes the new society. It is the latter, as much in the tradition of utopian socialism as it draws on Freud, which leads Marcuse to see society as a work of art, and which adds weight to the need to think again about what a process of social revolution would entail and how it might occur.

To begin with the problem of an affluent society:[36] while in Marxist theory the alienation and immiseration of a working class produces a revolutionary consciousness in response to the contradictions of bourgeois society, liberation is required now not from economic deprivation but from an affluence which impoverishes in a qualitative not quantitative way, masking this by quantitative improvements more than taken up by consumption:

> The problem we are facing is the need for liberation not from a poor society, not from a disintegrating society, not even in most cases from a terroristic society, but from a society which develops to a real extent the material and

> even cultural needs of man . . . that implies we are facing liberation from a society where liberation is apparently without a mass basis.
>
> (Marcuse, 1968b: 176)

The situation is new because the qualitatively better world, in which the quality of life has changed and not merely the level of income and consumption, is technologically possible. The advanced productivity of industrial society enables a possible abolition of all but the most minimal toil. What was utopian, a desire directed to phantasy, is now within reach and produces a potential 'leap into the realm of freedom – a total rupture' (Marcuse, 1968b: 177). The repressive Judeo-Christian work ethic becomes obsolete when work and play, and the realms of necessity and freedom, converge after the ending of scarcity. Then:

> This means one of the oldest dreams of all radical theory and practice. It means the creative imagination . . . would become a productive force . . . It would mean the emergence of a form of reality which is the work and the medium of the developing sensibility . . . And now I throw in the terrible concept: it would mean an 'aesthetic' reality – society as a work of art.
>
> (Marcuse, 1968b: 185)

In Berlin a few days later, Marcuse argues that a transformation of the technical and natural environments allows the world to turn into either hell or its opposite, and that this 'implies the necessity of at least discussing a new definition of socialism' (Marcuse, 1970: 64). The end of utopia of his Berlin title is, then, the end of a dream beyond realisation and its replacement by a real-possible world of plenty and leisure (as depicted in Seurat's *Bathers at Asnières* – see Chapter 1) and of work as play, a libidinal society.

Taking the talks in London and Berlin together lends the idea of society as a work of art a substance beyond aestheticisation, or retreat into the delusions of hedonism.[37] The difficulty is how to bring about the libidinal society, or whether it happens by itself under specific conditions. At the Roundhouse, building on *Eros and Civilization* and his critique of capitalism in *One Dimensional Man*,[38] Marcuse emphasises that the idea of society as a work of art entails a complete break from a repressive past, a negation that requires unconventional means of expression if it is to overcome conventional structures of repression.[39] Fusing social and psychoanalytic theories he maintains that such a break 'reaches into the depth dimension of the organism itself' so that 'qualitative change, liberation, involves organic, instinctual, biological changes at the same time as political and social changes' (Marcuse, 1968b: 184). The problem is how the new needs are articulated within an advanced capitalist society in which the contradictions are as dire as ever, while the potential opposition of a revolutionary class is diffused in consumption. The difficulty was aired in Berlin, and is restated in the *Essay on Liberation*:[40]

> This is the vicious circle: the rupture with the self-propelling conservative continuum of needs must *precede* the revolution which is to usher in a free society, but such rupture itself can be envisaged only in a revolution – a revolution which would be driven by the vital need to be freed from the administered comforts and the destructive productivity of the exploitative society, fred from the smooth heteronomy, a revolution which, by virtue of this 'biological' foundation, would have the chance of turning quantitative technical progress into qualitatively different ways of life . . . If this idea of a radical transformation is to be more than idle speculation, it must have an objective foundation in the production process of advanced industrial society.
>
> (Marcuse, 1969: 27)

The punctuation discloses, however, the speculative in an effort to put transformation beyond speculation. It shows, for me but I claim no authority, that in making liberation a drive Marcuse has not solved the difficulty of how transformation will occur. Affluence absorbs all social classes, so that the objective conditions for change exist but the subjective factor of human intervention does not materialise, and the notion of a revolutionary class is now obsolete. The 'instinctual need for a life without fear, without brutality, and without stupidity' (Marcuse, 1968b: 189) which Marcuse proposes at the Roundhouse still needs recognition, a role he entrusts (perhaps detracting from the idea of a drive) to an intelligentsia:

> Can we say that the intelligentsia today is a revolutionary class? . . . No, we cannot say that. But we can say . . . that the intelligentsia has a decisive preparatory function, not more; and I suggest that this is plenty . . . it can become the catalyst.
>
> (Marcuse, 1968b: 188)

This may be a response to the mainly intellectual Roundhouse audience, yet, though Marcuse stresses the need for self-organisation rather than the organisation of others, it seems like an avant-garde removed to the university.[41]

In Berlin, Marcuse similarly sees an intellectual class of radical students and academics as initiating liberation, in a situation in which utopia has become, no longer a dream, technologically viable:

> All the material and intellectual forces which could be put to work for the realization of a free society are at hand. That they are not used for that purpose is to be attributed to the total mobilization of existing society against its own potential for liberation.
>
> (Marcuse, 1970: 64)

Consciousness is now the site of intervention – to change how the world is perceived and received. The technological capacity which can end toil can release also 'a creative experimentation with the productive forces'; freed from capitalism it can 'become the concretely structured productive force that freely sketches out the possibilities for a free human existence . . . sustained and directed by liberating and gratifying needs' (Marcuse, 1970: 66). And Marcuse reiterates the idea of society as a work of art: 'if we are looking for a concept that can perhaps indicate the qualitative difference in a socialist society, the aesthetic-erotic dimension comes to mind almost spontaneously, at least to me' (Marcuse, 1970: 68). In 1967 this may have been illuminating, as the Hippies, Diggers, other radical groups and the mass of people wearing 'Make Love Not War' badges thought of themselves as making a new society.

In Paris in 1968, as recalled by Julia Kristeva in a way that coincidentally echoes Marcuse's essay on Aragon, the liberation of personal and social behaviour took on a political edge:

> Group sex, hashish, etc., were experienced as a revolt against bourgeois morality and family values. All of us from my generation went through it. This movement can only be described as political because it began by striking savagely at the heart of the traditional conception of love.
>
> (Kristeva, 2002: 18)

But, after the Summer of Love in San Francisco[42] and the summer of failed revolt in Paris, with Soviet tanks rolling into Prague, the moment passed. Another context for it was in any case a rapid expansion in art and fashion, and the market was not so inept as to ignore opportunities for consumption introduced by protest, as in the music industry.[43] In the affluent society, as Marcuse foresaw, an 'illusory bridging of the consumer gap between the rulers and the ruled' obscured the distinction between the real and perceived needs of the ruled to stifle the new imagination (Marcuse, 1969: 24). What remains, fifty years after the failure of 1918, is another withdrawal to an aesthetic dimension.

> where the miserable reality can be changed only through radical political praxis, the concern with aesthetics demands justification. It would be senseless to deny the element of despair inherent in this concern: the retreat into a world of fiction where exiting conditions are changed and overcome only in the realm of the imagination.
>
> (Marcuse, 1978: 1)

II MOMENTS

I turn now to Henri Lefebvre's idea of moments of liberation within routine.[44] This offers a speculative way out of the dilemma with which I began this chapter. I will return to it in Chapter 5 in relation to public monuments and the traces of urban occupation. Then, at the end of this chapter, I glance obliquely again at Marcuse's late aesthetic theory.

Rob Shields describes the idea of moments envisaged by Lefebvre as referring to 'those instants that we would each . . . categorise as "authentic" moments that break through the dulling monotony of the "taken for granted" . . . [to] outflank the pretensions of wordy theories . . . and challenge the limits of everyday living' (Shields, 1999: 58). In these vivid instants the potential of living is grasped, rather as Marcuse sees intimacy as transcending oppression. But whereas Marcuse sees the repressed memory of *bonheur* as retrievable in a dialectic informed by depth psychology, for Lefebvre the authentic is simply present in everyday life – and not the prerogative of an intelligentsia and more than of a revolutionary class. Authenticity may be a flash of insight, or an instant of joy when routine is interrupted. It is political and personal at once, and revolutionary euphoria as well as love.[45] Being in the everyday it belongs to everyone, residing in the paradox of the everyday.

In *Everyday Life in the Modern World*,[46] Lefebvre describes the everyday as restrictive but at the same time a realm of feelings, a realm of certainty but also an open-ended mutability:

> Everything here is calculated because everything is numbered: money, minutes, metre . . . Yet people are born, live and die. They live well or ill; but they live in everyday life, where they make or fail to make a living . . . It is in everyday life that they rejoice and suffer; here and now.
>
> (Lefebvre, 2000: 21)

This is the site of occupation and the emotive experience of spaces, dominated but not extinguished by the unified space of geometry and administration; it contains both carnival and lent, so to speak.[47] If planned space is reductive, lived space is heterogeneous and breathing, an idea developed a few years later in *The Production of Space*.[48] Everyday life is 'sustenance, clothing, furniture, homes, neighbourhoods, environment . . . Call it material culture if you like, but do not confuse the issue' (Lefebvre, 2000: 21, lacunae in original). It matters for the following reasons:

> The study of everyday life affords a meeting place for specialized sciences and . . . exposes the possibilities of conflict between the rational and the irrational . . . permitting the formulation of concrete problems of *production* (in its widest sense): how the social existence of human beings is *produced*, its transition from want to affluence and from appreciation to depreciation.
>
> (Lefebvre, 2000: 23)

So, Lefebvre makes space dialectical: it may change people but can itself be changed.

Lefebvre's writing is, like Marcuse's, a revision of Marxism; he is close at times to the Marx of the 'Theses on Feuerbach' (1845), which outline a philosophy of practice.[49] In his Vth Thesis Marx acknowledges Feuerbach's move from abstract thought to sensuous contemplation – to the experiential beyond the conceptual – but criticises his neglect of sensuousness as a practical human activity. Lefebvre gives a hint of what this implies: 'The way for physical space, for the practico-sensory realm, to restore or reconstitute itself is therefore by struggling against the *ex post facto* projections of an accomplished intellect, against the reductionism to which knowledge is prone', which he puts in brief as an 'uprising of the body . . . against the signs of the non-body' (Lefebvre, 1991: 201). So, if the tactics of revolution are those of intervention in the conditions of production, the conditions include the means of perception and conceptualisation, and the relation between them. This is not to say the concept is to be relinquished, any more than a conceptual space abolishes bodily spaces. Just as Adorno refuses to resolve the tensions of polarities such as the social and aesthetic, Lefebvre holds both lived and conceived space as having validity. The difficulty is that in modernity the lived is relegated to a marginal position, just as users are relegated to the role of passive consumers of spaces designed by experts.

This is why the 'architecture of pleasure and joy, of community in the use of the gifts of the earth, has yet to be invented' (Lefebvre, 1991: 379). However, in Situationism before 1968, Lefebvre saw a practice that was open to the unpredictable: of drifts (*dérives*) to reframe the city as lived space; of a psychogeographic imagination that created new ambiences within the city, and of the situation as 'a spatial/temporal event staged to catalyze liberatory transformation' (McLeod, 1997: 20). Or, as Simon Sadler puts it, 'The constructed situation would clearly be some sort of performance, one that would treat all space as performance space and all people as performers' (Sadler, 1999: 105).[50] The background to this is an urbanisation programme under de Gaulle in which Le Corbusier's influence is discerned in new towns such as Mourenx,[51] and the system-built *grands ensembles* to house workers on urban peripheries. The urge to re-encounter everyday life, for Lefebvre and the Situationists, can be understood as a rejection of the concrete form but equally of the power structure which builds all that.[52]

The point, though, of both the situation and the moment is that they, in their ways, provoke a new relation between change (in a Marxist sense of the production of history) and the dimensions of space and time. Traditionally, time has been the dimension of change, as in Hegel and still in Laclau; for Lefebvre, and for Doreen Massey in a feminist critique of Laclau,[53] it is space as a realm of occupation – or reclamation of a right to the city perhaps equivalent in materiality to the retrieval of a memory of *bonheur* in imagination for Marcuse – which is the ground of revolution. This is so even when time is the medium of the drift, because the way the drifting Situationists used time was, in effect, to waste it; the drift is a refusal of productivity and plan, the drifters like vagrants.[54]

In the *present* dimension of space, then, moments of insight occur when society is seen as the sham it is:

> Thus everyday life, the social territory and place of controlled consumption, of terror-enforced passivity, is established and programmed; as a social territory it is easily identified, and under analysis it reveals its latent irrationality beneath an apparent rationality, incoherence beneath an ideology of coherence, and sub-systems or disconnected territories linked together only by speech.
>
> (Lefebvre, 2000: 196–7)

But moments also enlighten, become moments of *presence* that, as Shields succinctly says, 'puncture the "everydayness" or banality of repetitive tasks like a ray of sunshine through clouds' (Shields, 1999: 60–1).[55] These are the transformative experiences alluded to by John Roberts, which he says are ignored by conventional political art (cited in Chapter 1, p. 14).

How, Lefebvre asks, can society not fall apart? He answers that it is by 'language and metalanguage, by speech kept alive under talk at one or two removes, under floods of ink', so that only in everyday life can a cultural revolution occur and be necessarily total (Lefebvre, 2000: 197). This is similar to Marcuse's idea of a chasm, a complete rupture of the existing structure. But for all radical theory the question is what is to be done. For Marcuse it was to identify a quasi-biological mechanism that would reintroduce objectivity into the process; but he accepts a need for recognition of this new drive, and aligns the intelligentsia with the task. Hence the old institutions will tend to stay and the new consciousness either be marginal or adapt to the dominant reality. In the end he withdraws to art as the safe house in the occupied land. For Lefebvre, the possibility is to redefine the problem: not as a trajectory assumed to be in time which requires the engine of a revolutionary class, nor as the reconstruction of a revolutionary class or avant-garde to carry out the mission; but as recognition of what is already present in everyday life. This still denotes that someone must do the recognising, but the difference is first that what is to be recognised does not have to be brought into being but exists; and second that the acts of recognition are heterogeneous and already occur. The problem then is how recognition of these authentic moments will impact on power, or decentre it. Moments of presence may be overlooked, just as Lefebvre's theory of them has been and as lived spaces are in his theoretical construction, but the moments irrefutably occur. They are perhaps what in an earlier period were called moments of grace, which, after god's death in the nineteenth century we must now describe in secular terms, and are immanent. From this, the problem of how transformation arises can be formulated in terms of a difference between an *immanence*, in which the new pervades and permeates the present like wonder or the morning dew, and an *imminence*, which is the soon-to-come but also the looming and engenders the end which is nigh, the sublimation of radical desire. In the imminent world, freedom is displaced in time or given form in the beautiful illusions of bourgeois art, while avant-gardes *interpret* the world *for others*; in the immanent world the transformative consciousness is every day.

But if the meek inherit the Earth in the landlord's lifetime, this is not allowed. Power leaks but the interests of capital use extraordinary violence to increase their accumulated wealth. For members of the affluent society, immiseration is reproduced at higher levels as higher incomes fuel increased debts. This depends on both a relentless promotion of consumption in every crevice of ordinary life, which is then no longer ordinary (any more than reality television is real except in a technical sense); and on a compartmentalisation of work and leisure so that productivity remains unchallenged in its sphere while the world outside toil, which is potentially playful, is the sphere of leisure time.[56] Moments of liberation recur but are forced into increasingly marginal locations, though the marginal, as in the emergent literature of an architectural everyday (see Chapter 7) and the increasingly radical literature of development studies, begins to rupture the centre-margin model – a spatial metaphor.

Perhaps Marcuse was not so far from answering the questioner in Berlin, but just not in the way he developed, through a drive for liberation. Perhaps he could have been more interested in the drop-outs, some of whom withdrew from the dominant society to make alternative, intentional communities which, in some cases, continue today.[57] Perhaps there is a correlation between the counter-culture of the late 1960s and Lefebvre's proposed rediscovery of carnival 'magnified by overcoming the conflict between everyday life and festivity' (Lefebvre, 2000: 206). Maybe carnival is society as a work of art. But none of this detracts from a feeling that the histories of European revolt are largely histories of failure, and this leads me to argue that, alongside a need as Lefebvre advocated to reinstitute a right to the city, there is a critical possibility in art which has some currency.

What can it do? At the least, the banality of the administered world can be thrown back, as Adorno saw in the plays of Samuel Beckett:

> At ground zero . . . where Beckett's plays unfold like forces in infinitesimal physics, a second world of images springs forth, both sad and rich, the concentrate of historical experience that otherwise, in their immediacy, fail to articulate the essential: the evisceration of subject and reality. This shabby, damaged world of images is the negative imprint of the administered world.
>
> (Adorno, 1997: 31)

The shabby, damaged world of Beckett's texts is also, strangely, a world of intimacy in which the semi-articulated feelings of the protagonist(s) are paraded.[58] Perhaps a case can be made that Marcuse's commentary on Aragon indicates an ordinariness in liberation, as in Lefebvre's everyday, which is both disruptive and delightful, dispersed while power tends to centralise yet in its diffuse way a force for the diffusion of power (I am not the person to make this case, the words running away with me). Or perhaps there is a parallel between Lefebvre's idea of moments of presence and Marcuse's of art which 'breaks open a dimension inaccessible to other experience' and is not bound by the reality principle' (Marcuse, 1978: 72).

Finally, then, I glance at Marcuse's late writing on an aesthetic dimension in which autonomy is a prevailing element: 'The work of art can attain political relevance only as autonomous work. The aesthetic form is essential to its social function. The qualities of form negate those of the repressive society' (Marcuse, 1978: 53). But autonomy distances, enshrining art in a world where there are no streets. There is no shortage of gloom, either, as Marcuse and Adorno posit the same aporia. This revolves around the inability of art to depict without aestheticising. Marcuse realised this in the 1950s:

> As aesthetic phenomenon, the critical function of art is self-defeating. The very commitment of art to form vitiates the negation of unfreedom in art. In order to be negated, unfreedom must be represented in the work of art with the semblance of reality. This . . . subjects the represented reality to aesthetic standards and thus deprives it of its terror.
>
> (Marcuse, 1956: 144)

Adorno writes in a similar vein:

> Every artwork today, the radical ones included, has its conservative aspect; its existence helps to secure the spheres of spirit and culture, whose real powerlessness and complicity with the principle of disaster becomes plainly evident . . . Artworks are, a priori, socially culpable . . . Their possibility of surviving requires that their straining towards synthesis develop in the form of their irreconcilability.
>
> (Adorno, 1997: 234)

But, again on Beckett, that art exposes contradictions:

> Even where reality finds entry into the narrative . . . it is evident that there is something uncanny about this reality. Its disproportion to the powerless subject, which make it incommensurable with experience, renders reality unreal with a vengeance. The surplus of reality amounts to its collapse . . . The more total society becomes, the more completely it contracts to a unanimous system, and all the more do the artworks in which this experience is sedimented become the other of this society.
>
> (Adorno, 1997: 31)

Perhaps this is what there is, given the failure of 1968 and that in the Summer of Love a new sensibility did not emerge to transform the world as a timeless pleasure-ground.[59] Marcuse admits in *The Aesthetic Dimension* that 'Art cannot change the world' though he goes on 'but it can contribute to changing the consciousness and drives of the men and women who could change the world' (Marcuse, 1978: 32–3). At least he has left behind the universalised masculine of his earlier writing; but still he insists on the drives. In 1967, The Doors sang on their album *Strange Days* 'we want the world and we want it now'[60] which is a refusal of the reality principle and claim for a world of immediate gratification of desires like that implied in the idea of society as a work of art.

And, in the end (which is not, any more than liberation has solutions), there is no need to wait for tomorrow (which tends to reproduce today) if . . . if . . . if only . . . if all, in the present, the immanence of freedom.

NOTES

1 Kellner sees Marcuse as turning to aesthetics in dark or personally difficult times after 1918–19, after the death of his first wife in 1951, and after 1968 (Marcuse, 1998: 29–30).

2 Those responsible for the murders, if tried, received light sentences: a *Freikorps* officer convicted of the murders of Luxemburg and Liebknecht was given four months, a private two years (Willett, 1982: 46).

3 Even the non-engaged Rilke had his apartment searched because he was a poet (Willett, 1982: 47).

4 Marcuse studied Lukács' *The Theory of the Novel* (1920) in Freiburg, citing a passage in his thesis (Katz, 1982: 33, 47).

5 'The dissolution and tearing asunder of a unitary life-form, the opposition of art and life, the separation of the artist from the surrounding world, is the presupposition of the *Künstlerroman*' (Marcuse, '*Der deutsche Künstlerroman*', p. 322 in original version, cited in Katz, 1982: 154). The thesis was published in 1978 in the first volume of Marcuse's German collected writings (Frankfurt, Suhrkamp).

6 In *Death in Venice* the mental state of the composer, von Aschenbach, disintegrates in the atmosphere of Venice; his drives are not assimilable to the bourgeois life in which his cultural status grants a privileged place: 'he belonged to a different order of humanity, to a different world, and in the face of the Dionysian forces which have their roots in that humanity and that world, no heroism or determination could protect him . . . if they break through only once, they demolish the bourgeois existence, shatter the harmony, bring to ruin all stability and order' (*Der deutsche Künstlerroman*, p. 326, quoted in Katz, 1982: 51).

7 '[A] travelling community of musicians and mimes, but in particular young clerics and students . . . whose assault shatters the stability of the established and ecclesiastical restrictions . . . They are totally outcast, permanently excluded; for them there is no place in the life-forms of the surrounding world' (Marcuse, *Der deutsche Künstlerroman*, p. 13, quoted in Katz, 1982: 42, lacunae as in Katz). O'Donoghue (1982) links the troubadours to heretical movements in the south-west of France.

8 'At least this part of the Hippies, in which sexual, moral and political rebellion are somehow united, is indeed a non-aggressive form of life: a demonstration of an aggressive non-aggressiveness which achieves, at least potentially, the demonstration of qualitatively different values, a trans-valuation of values' (Marcuse, 1968b: 190). On Romantic bohemians see Wilson, 2000.

9 Marcuse hoped to use his text on Hegel as his Habilitation (the necessary post-

doctoral thesis for an academic position), but: 'At the end of 1932 it was perfectly clear that I would never be able to qualify for a professorship . . . under the Nazi regime' (Marcuse, 2001: 3). Kellner cites Jansen, P.-E. (1990) *Befreiung denken – Ein politscher Imperativ*, Offenbach, 2000 Verlag, on Heidegger's blocking of Marcuse qualification. Katz draws on conversations with Marcuse to give a different account in which Heidegger's turn to Nazism is a shock to Marcuse, who learns of it only after he has left Germany (Katz, 1982: 85). Marcuse's letter to Heidegger (after a visit within the de-Nazification programme) points out that Heidegger never renounced his texts of 1933 (Marcuse, 1998: 263–4).

10 See 'Critical and Traditional Theory' (Horkheimer, 1972: 188–252), and 'The Present Situation of Social Philosophy and the Tasks of an Institute for Social Research' (Horkheimer, 1993: 1–14). See Hoy and McCarthy, 1994: 22–4 for a commentary on Marcuse's 'Philosophy and Critical Theory' (in Marcuse, 1968a: 134–58). Horkheimer argues that the material of philosophy 'can only be understood in the context of human social life: with the state, law, economy, religion – in short with the entire material and intellectual culture of humanity' (Horkheimer 1993: 1). Kellner sees the Institute's use of the term critical theory as 'a code for the Institute's Marxism during its exile period' (Marcuse, 2001: 9).

11 In Marcuse, 1968a: 3–42, 88–133. In 'The Affirmative Character of Culture', Marcuse asserts a separation in Greek classical thought of a realm of goodness, truth and beauty from one of utility, and in bourgeois society a separation of an aesthetic realm from ordinary life, fostered through the education of sensibility. The contradiction of a proclamation of universal Liberty and its denial by the mechanisms of exchange in capitalism produces a displacement of freedom and joy to the aesthetic: 'Bourgeois society has liberated individuals, but as persons who are to keep themselves in check. From the beginning, the prohibition of pleasure was a condition of freedom' and 'Only in the medium of ideal beauty, in art, was happiness permitted to be reproduced as a cultural value in the totality of social life' (pp. 115 and 118). He adds: 'That individuals freed for over four hundred years march with so little trouble in the communal columns of the authoritarian state is due in no small measure to affirmative culture' (p. 125).

12 *Hegels Ontologie und die Grundlegung einer Theorie der Geschichtichkeit* (1932), Frankfurt, Suhrkamp; in English as *Reason and Revolution: Hegel and the Rise of Social Theory* [1941] (1967), London, Routledge & Kegan Paul. See also Marcuse, 1972: 95–127 for a critique of Hegel and pp. 128–43 on Marx.

13 Marcuse, 1998: 200–14. Kellner notes that the title page of the manuscript is dated September 1945, that there may have been several drafts, and that elements of the various drafts may have been intended for insertion in *Eros and Civilisation* and *The Aesthetic Dimension*. This attests the importance of this somewhat overlooked text, reprinted in Kellner's editing of the collected papers from *Theory, Culture & Society*, vol. 10 (1993), pp. 181–95.

14 Marcuse cites Eluard's '*Les Sept Poèmes d'Amour en Guerre*' (from a clandestine edition, 1943).

15 'Sensuality as style . . . expresses the individual protest against the law and order of repression. Sensual love gives a "promesse du bonheur" which preserves the full materialistic content of freedom and rebels against all efforts to canalize this "bonheur" into forms compatible with the order of repression. Baudelaire's . . . "Invitation au voyage" is indeed, in the face of a society based on the buying and selling of labor power, the absolute negation and contradiction . . . and, at the same time, the utopia of real liberation' (Marcuse, 1998: 204). Cf.: 'When a love relationship is at its height no room is left for any interest in the surrounding world; the pair of lovers are sufficient unto themselves.' (Freud, *Civilization and its Discontents*, 1949 edition, London, Hogarth Press, p. 26, in Marcuse, 1956: 41).

16 'All the others live with and without their love . . . In contrast, Aurélien's and Bérénice's relationship binds itself to a "promesse du bonheur" which transcends the happiness of the others as much as a free order of life transcends all liberties within the established order of life' (Marcuse, 1998: 210). Similarly Marcuse writes in the *Aesthetic Dimension*: 'The sensuous substance of the Beautiful is preserved in aesthetic sublimation. The autonomy of art and its political potential manifest themselves in the cognitive and

emancipatory power of this sensuousness' (Marcuse, 1978: 66).

17 Marcuse writes of an inherent redeeming power in art, but: 'The aesthetic necessity of art supersedes the terrible necessity of reality, sublimates its pain and pleasure; the blind suffering and cruelty of nature . . . assume meaning and end . . . The horror of the crucifixion is purified by the beautiful fact of Jesus dominating the beautiful composition, the horror of politics by the beautiful verse of Racine . . . everything is in order again. The indictment is cancelled, and even defiance, insult, and derision – the extreme artistic negation of art – succumb to this order' (Marcuse, 1978: 49–50). Adorno, similarly writes in *Aesthetic Theory*: 'Artworks detach themselves from the empirical world and bring forth another world, one opposed to the empirical world as if this other world too were an autonomous entity. Thus, however tragic they appear, artworks tend a priori toward affirmation' (Adorno, 1997: 1). And, in a passage predating Marcuse's above: 'The artwork is not only the echo of suffering, it diminishes it; form, the organon of its seriousness, is at the same time the organon of the neutralization of suffering. Art thereby falls into an unsolvable aporia' (Adorno, 1997: 39).

18 Cf.: 'Form is the negation, the mastery of disorder, violence, suffering, even when it presents disorder, violence, suffering. This triumph of art is achieved by subjecting the content to the aesthetic order, which is autonomous in its exigencies' (Marcuse, 1969: 49).

19 Marcuse argues that while art in bourgeois society offers illusions its status is as a container of truths 'proper to art', divesting external reality of a claim to totality: 'Art's unique truth breaks with both everyday and holiday reality, which block a whole dimension of society and nature. Art is transcendence into this dimension where its autonomy constitutes itself as autonomy in contradiction. When art abandons this autonomy . . . art succumbs to that reality which it seeks to grasp and indict' (Marcuse, 1978: 49).

20 'The various phases and trends of anti-art or non-art share a common assumption – namely, that the modern period is characterized by a disintegration of reality which renders any self-enclosed form . . . untrue, if not impossible . . . This assumption is in flat contradiction to the actual state of affairs. Rather, the opposite is the case. We are experiencing, not the destruction of every whole, every unit or unity, every meaning, but rather the rule and power of the whole, the superimposed, administered unification. Not disintegration but reproduction and integration of that which is, is the catastrophe' (Marcuse, 1978: 50).

21 Marcuse, 2001: 95–106.

22 'Formulations Regarding the Two Principles in Mental Functioning' (1911), in Freud, 1991: 29–44.

23 Freud uses the terms *Es* and *Ich* as everyday terms.

24 This is not to say Freud lacked political awareness or affiliation. Fromm records Freud's identification with Victor Adler, leader of the Austrian Social Democratic Party, to the extent of renting the apartment at Berggasse 19 in which Adler had lived; he reads this not as unconscious memory of a visit to the apartment but an 'ambition to become a great political leader' (Fromm, 1959: 71). Marcuse sees Freud as setting his model of drives 'in a socio-historical world' (Marcuse, 1956: 12). Freud writes in 'An Autobiographical Study' (1935): 'I perceived ever more clearly that the events of human history, the interactions between human nature, cultural development and the precipitate of primeval experiences . . . are no more than a reflection of the dynamic conflict between the ego, the id and the superego, which psychoanalysis studies in the individual' (cited in Chasseguet-Smirgel and Grunberger, 1986: 34).

25 'we have become accustomed to taking as our starting-point the unconscious mental processes . . . the residues of a phase of development in which they were the only kind of mental process. The governing purpose . . . is described as the pleasure-unpleasure principle, or more shortly the pleasure principle' (Freud, 1991: 36). Freud explains that the reality principle is a gradual mediation of the pleasure principle in face of an external reality which is not always pleasurable, and summarises its momentum as (1) a heightening of sense awareness leading to a faculty of judgement; (2) a splitting-off of an area of mental

activity as phantasy; (3) realisation of delay in gratification; (4) a construction of myths of future reward; (5) education as imposition of the reality principle; (6) a reconciliation of the two principles in art; (7) a transition from auto-eroticism to procreation; (8) the freedom of unconscious processes from reality-testing (Freud, 1991: 37–42).

26 'The conflict between sexuality and civilization unfolds with this development of domination. Under the rule of the performance principle, body and mind are made into instruments of alienated labour; they can function as such instruments only if they renounce the freedom of the libidinal subject-object which the human organism primarily is and desires' (Marcuse, 1956: 46).

27 'For the middle-class thinkers of Freud's time, man was primarily isolated and self-sufficient' and although people need each other for profitable exchange and love 'they remain basically isolated beings, just as vendor and buyer on the market do; while they are drawn to each other by the need to satisfy their instinctual desires, they never transcend their fundamental separateness' (Fromm, 1959: 98).

28 Cf.: 'the substitution of the reality principle for the pleasure principle implies no deposing of the pleasure principle . . . A momentary pleasure, uncertain in its results, is given up, but only in order to gain along the new path an assured pleasure at a later time . . . The doctrine of reward in the after-life for the – voluntary or enforced – renunciation of pleasures is nothing other than a mythical projection of this revolution in mind' (Freud 1991: 41).

29 Marcuse follows Freud's definition of the *Es* as the largest and oldest layer of mental structure: 'free from the forms and principles which constitute the conscious, social individual. It is neither affected by time nor troubled by contradictions . . . all it strives for is satisfaction of its instinctual needs'; he continues that, conditioned by external reality, part of the *Es* gradually develops into the *Ich*, to mediate between the *Es* and the external world and preserve its existence, 'observing and testing the reality . . . adjusting itself to the reality, and altering the latter in its own interest' (Marcuse, 1956: 29–30).

30 'The memory of gratification is at the origin of all thinking, and the impulse to capture past gratification is the hidden driving power behind the process of thought' (Marcuse, 1956: 31). Marcuse cites Freud's *The Interpretation of Dreams*, 1938 edition, New York, Modern Library, p. 535.

31 For Freud '*Art* brings about a reconciliation between the two principles in a peculiar way. An artist is originally a man who turns away from reality because he cannot come to terms with the renunciation of instinctual satisfaction which it at first demands, and who allows his erotic and ambitious wishes full play in the life of phantasy. He finds the way back to reality . . . by making use of special gifts to mould his phantasies into truths of a new kind . . . valued . . . as precious reflections of reality' (Freud, 1991: 42). For Marcuse, 'The truths of imagination are first realized when phantasy itself takes form, when it creates a universe of perception and comprehension – a subjective and at the same time objective universe. This occurs in *art*' But the reality reflected is charged: 'behind the repressed harmony of form lies the repressed harmony of sensuousness and reason – the eternal protest against the organization of life by the logic of domination, the critique of the performance principle' (Marcuse, 1956: 143–4).

32 Coincidental similarities exist between part of *Eros and Civilisation* and Bloch's *The Principle of Hope*: Bloch seeks to establish hope as an extension of the self-preservation drive (Bloch, 1986: 51, 64); they are both interested in heretical groups in European history; and while Marcuse rejects the objectively given end of history which Bloch retains, Bloch (from Rosenzweig – see Levy, 1997: 177) adapts it as a redemptive freedom (Cf.: Benjamin, 1970: 255–66). But I can see the memory of Eden as only a memory of infancy (see Winnicott, 1986: 21–34; Fuller, 1988: 130–238).

33 See Nagel, 2002: 7–28 on play in the pre-Socratic world.

34 'Culture demands continuous sublimation; it thereby weakens Eros, the builder of culture . . . desexualization, by weakening Eros, unbinds the destructive impulses. Civilization is thus threatened by an instinctual de-fusion, in which the death instinct strives

to gain ascendency over the life instincts. Originating in renunciation and developing under progressive renunciation, civilization tends towards self-destruction' (Marcuse, 1956: 83). Stirk comments that the image of Eros as a culture-builder supports a reading of Freud as already integrating sociological and cultural factors into his theories; and that Marcuse was able to insert Freud's theory into a 'tradition of political thought that emphasized the erotic dimension of human community' (Stirk, 2000: 84).

35 Marcuse was fluent in German, French and English, and in 1967–8 participated in a UNESCO symposium on 'The Role of Karl Marx in the Development of Contemporary Scientific Thought' in Paris in May 1968, spoke to students at the Sorbonne, Ecole des Beaux-Arts and Nanterre (Katz, 1982: 185, n. 56), met Nguyen Than Le of the North Vietnamese delegation to the Paris peace talks, and visited student leader Rudi Dutschke in hospital in Berlin, as well as being at the Roundhouse and the Free University (Katz, 1982: 181–7). Other contributors at the Dialectics of Liberation Congress included R. D. Laing, Paul Goodman and Stokely Carmichael (Cooper, 1968). The five Berlin lectures are: 'Freedom and Freud's Theory of Instincts'; 'Progress and Freud's Theories of Instincts'; 'The Obsolescence of the Freudian Concept of Man'; 'The End of Utopia'; and 'The Problem of Violence and the Radical Opposition' (Marcuse, 1970 – German texts in *Psychoanalyse und Politik* (Frankfurt, Suhrkamp, 1968). These visits followed attempts by the American Legion to buy out his contract at San Diego, where he moved when his contract at Brandeis was allowed to lapse in 1965, the year of the first protest against the US war in Vietnam. He was pronounced unqualified to teach by then-Governor Ronald Reagan, and forced into hiding by a bomb threat (Katz, 1982: 169, 174–6).

36 See Galbraith (1958) *The Affluent Society*, in which he argues for a revision of the primacy of production.

37 See note 8 on Marcuse on the Hippies. On the English Diggers, see McKay, 1996: 18, 60, 134; on Diggers in San Francisco see Farrell, 1997: 219–22 and Braunstein and Doyle, 2002: 29–30, 78–9. On the historical Diggers of the English Revolution in the 1640s, see Petegorsky, 1999: 153–76.

38 The 1967 papers follow *One Dimensional Man*, published in 1964: 'Faced with the possibility of pacification on the grounds of its technical and intellectual achievements, the mature industrial society closes itself against this alternative. Operationalism, in theory and practice, becomes the theory and practice of *containment*. Underneath its obvious dynamics, this society is a thoroughly static system of life: self-propelling in its productivity and in its beneficial coordination' (Marcuse, 1964: 17). The book established Marcuse as a public intellectual, selling 100,000 copies in the USA within five years, translated into sixteen languages and reviewed in *Fortune* magazine (Katz, 1982: 168).

39 Of the utopian quality of the vision he writes: 'this is precisely the form in which these radical features must appear if they are really to be a definite negation of the established society: if socialism is indeed the rupture of society, the radical break, the leap into the realm of freedom – a total rupture', and cites Benjamin's anecdote of the shooting of public clocks during the Paris Commune (Marcuse, 1968b: 177).

40 Chapter 1 in the *Essay on Liberation* (1969) is titled 'A Biological Foundation for Socialism'. Marcuse argues that in an affluent society commodity production and exploitation permeate life; and that such resistance as is not suppressed by force is found in 'the diffused rebellion among the youth and the intelligentsia, and in the daily struggle of persecuted minorities' (p. 17). After an excursion into the obscenity of capitalist wealth accumulation, he asserts that consumerism has produced 'a second nature of man which ties him libidinally and aggressively to the commodity form' (p. 20). Class consciousness is then engineered by consumerism as servitude, so that complicity in the system is fostered 'in the instinctual structure of the exploited' so that the continuum of repression remains intact (p. 25). Liberation now requires 'subversion against the will and against the prevailing interests of the great majority of the people.' (p. 26).

41 Schnädelbach sees the blocking of revolt, when reform is seen as complicity but revolution distant, as an 'attractive argument for political abstinence' on the part of intellectuals, while the alternative of action leads to leftist terror, to which the reaction in the left is non-violence (Schnädelbach, 1999: 76).

42 The Summer of Love took place in San Francisco in 1967, after the Freedom Summer of 1964 (for civil rights). Hippies converted nineteenth-century houses in Haight-Ashbury into collectively run micro-communities. The Diggers were a presence here, declaring the death of money and birth of freedom (Farrell 1997: 219–27; Braunstein and Doyle, 2002: 94–5, n. 27).

43 See Seiler, 2000.

44 Shields observes the privileging of Lefebvre's spatial theories over his idea of moments: 'the Marxist geographer David Harvey is one of the only English-language writers to take notice in print' (Shields, 1999: 58); and adds in a footnote that Soja and Gottdiener know this aspect of Lefebvre, but that their professional interests as geographers marginalise it. Lefebvre first articulated a theory of moments in an essay '*La pensée et l'esprit*', *L'Esprit*, 1, May 1926, pp. 21–69. Shields sees the idea as a rejection of Bergson's duration (*durée*) which remains linear in the manner of progress. A moment for Lefebvre is not a point on a line towards a future, whether objectively given or not, but a variable sense of time passing in a present: 'In the moment, one does not feel the passage of time . . . we need to think of *lived time* qualitatively' (Shields, 1999: 59). This connects to the third category in Lefebvre's spatial theory, that of representational spaces (Lefebvre, 1991: 38–9). For a commentary on Lefebvre's *Critique of Everyday Life* (1947) see Gardiner, 2000: 74–86.

45 Shields cites the French (1961) edition of Lefebvre's *Critique de la view quotidienne, II: Fondements d'une sociologie de la quotidienneté*, Paris, L'Arche, pp. 341–57. Gardiner summarises: 'Lefebvre stresses that the everyday represents the site where we enter into a dialectical relationship with the external natural and social worlds in the most immediate and profound sense, and it is here where essential human desires, powers and potentialities are initially formulated, developed and realized concretely' (Gardiner, 2000: 75–6).

46 *La view quotidienne dans le monde moderne*, Paris, Gallimard; first translated into English in 1971 (New York, Harper & Row and Harmondsworth, Penguin) and reprinted (the edition used) in 2000.

47 McLeod summarises: 'Everyday life embodies at once the most dire experiences of oppression and the strongest potentialities for transformation' (McLeod, 1997: 14). Gardiner contrasts Adorno's quip that while there is no universal human history there is one from the catapult to the megaton bomb to Lefebvre's view of modern society as containing 'both repressive and emancipatory qualities' (Gardiner, 2000: 77).

48 Lefebvre sets out three spatial concepts: perceived space, in which every society develops characteristic (and ideological) ways of perception and ordering space; conceived space, which is generally the dominant space and employs a vocabulary of signs to represent space, as in the plan; and lived space, where meanings are constantly reconstructed for the spaces of material, bodily life. He defines lived space ('representational spaces' in the English translation) as 'the space of "inhabitant" and "users", but also of some artists and perhaps those, such as a few writers and philosophers, who *describe* and aspire to do no more than describe. This is the dominated . . . space which the imagination seeks to change and appropriate' (Lefebvre, 1991: 39). On conceived space he writes: 'Within the spatial practice of modern society, the architect ensconces himself in his own space. He has a *representation of this pace*, one which is bound to graphic elements . . . This *conceived* space is thought by those who make use of it to be *true*, despite the fact – or perhaps because of the fact – that it is geometrical: because it is a medium for objects, an object itself, and a locus of the objectification of plans' (Lefebvre, 1991: 361). But, as he earlier clarifies, the spaces of bodies are not abolished in the onslaught of the plan: 'This is not to say that during this period [of perspectival space] in Italy . . . townspeople and villagers did not continue to experience space in the traditional emotional and religious manner . . . by means of the representation of an interplay between good and evil forces at war throughout the world, and especially in and around those places of special significance for each individual' (Lefebvre, 1991: 79).

49 See Fischer, 1970: 152–62: 'By observing and interpreting changes as they occur, the philosophy of practice gains cognition which is never at a standstill; by the never ending endeavour to affect the development of the changing world, it continually transcends

pure observation and interpretation and becomes a motive force of practice (p. 157).

50 See Shields, 1999: 89–92, 103–7 on Situationism, and Harvey's Afterword in *The Production of Space* on Lefebvre's response to criticism from the Situationists, and the similar status within student movements in the USA and France of Marcuse and Lefebvre (in Lefebvre, 1991: 430). McLeod points out that Lefebvre was associated with the Surrealists in the 1920s, and the CoBrA group of Asger Jorn, Christian Dotrement and Constant Nieuwenhuis in the late 1940s. She sees the Situationists as deriving much of their critical apparatus from Lefebvre (McLeod, 1997: 21). See also Sadler, 1999: 19–20 on Lefebvre's impact on French sociology and planning.

51 Illustrated in *Internationale situationiste* 6 (1961) – see Sadler, 1999: 53, fig. 1.23.

52 Sadler finds resonance between Situationism and Lefebvre's recognition of a need for praxis in urban studies to engender an experimental utopia and deployment of an imaginary in the production of new concepts of urban life (Sadler, 1999: 47). Starr notes a parallel move on the part of Lefort, after 1968, towards cultural and everyday activities as an exit from the impasse of political debate (Starr, 1995: 25).

53 Massey takes Laclau to task in *Space, Place and Gender* on the grounds that he reproduces dualism in polarising space and time, to which she compares currents in radical geography and gender studies: 'the dichotomous characterization of space and time . . . may both reflect and be part of the constitution of . . . masculinity and femininity of the sexist society in which we live' (Massey, 1994: 259). Douglas also points out the tendency for one term in a duality to be privileged, and affirmed in a gendered division of labour: 'From simple complementarity a political hierarchy has been derived' (Douglas, 1987: 49).

54 'Situationists uncovered the social body of "the naked city" by becoming streetwise. Drifters were effectively vagrants . . . The passages of the drift were lined with cheap shops and cafés; the ghettos offered not only an "ambient other" but also nonbourgeois, nontourist cost of living' (Sadler, 1999: 93). The idea of the drift as vagrancy has, too, a resonance with Foucault's history of the *Hôpital Général* from 1656 as containment of the non-productive, the vagrant and insane (Foucault, 1967: 38–64).

55 Shields quotes Greil Marcus on the 'tiny epiphanies' of moments in everyday life 'in which the absolute possibilities and temporal limits of anyone's existence were revealed' (Marcus, 1989: 144, quoted in Shields, 1999: 61). The millenarian society of immanence is described by Bloch in his reference to Joachim of Fiore – Bloch, 1986: 509–15.

56 In 'The Stars Down to Earth', Adorno comments on the split between a realm of work and a realm of leisure: 'The idea is that by strictly keeping work and pleasure apart, both ranges of activity will benefit: no instinctual aberrations will interfere with seriousness of rational behaviour, no signs of seriousness and responsibility will cast their shadow over the fun. Obviously this advice is somehow derived from social organization which affects the individual as much as his life falls into two sections, one where he functions as a producer and one where he function as a consumer' (Adorno, 1994: 71).

57 See Schwartz and Schwartz, 1998: 269–90.

58 See Keller, 2002: 133–71 on Beckett's *Waiting for Godot*.

59 'The new sensibility has become . . . *praxis* . . . negation of the entire establishment, its morality, its culture; affirmation of the right to build a society in which the abolition of poverty and toil terminates in a universe where the sensuous, the playful, the calm, and the beautiful become forms of existence and thereby the *Form* of society itself (Marcuse, 1969: 33).

60 From 'When the Music's Over', *Strange Days*, New York, Electra 7559-62548-2 (first issued 1967, various re-issues in CD).

5

1989

AFTER THE WALL

•

In Chapter 1 I took the destruction of the Vendôme Column during the Paris Commune as a point of departure for discussion of the Realist avant-garde. In this chapter I begin with the dismantling of the Berlin Wall in 1989. Both chapters reconsider monuments but whereas in Chapter 1 the destruction of a monument denotes the counter-narrative of a revolt that failed soon after, in this chapter the dismantling of the Wall marks the encapsulation in history of the political system engendered by a continuation by other means and with other strands of that counter-narrative. But the Wall was and is perceived ambiguously: from the side of the German Democratic Republic (GDR) it was a border security system; from the West a surface for graffiti, which lent it the semblance of a monument to free expression. These ambiguities were enhanced when sections of the Wall were resited in the free-market world. In this chapter, then, I take the dismantling of the Wall as a point of departure from which to examine ambiguities of meaning in public monuments. I see these, in both their implicit and explicit contents, as shaping a society's representation of itself, and as such sites of imposition and intervention. Because there are vicissitudes in representation, I ask to what extent the meanings of generic forms such as the memorial can be renegotiated; and whether it is viable to democratise the genre of the monument or subvert it from within. Finally, I add a brief note on the ambivalent attraction of demolition, which connects to the theme of ruins that runs through Chapter 6, where counter-monuments in Germany are discussed in relation to intervention in narratives.

I THE BERLIN WALL

> The East Germans voted with their feet. Since the beginning of that year the number of people seeking refuge in the Federal Republic had risen spectacularly. In early September Hungary's relatively liberal leaders, breaking

5.1 • A section of the Berlin Wall in Manhattan

> the Communist rules of the game, allowed those refugees to freely cross the Austrian border. Prompted by Gorbachev to reform, pushed from below by mass demonstrations, particularly numerous in Leipzig, the German Communists dropped their old leader . . . on October 18. The new leadership . . . did not know how to deal with the mounting tide of protestors. On the night of November 9 it yielded and promised to grant everybody a permit to go west. West Berlin was then invaded by a human flood. The day after, the bulldozers went into action. The crumbling of the Wall, erected back in 1961, was the symbolic climax of this East European movement.
>
> (Singer, 2001: 12)

Daniel Singer sees the dismantling of the Wall in terms of a mass movement. The people moving into West Berlin might almost have been singing revolutionary anthems. Seventy years after the failure of the German Revolution they ran, however, not towards the dawn of a new society but to what would become for them the uncertainties of a market economy designed to ensure the triumph of a new world order. Although the end seemed sudden, it had begun in the previous decade when the leadership of states in the East block looked to a western-style consumerism without the means of delivery.[1] There was the obsessive security apparatus and the constant informing by one section of the population on the rest so that most felt insecure (which produced, particularly in the GDR, such a quantity of data that its handling constituted a growth industry). There were casualties, such as those who were imprisoned and those who were caught or killed fleeing across the wire, and even in the relative comfort of the academic sphere, for instance, Ernst Bloch (who returned to accept a chair in philosophy at Leipzig in 1949 because he believed in the Marxist state) was declared a revisionist under the Ulbricht regime.[2] But although much went badly wrong, citizens were protected from economic swings and production was organised to provide work, prices were controlled and in the 1970s there had been an economic upturn (which did not last, fomenting the dissent that erupted in 1989).

I draw attention to variance in possible interpretations of these events not to take a position – I am not a German – but because it opens other areas of ambiguity, in readings of the Wall, and from that of other public expressions of values: the informal (graffiti) and the formal (monuments). But back to the Wall: the Border Security System for the National Boundary West, to use its name, first appeared as a barbed-wire fence on August 13th, 1961 in Berlin. Blocks of reinforced concrete were added at street crossings and the fence soon became a wall 30-cm thick, topped by breeze-blocks and barbed wire. Beyond the city a fence of several layers of wire ran through the forests, guarded by watch-towers. The next Wall was produced industrially, using prefabricated sections. Buildings beside it were either cleared or their windows bricked-up, while listening devices were used to detect tunnelling. On the west side, someone painted 'EIN DEUTSCHLAND' at Potsdamer Platz. In its final form the Wall was 4.1 metres high and 1.8 metres thick, and a site of artistic activity.[3] In 1986, New York graffiti artist Keith Haring painted a 200-metre section in pictographic style; Christophe Bouchet's *Hommage à Duchamp* fuses New York graffiti with avant-garde art, combining spray-can figures in vivid colours with a ceramic urinal.[4] There was even a competition for designs for the Wall, with a first prize of 3,000 DM.[5] Winfrid Hagendorf's entry adapts the visual code of a standard German road sign: Berlin with a vertical arrow in the top box, and Berlin crossed by a red diagonal in the lower. Many people simply wrote on the Wall in German or American: 'MAUERKUNST LEBENKUNST' and 'PIRATE ART Bye Bye Berlin'.

When restrictions on movement to the West were lifted, the Wall ceased to have a function. Individuals hacked at it and took pieces away to prove they had been there while a whole section was removed to Texas A&M University and sited next to a burial ground for University mascots; on April 21st, 1993 George Bush (senior) gave a speech in front of it.[6] Another section with characteristic graffiti is sited in a plaza in Manhattan, near the Museum of Modern Art. White garden furniture has been placed in front of it; coffee and bagels are available. To resite the Wall is to recontextualise it, but its recoding is ambiguous: it is in one way a monument to victory, brought as the spoils of war like an enemy's standard; in another it is almost decorative, like a piece of street furniture in a post-Cold War climate in which it recedes into a past.

As the latter, it is added to a genre conventionally value-free, which makes its recontextualisation a decontextualisation of sorts. But there is a further more fracturing ambiguity: graffiti on the West face of the Wall signifies a freedom of expression prevented in the East. The *New York Times* printed pictures of graffiti appearing for the first time in the East in 1989: 'another sign of the newfound freedom springing up next to McDonald's and the polling booth' (in Cresswell, 1996: 45). In the 1980s graffiti appeared also on the New York Subway, using techniques like those of the Wall and with equal invention (a key quality of western art); and if subway graffiti emphasised tagging rather than ideology, this takes the signature (also important in art) as image. But subway graffiti appeared by night to denote the presence of an underclass. Richard Sennett writes in *Conscience of the Eye*: 'These endless monster labels might be rather grandly described as the making of a theatrical wall, the subway cars treated as a neutral backcloth to be brought to life by dramatic gestures' (Sennett, 1990: 205). He cites passengers' reactions to graffiti as indicating 'the subway is a dangerous place' (Sennett, 1990: 206),[7] yet his analogy of a theatrical backdrop would suit the white-painted surface of the Wall. There is an ambivalence reminiscent of the use of abstract art in the 1950s: some US Senators fulminated against it as a communist plot, but paintings consisting of non-referential drips and smears were toured internationally to state a western aesthetic and antidote to Socialist Realism.[8] This divergence of readings of graffiti as threatening disorder, or as licensed disorder,[9] turns attention back to the Wall. Hilary Lawson, for example, refuses the conventional term 'fall':

> The Wall did not fall, and alternative descriptions such as: the abandonment of the border by the East German guards, the euphoric meeting of East and West Germans on the border in Berlin, the mass chipping away at the concrete wall, are no closer to some supposed real event that lies beyond each and every description.
>
> (Lawson, 2001: 137)

This begs the question as to whether public monuments in general are open to recoding, and whether there is scope to intervene in the process. The destruction of the Vendôme Column was an attack on a sign of oppression, its toppling by allusion a toppling of the oppression it signified. Something like that could be said of the Wall (which has alternative readings). The abuse expressed towards the statue of Napoleon on top of the Column was a reversal of power, its enactment a necessary demonstration of this. Similarly, in Budapest in 1956, a statue of Stalin was abused by spectators.[10] But meanings can be shifted through irony, too, as in Krzysztof Wodiczko's projection onto Nikolai Tomsky's *Lenin Monument* (1970) in Berlin in 1990: Lenin is now a Pole in a red and white striped T-shirt, pushing a shopping trolley of electrical goods to sell in Warsaw.[11] The counter-reading, in which both Lenin and the proletariat assume new roles, depends on the monument's survival as foil for the projection. There were informal adaptations, too, around this time: 'No Violence' on a sash on the *Lenin Monument*; and in 1990, '*Wir sind unschuldig*' (we are innocent) on the base of Ludwig Engelhardt's *Marx-Engels Monument*.[12] I return to Wodiczko's projections as counter-monuments below but first want to look at what happened to the statues.

The *Lenin Monument* was destroyed in 1991. Some of its stones were taken to the Friedrichshain cemetery and placed on the graves of Luxemburg and Liebknecht.[13] In Moscow, a statue of Felix Dzerzhinsky was craned out of a square at the Lubiyanka prison in front of the Western news media on August 22nd, 1991, the second day of Yeltsin's coup. The *New Yorker* reported 'monuments all over the country – fierce icons of the longtime socialist-realist hegemony – were being toppled and carted off' (in Levinson, 1998: 13–14). This had happened before when Khrushchev revised Stalin's place in history, but after 1989 was more indiscriminate. For a mix of reasons, however, the statues in some cities were not destroyed but re-presented, rather as sections of the Wall were recoded in New York and Texas. When, for instance, the price of bronze fell so that reintegration into the cycle of production ceased to be viable,[14] a bronze foundry at Gliwice in Poland started to take the statues back with the idea of making a museum for them. In Moscow the statues were put in a Temporary Museum of Totalitarian Art. In Budapest they were removed to *Szóbórpark* (Statue park) on the city's outskirts, designed by Akos Eleod;[15] only the Soviet War Memorial on the graves of Soviet soldiers killed in the liberation of 1945 remains in its site opposite the US Embassy. And in Lithuania in 1999 the Soviet-period statues were collected in a park at Grutas, set in a peaceful landscape but in order to remind the next generation of a period seen not only as of Stalinist repression but also, in a newly nationalist culture, of Russian occupation.[16] In *Disgraced Monuments* (1992, Channel 4), Laura Mulvey and Mark Lewis see the monuments in store as having an uncanny presence. Mulvey argues that their meanings are destabilised:

> their disgrace and removal may encapsulate, as image and emblem, the triumphal overthrow of an *ancien regime* for which they had presented a public face . . . [or] their ultimate fate raises questions about continuity and discontinuity, memory and forgetting, in history; about how, that is, a culture understands itself across the sharp political break of revolution.
>
> (Mulvey, 1999: 220)

Similarly, Judith Rugg writes that in creating *Szóbórpark* 'the authorities divested themselves of any obligation to remember and also relieved the viewer of the burden of memory' (Rugg, 2002: 8). But Renata Salecl sees Mulvey's position as a view from the West which assumes 'the current and former rulers do not differ in how they deal with historical memory' (Salecl, 1999: 99). She accepts that to erase monuments allows the past they represent to be romanticised, but argues that after major upheavals such erasures are necessary: 'If we take the case of post-Hitler Germany, one does not expect to see the Fuhrer's pictures in public places' (Salecl, 1999: 99). The removal of objects from public spaces was not, as it happens, confined to the ex-East bloc. In 1989, Richard Serra's *Tilted Arc*, a wall of Cor-Ten steel plates, was craned by night out of Federal Plaza, New York after orchestrated protests from users of the space and a defence by the art world. I wonder, simply, what kind of park would have been made for the western equivalent of the disgraced monuments had the crowds poured in the opposite direction – from West to East. What would have become of Jonathan Borofsky's *Hammering Man* (1992) sited outside the European Bank in Frankfurt (and outside the Fine Arts Museum in Seattle)?[17] Or Raymond Mason's *Forward* in Centenary Square, Birmingham? This resin sculpture of the city's tradespeople seeming to march to the bright dawn represented by the new Convention Centre was installed while western newsreels showed statues being removed in Moscow; it was described as part of an urban renaissance, but in a local paper as Red Square coming to Birmingham. As it happens, *Forward* (in resin not steel) met a more arbitrary fate than the disgraced monuments in 2003.[18]

I ask the question above, if fancifully, to draw attention to a discrepancy in conventional readings of public art. Of course, the removal of *Tilted Arc* cannot be compared to that of the statues of Dzerzhinsky in Moscow and Warsaw. As founder of an oppressive security apparatus Dzerzhinsky was a hated figure; the removal of his statues was, like the destruction of the Vendôme Column, an act of the kind Salecl sees as necessary. *Tilted Arc* ruined people's lunchbreaks, not their lives. But public edifices are where and as they are not simply by regulation in a technical sense. As Kirk Savage writes, they 'do not arise as if by natural law to celebrate the deserving; they are built by people with sufficient power to marshall (or impose) public consent for their erection' (cited in Levinson, 1998: 63). So, if graffiti on the West side of the Wall expressed autonomy, the genre of abstract, steel sculptures typified by Serra (or its figurative equivalents) expresses the implicit values of the West – in London's Broadgate, Serra's *Fulcrum* functions as a piece of blue-chip art affirmative of the status of a blue-chip development site. *Tilted Arc* and *Forward*, or Antony Gormley's cruciform iron men on the walls of Derry,[19] or the bronze likenesses of Cary Grant, Thomas Chatterton, and John Cabot in a new public space in Bristol, are as ideological in their way as the disgraced monuments. The term 'sculpture' may be used to set them apart from the non-art genre of statues, but do they not as much, if sometimes obliquely, reproduce the categories of a system of social ordering and thereby play the part of public monuments in maintaining a regime?

II (EXTRA)ORDINARY MONUMENTS

When post-modern culture entails fusions of art, mass media and consumption, the purist aesthetic of late modernism (as expressed by Serra) begins to seem past its sell-by date. The universality this art claimed was in any case undone from within in Clement Greenberg's call to keep art moving.[20] Style change is modern art's equivalent of regime change, and regime change destabilises the notion of power as universal. Although scale and grandeur in monuments are used to state permanence, implying that each regime is a culmination of history, it may be that such monuments conceal the instability of the situation. The *Millennium Monument* in Budapest was conceived in 1881 (and built between 1896 and 1929) to commemorate a thousand years of the Hungarian kingdom, yet in effect to state the merger of Hungary into the Austro-Hungarian Empire. A central column supports the Archangel Gabriel and is flanked by two colonnades within which were placed statues of Stephen, first Christian King of Hungary (crowned in 1001), and Franz-Joseph, Emperor of Austria-Hungary (crowned in 1848). While the monument was completed, the dual-monarchy drifted to the war with Serbia which undid it. In 1918, the short-lived socialist regime removed the Hapsburg figures; then the counter-revolutionary regime redesignated the monument as a memorial to the Hungarian dead of the war. The site was renamed Heroes Square by the fascists. In 1945 statues of the revolutionary Istvan Bocksay and leader of the 1848 insurrection Lajos Kossuth were placed where Franz-Joseph and King Stephen stood before,[21] and Andrásy Avenue, which leads to it, renamed after Stalin. The monument survives, but its meaning is clearly mutable. This suggests a possibility to interrupt perceived meanings and nudge them in a new direction, exposing a narrative as sham or countering it. One form of interruption is the renaming of statues: Marx and Engels as the pensioners, or sculptures of female deities in fountains in Birmingham and Dublin as floozies in jacuzzis.[22] A direct approach was the IRA's attempt to blow up statues of English kings in Dublin.[23] But in the context of this chapter the question is whether the genre of the monument can be renegotiated (democratised or subverted from within), using the tensions of meaning created in Wodiczko's projections or by adding to it in ways expressive of counter-meanings. A set of cases relevant here comprises counter-monuments referencing the Holocaust in Germany, discussed in the next chapter. Here I look to efforts to recode street-level figurative sculpture and at public memorials, to Wodiczko's projections again, and to monuments of a different kind – to industrial history – by Herman Prigann (one of whose works, also, occupies a site on the ex-border, in the Harz mountains in Germany).

In the nineteenth century, statues were put on plinths to make people look up to them. This confirmed the privileges of those represented, their status as model citizens (or colonialists), and the power of the establishment to put them there. In the early twentieth century, avant-garde sculpture was taken off the plinth as a gesture against the privileged status of the museum object. Sculpture on the floor is now commonplace in galleries; in public streets life-size likenesses of historic, folkloric or literary figures similarly stand at street level. For example, James Joyce and Molly Malone stand on street corners in Dublin, Fernando Pessoa sits outside a café he used to frequent in Lisbon and I. K.

Brunel sits under the arches at London's Paddington Station. Everyone can touch these figures; children can climb over them because the street is not a museum, while tourists like to be photographed next to them. But the ordinariness of the figures does not bring to earth the individuals depicted, who remain celebrities in as much as those who seek vicarious status by being photographed next to them are not. In Bristol, the bronze figures from the city's history are all white men; they reproduce the usual exclusions while affirming the dreams of the culture industry, as Adorno calls it.[24] Perhaps street-level likenesses are, far from egalitarian, successors to the bronze figures from military, commercial and cultural histories found in, say, New York's Central Park.[25] I would say the same of any recruitment of citizens described as ordinary to the peripheries of stardom, as in *Path of Stars* (1993), a paving design by Sheila Levrant de Bretteville.[26] Griselda Pollock argues that adding women into men's art history is an inadequate response to cultural sexism because to demand women's consideration 'not only changes what is studied and what becomes relevant to investigate but it challenges the existing disciplines politically' (Pollock, 1988: 1); this can be applied, too, to other categorisations of exclusion.

If, then, naturalism naturalises social stratification, what are the alternatives? Is the genre useless to radical sentiment, or can it be turned? I look now to Kevin Atherton's *Platforms Piece* (1986) at Brixton Station, south London

5.2 • Kevin Atherton, *Platforms Piece*, Brixton

as a work which departs from naturalism while retaining the traditional medium of bronze. It consists of three life-cast figures of commuters. The models were approached by the artist during the morning peak hour, given money for clothes and a bag from the market, and cast using the lost-wax process. Two of the figures are black, one white; two are female, one male. But why does Atherton use bronze, a medium traditionally associated with the representation of privilege, for these clearly non-privileged figures? Is it an attempt to redistribute the privileged status of representation (which could be an analogy for a redistribution of wealth, possibly)? If so the figures could be compared to Socialist Realism; yet there is no heroic dimension, as epitomised in Vera Mukhina's *Worker and Collective Farm Girl* (1937) for the Soviet pavilion at the Paris World Fair.[27] No, the particularity of the life-cast separates these figures from both naturalism and Realism. The figures at Brixton Station do not stand for a class, though they reflect the mix of gender and race of users of the platforms. Atherton sees the process as allowing 'a resonance between the figure and the viewer', and the work as site-specific (Serpentine Gallery, 1988: 13). The latter is in the specificity of the publics represented as well as the vital participation of the people represented in the process of their production (something more recently taken up in various global works by Antony Gormley). This opens, figuratively, the space in which a society's self-representation is determined: 'The ambiguity of urban forms is a source of the

5.3 • Michael Sandle, *St George*, London

city's tension as well as of a struggle for interpretation. To ask "Whose city?" suggests more than a politics of occupation; it also asks who has a right to inhabit the dominant image of the city' (Zukin, 1996: 43).

A similar argument can be made for the photographic image of tenants of social housing blocks displayed in posters on bus stops, advertising columns, and on the sides of their blocks by José Maças de Carvalho, in the Marvila district of Lisbon during the project *Capital do nada* in 2001 (see Chapter 7). In both cases, images of the disenfranchised gain visibility. While the question lingers as to whether visibility is enough, what emerges when alternatives are attempted is the ideological content of naturalism and a possibility for new, critical kinds of realism. Writing on *Tilted Arc* as an ideological insertion in public space, Rosalyn Deutsche argues that

> Opening the question requires that we dislodge public art from its ghettoization within the parameters of aesthetic discourse, even critical aesthetic discourse, and resituate it, at least partially, within critical urban discourse. More precisely, such a shift in perspective erodes the borders between the two fields.
>
> (Deutsche, 1992: 161)

But I need to add a further dimension: not only urban spaces but the spaces of historical narrative can be contested. If how we are is articulated in stories we construct about ourselves, which we enact, these stories are also constructed for us, not least in public monuments and memorials. The role of memorials in the construction of national identity is rehearsed elsewhere, notably by Jon Bird and Joe Kerr.[28] I look now at only two examples which seem radical restatements of the genre in their very different times. The first is Charles Sargeant Jagger's *Artillery Memorial* at Hyde Park Corner, London (1922); the second is Maya Lin's *Vietnam Veterans Memorial* in Washington, D.C. (1982).

Jagger's memorial was unveiled when my parents (who were the same age) were one year old, and it is necessary to make a historical adjustment when reading it. Its context includes the predominance of neo-classicism in memorials and public buildings at the time, as in the naked youth of F. Derwent Wood's *Machine Gun Corps Memorial* nearby and in a different way in Lutyens' *Cenotaph* in Whitehall; and the aftermath of the 1914–18 European war, which produced a widespread distrust of the officer class who ineptly led the slaughter, and a difficulty in dealing with loss on an unprecedented scale (addressed as an impossibility of utterance in the blank spaces of the *Cenotaph*). I assume, but do not know, that Jagger worked from his own recollections; he was an officer, wounded at Gallipoli and in France, and awarded the Military Cross. But the memorial does not commemorate the officer class any more than it follows the conventions of neo-classicism; neither does it forego utterance. Four bronze soldiers surround a plinth that supports a stone 9.2-inch Howitzer; all are from what officers call other ranks, and the fourth is dead, laid out

5.4 • Maya Lin, *Vietnam Veterans Memorial*, Washington, D.C.

under his coat with his helmet resting on his chest to represent the 56,700 dead or missing from this Regiment. There is a Latin inscription, 'UBIQUE', above a list of places – and the euphemism (to me but perhaps not Jagger) of 'Here was a Royal Fellowship of Death'.[29] It would be easy, strolling through the leafy traffic island, which has become a kind of memorial park, to see this as an aestheticisation. But it seems to me now a realism.

Moving on sixty years, the *Vietnam Veterans Memorial* is, too, well covered elsewhere.[30] I introduce it here because it seems to be an inversion of monumental form that has attracted an unusually diverse public while stating a complex relation to its site and a tension between the emptiness of a cenotaph and the awe of a recitation of the names – a traditional memorial device – of the 58,476 US dead and missing presumed dead in a war that produced no victory (for this side). What is striking is not only that the names are inscribed on the memorial's black, polished stone in order of the date of death, but that the list begins in the centre of its shallow V-shaped cut in the grass of the Mall. This makes a reading of the list from first to last always a return to the first, while the list is palpably finite. That walk along the names, through the sight-lines established by the siting of the memorial, also offers views of the obelisk of the *Washington Monument* and the colonnaded enclosure of the *Lincoln Memorial*. These two focal points in a rhetoric of the 'free' world are, I suppose, what the war was fought to preserve and in effect broke when a significant part of the population opposed it, and when the body-bags arrived with an unsupportable frequency (against which US foreign policy still reacts). Probably, this memorial, like Jagger's, aestheticises its content as much as it faces it, but the difference between it and the *Washington Monument* draws out a quality which may be interesting. Charles Griswold notes that the obelisk appears the same from any side, an indifference (one might say a Kantian disinterestedness?) matched by its lack of inscription:

> The Washington Monument does not carve out a space particular to itself . . . into which the beholder is drawn and thus disconnected from the surroundings. It is not an absorbing monument in the way the VVM is. . . . although people look and refer to this monument, they rarely sit and contemplate it and infrequently celebrate or demonstrate at its base. None of this contradicts the fact that the Washington Monument also serves as the centre of the Mall, if not the city. It is a space-defining, orienting structure even as (or perhaps because) it is indifferent to this or that perspective.
> (Griswold, 1992: 88–9)

The indifference of the bird's eye (or god's eye) contrasts with the immediacy (or immanence) of the names, as of the life-cast figures in Brixton and Jagger's realism. One states power, the cold eye which sweeps over the field, which subsumes personal grief in national mourning; the other experience, which is always personal and political. John Beardsley sees the *Vietnam Veterans Memorial* as having provoked a national drama reproducing the conflict engendered by the war, and 'a much needed catharsis' (Beardsley, 1989: 124–5). Meanwhile discussions continue at the time of writing on an appropriate form of memorial for the dead (the US dead, or all the dead?) of September 11th. Following a lecture at Staten Island College by Mierle Ukeles, whose work I discuss in Chapter 7, a participant suggested a solution which would unite those in power with those who had never experienced power except as its objects (p. 162). Whatever is resolved, however, will probably be less radical after a return to normality.[31]

I want to look now at a subversion of the monument through irony and acidity in the work of Michael Sandle, and then end this section by returning to the creative tensions of meaning in Wodiczko's projections. The link to memorials is not obvious, though Sandle has designed a (non-ironic) war memorial for the island of Malta and Wodiczko has made many projections onto them. But the link is in the assertion of subjectivity with which both contend, as do Lin, Atherton and Jagger in their ways. In Sandle's work the identity of the subject tends to be stated as an absence, or erasure. *Der Trommler* (1985), for instance, fuses the type of the drummer-boy seen in nineteenth-century military pictures with the faceless soldiers in Goya's *Executions of 3rd May, 1808* (1814, Prado), and perhaps the sculptural language of Epstein or Boccioni, to become a robotic techno-parody. It is neither naturalist nor realist, and uses the familiarity of its genre to destabilise its references. *St George* (1987), outside a modern office block in London, adapts heraldic representation and the traditional material of patinated bronze; the saint is engaged in thuggery: 'my St George isn't your usual "officer" type killing dragons with an air of insouciance . . . he's actually working very hard, and is a nasty piece of work' (in Bird, 1988: 39). And *A Twentieth Century Memorial* depicts a skeletal Mouse who sits operating a brass machine gun. It is not an explicitly anti-war image but alludes to a militarisation and Americanisation of life.[32]

Life in North America, too, was subjected to Americanisation as homeless people were cleared from the streets of New York and Los Angeles.[33] In the 1980s, Wodiczko used projections to draw attention to their presence – on the *Civil War Memorial* in Boston in 1986–7, and on the *National Monument* at Calton Hill, Edinburgh in 1988.[34] The manacled legs of the dispossessed were projected onto the Westin Bonaventure Hotel, Los Angeles in 1987:

> Dominant culture in all its forms and aesthetic practices, in what it says and does not say, remains in gross contradiction to the lived experience, communicative needs and rights of most of society, whose labour is its sole base. Transmitted not only by the media but also by the Built Environment, and controlled by its commercial and political sponsorship, it creates miscommunication, alienation, misrepresentation and life-in-fantasy while holding a monopoly over public life, education, and the development of a communicative experience.
>
> (in Freshman, 1992: 139, first published (1989) in *Matrix 103*, exhibition brochure, Hartford, Wadsworth Athenaeum)

Wodiczko's projections foreground the military associations of public monuments, too: missiles on the *Victory Column*, Stuttgart (1983) and *Nelson's Column*, London (1985); Soviet and US warheads linked by chains on the *Memorial Arch* at Grand Army Plaza, Brooklyn (1984–5) and an assault rifle and petrol pump nozzle in skeletal hands on the *Arco de la Victoria*, Madrid in 1991, three days after the outbreak of the first Gulf War.[35]

The projections need special equipment and permissions, and last a short time. Nonetheless, they may, as Wodiczko claims, create new readings which endure in public memories:

> The aim of the memorial projection is not to 'bring to life' or 'enliven' the memorial nor to support the happy, uncritical, bureaucratic 'socialization' of its site, but to reveal and expose to the public the contemporary deadly life of the memorial. The strategy of the memorial projection is to attack the memorial by surprise, using slide warfare, or to take part in and infiltrate the official cultural programs taking place on its site. In the latter instance, the memorial projection will become a double intervention: against the imaginary life of the memorial itself, and against the idea of social-life-with-memorial as uncritical relaxation.
>
> ('Memorial Projection', *October*, 38, Winter 1986, in Freshman, 1992: 115)

Using a replica of Verrocchio's *Colleone Monument* (on to which he had previously projected in Venice) in Poland during the military emergency of 1986, Wodiczko projected a skeletal horse with chains for a bridle, and a rider carrying a police riot stick and wearing a swastika armband.[36] The message is obvious, as are references to traditional types of doom such as the Horsemen of the Apocalypse (here fused with Nazism's final solution) but the image is nonetheless startling. Perhaps it trades on associations that have currency in public imaginations; perhaps it is in playing with what is partly known that unexpected meanings are produced which make incisive rents in the fabric of the structures of social ordering. But words can run away, and I must be careful not to be drawn into the adrenalin of high drama. I need to find instead a renegotiation of rationality beyond the disinterested stance of the *Washington Monument* and the illusion of objective knowledge which is integral to the view from above (the view of power). The above reconsideration of public monuments and memorials leads me to this. As does Primo Levi.

> It is, therefore, necessary to be suspicious of those who seek to convince us with means other than reason, and of charismatic leaders: we must be cautious about delegating to others our judgement and our will. Since it is difficult to distinguish true prophets from false, it is as well to regard all prophets with suspicion. It is better to renounce revealed truths, even if they exalt us by their splendour or if we find them convenient because we acquire them gratis. It is better to content ourselves with other more modest and less exciting truths, those one acquires painfully, little by little and without short cuts, with study, discussion and reasoning, those that can be verified and demonstrated.
>
> (Levi, 1987: 397, in Griffin, 1995: 392)

Levi frames his caution in the language of his profession as a chemist. But verification is also in insights gained through personal experience, and demonstration possible in a critical investigation of conditions.

III DEMOLITION AND ATTRACTION

After the Wall was dismantled the rebuilding of Berlin was a great opportunity for architects.[37] Sony Plaza in Potsdamer Platz is bigger than that in Manhattan; old cheap areas in the East have become districts for designer bars and art-hotels. The new building projects erase the spaces of public memories where the Wall was like a fault-line. Hilary Lawson sees the narrative of the end of the Cold War as a closure:

> We may imagine the fall of the Berlin Wall as a specific event, remembering perhaps the images of the night when the Wall was wrapped in people.

> There are other equally effective closures for that night. It was for idealistic party members of East Germany the end of a dream; for some Turkish families who were shortly to be forced to leave it was the rise of a new German nationalism; for American hawks it was the defeat of communism.
>
> (Lawson, 2001: 137)

And if the opportunities for architects are a collective closure of one difficult history which followed another even more so, might there be other reactions, and might they be more informal and troubling?

If the dismantling of the Wall was exciting as an opportunity for a release of emotions pent up in the atmosphere of an authoritarian state, it may be appropriate to ask what those emotions were (apart from the assumed reversal of power and entry to the free-market world). Was there an excitement in the act of demolition itself? Thinking back to Chapter 4 and references to Freud's *Civilization and its Discontents*, is there not a quasi-joy in destruction, which runs against the requirement to set aside joy today in favour of a superior good? In *Fragments of the European City* Stephen Barber draws attention to the pleasures of demolition sites:

> Demolition: the transformation of the city is a restless process of negation. When the city is settled, an atmosphere of congelation rises to the surface, tempting acts of aggression against the city. The city is perpetually invested with a dynamic jarring and upheaval of its configuration. Demolition of the city's elements strengthens what remains, and also strengthens the sense of a vital damaging through which the city takes its respiration. Demolition by exterior forces exerts a particular force of poignant dislocation which remains vivid over decades.
>
> . . . it is the infliction of damage by the city upon the city which accentuates the vision of transformation. The visual arena of the city *must* move through concurrent acts of construction and obliteration, extrusion and intrusion, incorporation and expulsion. The periodic demolition of entire areas of the city make its perspectives swing crazily, imparts a sense of exhilaration which is compounded from anticipation of a new 'coming into being', and from a lust for raw destruction.
>
> (Barber, 1995: 29)

But it works in more than one way: the prospect of ruins can be aestheticised like anything else to conjure a happy ending; or perhaps it is more accurate to say a happy-enough ending, an awareness of mortality and an accommodation of destruction's totality as the timescale of architecture slides into that of geology, though there is a possibility to hold out:

> The ideas that ruins awake in me are grand. Everything is annihilated, everything perishes, everything passes, there is only the world that remains, only

> time which endures. How old it is this world! I walk between two eternities. Everywhere I cast my eyes, the objects which surround me announce an end and make me yield to that end which awaits me. What is my ephemeral existence in comparison with that of the rock which is effaced, this valley which is forged, with this forest which tremble, with these masses suspended above my head which rumbles. I see the marble tombs crumble into dust; and I do not want to die!
>
> (Diderot, 1963: 228–9, quoted in Roth, 1997: 59)

Or, there is a beautiful disintegration which appeases the denial of mortality and subsumes mortality in cultural duration, as in a drawing given by Albert Speer to Hitler:

> after generations of neglect, overgrown with ivy, its columns fallen, the walls crumbling here and there, but the outlines still clearly recognizable. In Hitler's entourage this drawing was regarded as blasphemous. That I could even conceive of a period of decline for the newly founded Reich destined to last a thousand years seemed outrageous . . . But he himself accepted my ideas as logical and illuminating. He gave orders that in the future the important buildings of his Reich were to be erected in keeping with the principles of this 'law of ruins'.
>
> (Speer, 1970: 56, quoted in al-Khalil, 1991: 39)

The energy of destructiveness is repressed to become the vista of a heroism in ruins, organised and controlled as a fragmentary but still longed-for utopia. In this way even the totally authoritarian state can become picturesque, projecting onto a spectral future aspects of a gothic (and ghostly) past of references to Emperor Frederick asleep in a cave in the Harz mountains.[38] The ruin is the counter-weight to the halls of marble of the new Chancellery, also designed by Speer, in which visiting diplomats were forced to walk on slippery marble floors.[39] Both are essays in the sublime.

But I want to suggest that there is a counter-possibility: that (just as there are anti-monuments to fascism, discussed in the next chapter) there can be a positive remembering of modernity, and the industrialisation that was one of its main achievements. What I have in mind is how the industrial past is remembered, not in heritage museums and the conservation of old mills and mine- or iron-workings (which have a place in cultural tourism and may be educative), but in a transformation of its sites as a new form of monument.

This leads me to the work of Herman Prigann in Germany, most of whose projects are an adaptation of post-industrial sites. Prigann was born in 1942 in Gelsenkirchen, and recalls playing in bomb sites as a child. He speaks of 'a personal story, this fascination with the landscape of ruins', but also of a tendency to sanitise industrial landscapes as parks, which is to him like making follies: 'people like to make artificial ruins, but I make a cut. That kind of answer is not useful anymore, but to use the industrial culture of the recent past as traces, that could be really important' (conversation, September 2001). For example, near Cottbus in the ex-GDR, where a village stops and the road disappears into a vast open-cast mine which consumed several of the houses before operations for brown coal extraction ceased, Prigann worked with a local construction team (providing work) to make a large earth ramp planted with broom – from which came the work's title, *Gelbe Rampe* (*yellow ramp*) – on top of which were placed several concrete slabs, as found, detritus from an industrial site. In a conversation (September, 2001), Prigann stated the key elements of his work: material, place, use by people and work provided locally in construction (also a means of dialogue). Seen from afar the earthwork and slabs have a resonance of neolithic mounds and henges, though the combination is specifically modern. Sheltering from wind and rain behind the slabs, which are about 10 metres high, I began to sense the explicit industrial quality of the site – putting aside memories of hill sites in the south-west of England. The monument restates an industrial past to say both that it is over and that it was made by people. It is there, the good and the bad together, as what was made then. Far from a Kantian statement of disinterested judgement, this appears to be a statement that is interested (engaged) but non-judgemental. It stands as tension between the picturesque and the blatant, between the attraction of ruins and their actuality.

To take another case, at Marl, in a zone of chemical industries in the Ruhr, Prigann and a local construction team introduced mounds of earth to a derelict water purification plant (the concrete walls of which cracked as water levels rose after the end of deep mining and shutting down of the pumps which kept the mines from flooding). The brick casing of the concrete structure has been removed except at the corners, where it rises pyramid shaped and rough edged, coming close to the picturesque, a ditch dug around the building (to which entry will become unsafe as it collapses further), and planting of species native to the area selectively introduced. The scale of the central concrete structure, which is filled with pipes and dark chambers, is like that of an Egyptian hypostyle hall. The ruin can be seen as a whole from an adjacent building, which will remain on site, its flat roof used in local events such as firework displays. The proportions of main building to a subsidiary structure reduced to a steel skeleton, the reflections in water, the asymmetry of the main structure and the skeletal girders of a small adjacent building, fit a traditional sense of pictorial regulation. But this is, again, industrial Germany and the materials are those of heavy plant. It will be centuries before a structure as massive and reinforced as this falls down, but in time the plants, both those introduced and already there, will take it over.

Prigann sees his projects as interventions that are both social and aesthetic. He normally begins with walks through and around the site, gleaning local knowledge in informal encounters; this leads to detailed proposals and visualisations which are put to public consultation as well as negotiated with public

5.5 • Herman Prigann, industrial ruin at Marl, Ruhrgebeit

authorities (who in most cases are the owners of the site). He sees the development of each project as a dialectic involving site, material, work and use, leading to a set of ruins that are staged, and adapt as natural processes of decay and growth become more visible than the artist's reconstruction (though the shaping of the site by succession growth is initially shaped by the artist's intervention and retains traces of it). To use the term 'deterioration' to describe this process would be to see only one half of an integration of art (insertion) and decay (deletion). When I first encountered these works I was perplexed.[40] There seemed little more than piles of earth, broken walls, bits of concrete, industrial detritus strewn around amid weeds, brambles and dog roses. As I spent time walking through the sites, however, each took on a particular scale, as happens in archaeological sites. I began to realise, too, that the weeds and bushes were part of the work – arriving at each site Prigann would say: 'My collaborator is doing well.' But, equally, he emphasised that concrete is not a material which should be imbued with guilt. The industrial past had good things, too, was constructive as well as at times destructive. In this way it is like the state which built and eventually dismantled the Wall.

One of Prigann's projects has a different kind of site and set of referents: *Ring der Erinnerung* (*Ring of Remembrance*) is a circular rampart of earth with four openings, overgrown by brambles and other wild vegetation, on the site of the ex-border in the Harz mountains. It is approached through conifer forest,

5.6 • Herman Prigann, *Ring der Erinnerung*, near Braunlage

on a path that passes a short section of the Border Security System for the National Boundary West, and further on, on the edge of what was a wide no-one's land, is a watchtower in a clearing. The area was affected by acid rain in the 1970s, and dead trees from the site were used in making the structural element of the work. The forest is strangely silent; it is a monoculture that supports few species of bird, though deer occasionally run through a clearing.

Prigann conceived the *Ring* in 1989 when it appeared that the two Germanies would be reunited. His proposal combines two themes:

> The wooded landscape of the Harz shows considerable evidence of being a dying forest, and so the material from which the rampart is built will be dead wood from this region . . . The dead wood of the rampart will rot in the course of time and become soil. Planting with brambles and other climbing plants means that the dead wood become overgrown. Thus the rampart make it possible to reconstruct and comprehend metamorphic events from two points of view: decay and growth.
>
> The circular rampart itself is placed in the landscape in such a way that it covers the old German–German border as a circular sign and in this sense stands as a symbol for this historical phenomenon. Nine of the old border posts will be left in the middle.
>
> (Prigann, 1993: 49)

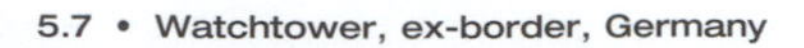

5.7 • Watchtower, ex-border, Germany

The work combines processes of growth and decay in which the decay of one material supports the growth of another, with a seal set on the linking of the two sides of a border which will also slowly decay as its histories recede when those who lived through them have died. But the line between the sides of the border is not erased; it is marked by a line of tall concrete posts which formed part of the border security apparatus and which Prigann has retained in their positions. Earlier in this chapter I mentioned Salecl's perception that after an upheaval there is a need to remove the old monuments, while western views tend to favour preservation to attest what might otherwise be forgotten. I broadly agreed with Salecl, but in this case I see the concrete posts as a vital counter to the rampart: line against circle; industrial object against natural decay and growth; and objects that will take a long time to decay and be detectable long after that.

At each of the four entrances to the circle, and in its centre, flat stones are set in the ground, inscribed 'FAUNA', 'FLORA', 'LUFT', 'WASSER' and 'ERDE'. Prigann sees these descriptive categories of the natural world as setting historical time beside biological or geological time. In 2001 when I saw the *Ring*, the brambles had already overgrown much of the structure. Inside the circle there was an uncanny quietness, not of a meditative quality but more an absence of the usual sounds of nature – but it is impossible to visit such a charged site without carrying into it baggage of feelings about the histories of which the site still has visible reminders. Prigann sees a contemplative aspect:

'the circle is and was a place of contemplation' (Prigann, 1993: 49); but in some cases the old circles were places of sacrifice. Prigann once found a dead fox in the centre of the circle. The area is remote and has always been poor, with its own folklore (and some neolithic circles, though the *Ring* is not a reference to them).

Prigann is adamant that he is not making a monument to German history, nor a history of all the Germans as a local functionary tried to characterise it. The *Ring* is an ecological monument stating an interconnectedness of processes of decay and growth in contrast to the fixity of the border posts; and perhaps sees human history in those terms of forces in tension (which is one way of saying connected). The method of construction, however, reveals a possible additional layer of meaning in its derivation from an early work by Prigann, *Hanging Tree* (1985, Vienna), in which an upended tree trunk is suspended from the apex of three poles leaning together, like a pendulum. Prigann writes of the symbolism of the Tarot hanged man, and the image of the inverted world-tree in the Katha Upanishad: 'This . . . is purity, is the Brahman, is what is called Non-Death. All worlds rest in it' (Prigann, 1993: 20–1, citing M. Eliade, *Religion and the Sacred*).[41] But my reflection on it is in terms of two kinds of time, the cycle of growth, and the certainty of the border (its time run out). Jochen Boberg writes, from the work in progress:

> Herman Prigann is pacing a monumental ring made from remains of dying nature on the violently cleared area between Germany and Germany, in a sector through which the markings of former border defences cut: a lunatic figure that defines the field of fire. Twin stones in the East and the West determine direction. In the centre, 'Terra Stein', a marker. This is a dangerous game. These are the signs of a sacred precinct; but what collection is coming together here, what sacrifice is being presented, what instruction issued? There is no community that would have chosen this place, no faith that determines it, no hand that takes care of it for ritual purposes. Taken from nature by a violent intervention, shaped by knowledge of our history, this work is once more abandoned to nature, and thus becomes a remembered 'Ring der Erinnerung' – a circle of remembrance. This opens up an enormous dialectic . . . He has chosen a place that hurts for this work, a form that is profoundly correct, dimensions large enough to be perceived as an assertion.
>
> (in Prigann, 1993: 64–5)

The *Ring* occupies a site of unusual difficulty. It could be forgotten, yet . . .

Another is Dresden. In the quotation from *Fragments of the European City* above, I extracted (in the lacunae above) this remark in parenthesis: '(in the heart of Dresden, the black ruins of the blaze are still evident and displayed after fifty years)' (Barber, 1995: 29), because it seemed to point to an emotion more complex than the thrill of demolition sites. The ruin of the Frauenkirche to which this refers was left in its black state after the fire-storm caused by allied bombing on February 13th, 1945, which killed perhaps as many as 100,000 people. The heat was so intense that some victims were vaporised; others were found intact in rooms, sitting at table, from which all air was suddenly extracted. The numbers are meaningless; it was annihilation to which no monument could be adequate, any more than to Auschwitz.

In Dresden – perhaps, I cannot say – the black shard of the church is a statement that has a function of witness, requiring no translation and experienced personally by each observer. But Dresden is being rebuilt economically and has its Hilton Hotel now as well as the Soviet-era concrete blocks; the Hilton is in the centre, near the Frauenkirche. The black ruin is being rebuilt piece by piece from the catalogued and stored fragments. Stefan Hertmans writes on the restoration in *Intercities* as 'still a difficult topic of conversation with people from the area'; the question is whether the ruins of 'what was once the most beautiful Baroque church in Europe' should be left in their destroyed state or whether those who see them left that way as 'a protest against the English' risk being taken as 'nostalgically conservative Germans' (Hertmans, 2001: 57–8).[42] They might be; which is not an argument for restoration. So I remember the debate over disgraced monuments, and look to possible tensions between presences and absences, memories and forgetting, and mutable meanings. I still wonder what would have happened had history been different, if the revolution had not failed. But it did. Now? As Samuel Beckett writes at the beginning of *The Unnamable*: 'Where now? Who now? When now? Unquestioning. I, say I. Unbelieving. Questions, hypotheses, call them that. Keep going, going on, call that going, call that on' (Beckett, 1975: 7).

NOTES

1 'From the point where the system proved incapable of satisfying the consumer expectations which it was itself calling forth, it ran up against growing political dissatisfaction' (Kagarlitsky, 2001: 55). Economic colonisation was in any case relentless: 'A capitalist exchange economy begins by breaking down local hierarchies and traditions and substituting a production-consumption system with an inherent mechanism of accumulation. The apogee of such an expansionist economy is a universal world-system which regulates and replaces all other local and particular sources of cultural identity' (Angus, 2000: 78). Yet the immediate impact of the change outside its epicentres such as Berlin – Leipzig had also been a hub of growing dissent within the GDR – was felt incrementally. In 1990 the citizens of Croatia received postcards from the Croatian Post Office and Telecommunications service: 'The postcards were a double miracle because of what was written on them: from now on, when waiting in a post office, one must stand behind a yellow line of the floor. This yellow line will indicate a so-called "space of privacy," so that every citizen from now on will be able to do his or her business alone at the window' (Drakulic, 1993: 940–5).

2 For an account of Bloch's relations to the GDR regime, and shifts in perceptions of his work after the *Wende* (turn) of 1989, see Ernst and Klinger, 1997.

3 My description of the Wall is derived from Hildebrandt (1988). He comments: 'Barbed wire awakes bitter memories, especially in Germany'; and quotes the following from a dictionary: '"*Mauer*" is a Germanic loan-word from the Latin. The English word

"wall" is based on the medieval derivation of the latin *vallum*, *vallus*, and originally indicated a pole, stake or palisade. It could also be used in a literary sense for "protection". This expression does not exist in, for example, Chinese or Japanese, where the character for "Wall" means earth or clay. The material, not the function, is expressed' (ibid., n.p.). The term 'Deutschland' in the Wall's graffiti indicates reunification, a contentious issue: Jaspers opposed it to freedom in *Freiheit und Wiedervereinigung* (freedom and reunification), and Grass wrote: 'whomever today reflects on Germany and seeks an answer to the German Question, must also think about Auschwitz' (G. Grass, 1990, *Deutscher Lastenausgleich: Wider das dumpfe Einheitsgebot*, Frankfurt, Luchterland, p. 11, quoted in Rabinbach, 2000: 163).

4 For illustrations of New York graffiti using spray-can paints, see Chalfant and Prigoff (1987) and Robinson (1990). For a critical discussion of art both on and of the Wall, see Deutsche, 1996: 139–41. Deutsche notes that artists painted on a section of the Wall between Martin Gropius Bau and Künstlerhaus Bethanien, near art institutions; she categorises most of the Wall paintings as neo-Expressionist, and unable to 'confront the specificities of their site' (p. 140). See Cresswell, 1996: 36 on Haring in New York. Dovey notes that the West side was re-whitewashed sometimes, once to remove anti-Reagan graffiti during an official visit; and that once graffiti-covered fragments gained exchange value after 1989 a cottage-industry appeared to manufacture them, complete with certificates of authenticity in English (Dovey, 1999: 66–7).

5 The competition was organised by Museum House at Checkpoint Charlie in 1984, and received 288 designs (Hildebrandt, 1988).

6 The Holocaust Memorial Museum opened in Washington, D.C. on April 22nd, detracting from news coverage of Bush's speech; Elie Wiessel, returning from a visit to Bosnia, told Clinton: 'I cannot sleep since what I have seen' (Mestrovic, 1994: 32). In Texas, a Ku Klux Klan rally was planned for April 24th.

7 Sennett quotes N. Glazer (1979) 'On Subway Graffiti in New York', *Public Interest*, Winter, n.p. The account denies a direct connection between graffiti and crime but sees both as 'part of one world of uncontrollable predators' (Sennett, 1990: 206–7).

8 'While Abstract Expressionism was being exported around the world to showcase American freedom (and implicitly to condemn Soviet totalitarianism), individual Abstract Expressionist artists were being targeted and their patriotism called into question' (Rich, 1994: 227).

9 Cresswell argues that the Western city is a product of reason and progress, while the graffiti artist 'is the insane spoiler who resists reason and introduces chaos' but in another context graffiti becomes a 'sign of a free spirit closing the curtain on the stifling bureaucracy of Communist authoritarianism' (Cresswell, 1996: 45–6).

10 Michalski illustrates a crowd abusing the head of Sandor Mikus' *Stalin* (1951) in Budapest, noting 'especially vivid acts of public denigration' which annoyingly he does not specify (Michalski, 1998: 140–3, figs 95, 96).

11 Wodiczko is quoted as seeking to expose the difference between the idealism of the statue and the lives of those living in the cities of the ex-East block, arguing that such statues should not be demolished but retained as historical witnesses (Freshman, 1992: 158–9).

12 Michalski notes that Marx and Engels were sited in a garden behind the Palace of the Republic, and sees the monument as denoting a relaxation of ideology: 'The realized work shows the two thinkers standing or sitting demurely on a low, almost indiscernible pedestal. They are available to anyone to touch, and children were permitted to play around them – a trademark of public sculpture' (Michalski, 1998: 147, fig. 99) He adds that they were known as 'the pensioners'.

13 Michalski, 1998: 150 citing 'Getrennt zu Karl und Rosa', *Frankfurter Allgemeine Zeitung*, January 11th, 1997. The original monument to Luxemburg and Liebknecht at the cemetery was designed by Mies van der Rohe and destroyed in 1933 (Harbison, 1991: 48).

14 Michalski cites Pétain's decree of 1941 in which French monuments, such as

Bartholdi's *Balloon of Ternes* (1906), were thus reintegrated (Michalski, 1998: 151).

15 Levinson, 1998: 70–4; Trowell, 2000; and Rugg and Sedgwick, 2001. In the catalogue (1995, Budapest, Akas Reithy) Eleod states that he tried 'to treat this terribly serious theme with the proper amount of seriousness' (in Levinson, 1998: 71–2). Rugg and Sedgwick note that the park is surrounded by billboards and electricity pylons: 'In its isolation from the city, its hollow objects, its silence and its emptiness, the Park evokes the absence of the ideology that the objects were originally intended to monumentalise' (Rugg and Sedgewick, 2001: 95). Trowell remembers it having 'a distinctly eerie and unsettling air, whatever your politics' while the tourists bought tins labelled 'The Last Breath of Communism' (Trowell, 2000: 99).

16 A tourist web site states: 'The number of ideological sculptures in one exposition is very rare and unique . . . It is [the] heritage of several decade of Lithuanian monumental art, despite its ideological context. These symbols of [the] Lithuanian national tragic time . . . enclose truth about soviet occupation times to us and especially our children' (http//:www.travel-lithuania.com/grutas/general_info.htm). See http//:www.muzeijai.lt/Druskininkai/gruto_parkas.htm (in Lithuanian) for images of individual works. I am grateful to Laima Kreivyte for telling me about Grutas.

17 There is a small version in Bâle, Switzerland. Borofsky is quoted as saying that the image represents 'the worker in all of us' and that the aim of siting editions of the work around the globe is 'to connect all of us together' (in Rupp, 1992: 95).

18 See Hall, 2003 on public art in Birmingham. The statue was set on fire on April 17th, 2003: 'A teenager has been charged with arson after an attack which destroyed a £200,000 city centre statue . . . known locally as the Lurpak sculpture because it looks almost as if it is carved out of butter' (BBC news bulletin, April 18th, 2003).

19 MacAvera, 1990: 111–13; after various criticisms, not least the inadequacy of the work to its stated task of reconciliation in a divided city, MacAvera writes: 'The whole point about Gormley's work is not that it is site-specific but that it is site-general' (p. 113).

20 In his essay 'Avant-Garde and Kitsch', Greenberg writes that the 'true and most important function of the avant-garde' is to 'keep culture *moving* in the midst of ideological confusion' (Greenberg, 1986: 8) – see Chapter 2.

21 Levinson, 1998: 5–9 citing András Gero (1995) *Modern Hungarian Society in the Making: the Unfinished Experience*, trans. J. Patterson and E. Konez, Budapest, Central European University Press, pp. 203–22.

22 Miles, 2000: 21.

23 A statue of William III in Petersfield, Hampshire, by John Cheere (1757), also was tarred and feathered in an election campaign, and restored but not regilded by public subscription, in 1913; and a thistle under the horse's foot of Peter Sheemaker's statue of William in Hull (1734) was stolen by Jacobites (Darke, 1991: 80, 215). The victor of the Boyne's prowess is also dented in a popular rhyme: 'Here's King Billy with his ten-foot willie; he showed it to the Lady next door. She thought it was a snake so she hit it with a rake, and now it's only two foot four' (for which I am indebted to Elizabeth McFall from Carrickfergus).

24 Adorno uses the term instead of popular culture, and suchlike, because mass culture is not made by but to control the masses: 'The dream industry does not so much fabricate the dreams of the customers as introduce the dreams of the suppliers among the people. This is the thousand-year empire of an industrial caste system governed by a stream of never ending dynasties' (Adorno, 1991: 80). See also Adorno and Horkheimer (1997: 120–67) for the 1947 formulation.

25 Of 51 monuments erected in Central Park from 1859 to 1977, 28 represent male figures, 7 show animals, 12 are abstract, composite or allegorical, and 4 are female; but of those one is Mother Goose and another Alice in Wonderland (who appears twice) while the subject of the *Frances Hodgson Burnett Memorial* (1936) is a girl holding a seashell. *The Indian Hunter* (1866) by J. Q. A. Ward is the only non-white presence, described in a guidebook as typifying interest in native Americans after the Civil War and portraying the type as a noble savage (Gayle and Cohen, 1988: 187–246). A comparison could be made

with the native American in the stories of Karl May (Bloch, 1991: 154–7); Gandy sees Olmsted, the park's designer, as creating 'a unifying national culture in the midst of sweeping social and economic change' (Gandy, 2002: 99).

26 Lippard, 1997: 242–3. The project, in the Ninth Square district of New Haven, Conn., takes the names and details of past and present residents as material for a series of stars set in terrazzo paving, presumably to walk over. De Bretteville has also worked as lead-artist in projects in Los Angeles organised by The Power of Place, producing *Biddy Mason: Time and Place* (1989), an 81-foot documentary wall, and another paving scheme for Little Tokyo (Hayden, 1995; 2001).

27 Wilson, 1993: 106–7. On heroism in everyday life in the Soviet Union in the 1930s, see Fitzpatrick, 1999: 70–75.

28 Bird, 1988; Kerr, 2001. Bird writes: 'Into this public space were inserted devices for social subordination, particularly for the recognition and celebration of hierarchical authority under the aegis of ritual and commemoration' (Bird, 1988: 30). Kerr notes that a statue of Lenin commissioned in 1942 during Anglo-Soviet collaboration in wartime was not only frequently attacked by English fascist gangs but also removed by the authorities in 1948 (pp. 72–3).

29 Darke, 1991: 63. Jagger illustrates the work's production in *Modelling and Sculpture in the Making* (number 5 in the *'How To Do It' Series* (1933)). Plates XIX and XX show a drawing and relief of a soldier draped on barbed wire – *No Man's Land* – but his text is about technique: 'It is an interpretation on the flat of objects of three dimensions . . .'. He adds cases of fine sculpture: *La Vierge d'Alsace* by A. Bourdelle is 'an outstanding masterpiece of modern art' (plate XXX); F. Metzner's architectural sculpture at the Rheingold Restaurant, Berlin is 'one of the most successful achievements produced by the modern school of sculpture' (plate XXXII).

30 Griswold, 1992 (in both Mitchell 1992: 79–112 and Senie and Webster 1992: 71–100); North 1992; Lopes, 1987.

31 I attended the lecture, on February 5th, 2002. Ukeles was working on the site of Fresh Kills landfill where New York's garbage was taken by barge for many years until 2001 when the site was closed to be reclaimed as a nature reserve. After September 11th, the debris, a euphemism for the ash of the site including that of people, was deposited on the landfill. See p. 162.

32 Sandle says in an interview with Bird that 'although it's about war, it's not necessarily antiwar, it's to do with my rather ambiguous thoughts about my own aggression, and war as an historical constant' and 'I want the spectator to feel the chill of recognition, that there is something about the work that i not just decorative . . . I think that the work is based upon something real . . . a neurotic anguish . . . But possibly it's also the sign of a correct, objective estimate of the world as a pretty dangerous place' (in Bird, 1988: 35).

33 Deutsche, 1991 [reprinted, pp. 49–108 in Deutsche 1996]; Rosler, 1991; Davis, 1990.

34 Freshman, 1992: 136–7 and 148–9; On Wodiczko's homeless person's vehicles see Freshman, 1992: 54–73; for critical discussion, see Deutsche, 1996: 3–48.

35 The projection in Stuttgart took place during the German Federal election of 1983, in which the Christian Democrat Party endorsed the siting of Pershing missiles in Germany (Freshman, 1992: 100–1). During the projection in London an unauthorised projection was made of a swastika onto South Africa House. Wodiczko write: 'Many people told me that even though they hadn't seen the actual projection . . . somehow when they look at the pediment the swastika is seen as missing, as a kind of afterimage' (R. Gilroy, 1989, 'Projection as Intervention', *New Art Examiner*, February 16th, cited in Freshman, 1992: 117). On the projection in Brooklyn, Wodiczko notes a contrast between the monument's Beaux-Arts style, and function in commemorating the Union Army's victory over the Confederacy, and two naturalist bas-reliefs by Thomas Eakins mounted on it showing returning soldiers: 'It was a time when the public was being prepared for impending peace talks between the U.S. and Soviet governments. . . . this social and auditory interaction helped the visual projection survive in the public's memory as a complex experience' (in Freshman, 1992: 112–13). On Madrid, see

Freshman, 1992: 162–3. '¿ CUANTOS ?' is sited on the upper face of the arch, which commemorates Franco's victory in the Civil War.

36 The projection marked the 47th anniversary of Poland's invasion by Germany; see Freshman, 1992: 132–3.

37 'In 1989, Berlin's reunification brought with it heady aspirations and a great building boom. As capital of a reunified Germany, Berlin was once again to become a thriving metropolis, a world city, a global city. Again, architects and planners saw Berlin as a great opportunity: this time to remedy not only the lingering effects of war, but also those of modernist planning and the politics of division. This, of course, meant further demolition' (Siegert and Stern, 2002: 118).

38 Frederick I Barbarossa sleeps in a cave in the Kyffhäuser Mountain in the Harz region; more orthodox histories have him drowned in Anatolia in the Third Crusade (1189). Bruno Schmitz' *Monument to Wilhelm I* (1896) is situated on this mountain, and includes a statue of the Emperor awakening: 'The victory of 1870/71 [Sedan] and the unification of Germany gave a new impetus to this old myth, the white-bearded Wilhelm having seemingly fulfilled the hopes and longings of the old red-bearded Emperor' (Michalski, 1998: 62; illus. p. 63). Bloch writes: 'The emperor was not allowed to remain dead for the excited imagination . . . But the legend transported the emperor to a mountain, first into Etna . . . then, proceeding northward, into the Kyffhäuser. Ancient, cthonic images were associated with this figurative grave: in pre-Christian times a mountain cult was at home on the Kyffhäuser . . . Frederick II took his place and only much later did the heretical emperor change places with Friedrich I Barbarossa, the pious, insignificant rule, the romantic epitome of banal imperial glory in the style of Wilhelm "the Great"' (Bloch, 1991: 120–1; see also Bloch, 1991: 57, n. 33).

39 'He [Hitler] ordered the marble floor left exposed since "diplomats should have practice in moving on a slippery surface"' (Speer, 1970: 113, in Dovey, 1999: 61).

40 I visited the sites about which I write here in May 2001 in company with the artist, Heike Strelow (an independent curator from Frankfurt) and Annette Berger (a designer working on a book about Prigann's work); this allowed extended conversations and wide-ranging discussions. Quotations from Prigann below are from this visit, and a subsequent meeting in his studio in Barcelona in September, 2001. The visits were supported by the AHRB.

41 Prigann has also originated a book on trees in which a much wider set of historic and symbolic derivations is detailed (Prigann, 1984 – see especially pp. 34–9).

42 Hertmans concludes: 'But at the same time these are banal speculations because all the ruins in the world are of course a permanent indictment of the insuppressible urge to destruction that from time to time sweeps through even the most "civilised" society' (Hertmans, 2001: 57–8).

6

1993 (I)

IN MEMORIES OF DARK TIMES

•

In the previous chapter I asked whether the form of the public monument could be democratised or subverted from within, ending with reference to the restoration of the Frauenkirche, Dresden. In this chapter I am still concerned with narratives – that is, with constructions of history and the possibility to de- or reconstruct them – but the focus is on how, if they can, narratives deal with the extreme history of the Holocaust (as remembered in the 1980s and 1990s), and how art and architecture mediate this history which is many histories experienced or encountered in different ways. I begin in 1993 with the opening of the Holocaust Memorial Museum in Washington, D.C., which was designed by James Inigo Freed and is described as a narrative museum, and compare it with the Jewish Museum in Berlin, which was designed by Daniel Libeskind. This leads me to Adorno's aesthetic theory, and the problem of aesthetics and suffering. I refer briefly to the writing of Samuel Beckett and Paul Celan, and suggest that, if it is possible to write poetry after Auschwitz, representation fails, leaving the option of oblique allusion and a language as fractured as the history sedimented in it. In the second section of the chapter, moving to contemporary art which references the dark times, as it were, of European history, I look at the counter-monuments of Jochen and Esther Shalev Gerz, and a developing performance work by Dan Gretton. Finally, I reconsider the possibility of critical reflection in John Goto's renegotiation of the tradition of history painting through a digital manipulation of images. I leave open the question as to whether such work contributes to a critical consciousness in society or is confined to the limited spheres of cultural and academic life – though I take up this question in Chapter 7.

The following quotation concerns the Holocaust Memorial Museum.

> Human beings give testimony. Testimonies are also given by objects. In Freed's Holocaust museum, hundreds of photographs of the prewar Lithuanian shetl Ejszyszki near Vilnius and its inhabitants, placed in a

> tower-like structure, are made to tell the tory of a destroyed community. Another room offers not photographs, but objects, hundreds of shoes of the victims of the Majdanek concentration camp near Lublin, Poland. Metonymically, they give evidence for people about whom little more is known today.
>
> (Weissberg, 2001: 21)

I A NARRATIVE MUSEUM

The Holocaust Memorial Museum was opened by President Clinton on April 22nd, 1993. It is situated adjacent to national historical institutions and the *Washington Monument* and *Vietnam Veterans Memorial* on the Mall. The history it exhibits took place in Europe (though many of those afflicted fled to North America, or sought to). Liliane Weissberg calls it 'the U.S. Holocaust Memorial Museum' to differentiate it from the Jewish Museum in Berlin (Weissberg, 2001: 20), and this might be a more accurate name. The Holocaust Memorial Museum was established by Act of Congress, has Federal funding, and has been described as adapting the history of the Holocaust for the American public.[1] In that context, Michael Berenbaum states that the narrative should be told 'in such a way that it would resonate not only with the survivor in New York and his children in Houston or San Francisco, but with a black leader from Atlanta, a midwestern farmer, or a northeastern industrialist' (quoted in Young, 1993: 337). But the museum can be situated also in a context reflecting liberal humanism rather than national identity (though one may claim the other), and is seen by James Young as setting 'an ideal of catastrophe against which all other destructions will be measured' (Young, 1993: 338). As a non-Jew with no personal connection to this history perhaps I should hesitate to comment, yet question such universalism and wonder what could be said of other genocides – those of native peoples in North America and Australia – and why a benchmark of annihilation would be useful. I also wonder whether extreme histories, like apocalyptic myths, produce adrenalin and are thus addictive. There is also a specific question as to how such a history is institutionalised in a museum, and how this differs from the less concretised acts of witness which began among Jews in the US in the 1940s.[2] Was the Holocaust an exception to the history of modernity, or its product?

Zygmunt Bauman sees the Holocaust as having a defining presence in European history alongside the French Revolution and discovery (as he puts it) of America, which requires a particular effort of understanding because, since the 1930s, 'nothing much happened to those products of history which in all probability contained the potentiality of the Holocaust' (Bauman, 1989: 86). If the Holocaust was not an aberration but an outcome of ideational and technical elements in modernity which are still operative, he continues 'there are reasons to be worried because we know now that *we live in a type of society that made the Holocaust possible, and that contained nothing which could stop the Holocaust from happening*. For these reasons alone it is necessary to study the lessons of the Holocaust' (Bauman, 1989: 88, emphasis in original). He notes that the technical capacity and hierarchic organisation of an industrial society achieved in four years what pogroms would have taken two centuries or so to accomplish.[3] If he is right, then we could say the Holocaust is not a regression to barbarism, figuratively an eruption of hate as non-organised as the (Freudian) drives in the *Es*, but a product of the ordering mechanism effected in the *Ich*, or civilization. The Holocaust was, then, a machinery oiled and kept working by a large number of people in visionary, managerial, and menial capacities. The coldness of the operation is seen in a memo on adaptations to a fleet of Saurer vans, dated June 5th, 1942: 'Since December 1941 97,000 have been processed by the 3 vehicles in service with no major incidents . . . The normal load is 9 per square metre.' Saurer vehicles are spacious, and loading to full capacity destabilises them; so 'A reduction in capacity seems necessary. It must be reduced by one metre instead of attempting to solve the problem, as hitherto, by reducing the number of items loaded.' It could be a transport of goods or animals, but the clue is in a second reason for not filling the space – it extends the operating time 'as the void must be filled with carbon monoxide', while 'the merchandise' tends to push to the rear doors to be found there at the end.[4]

In the film *Shoah* (1985) Claude Lanzmann interviews a man who removed bodies to the furnaces after the van made its tour through the woods. Lanzmann refuses the role of narrator, putting one interview after another without comment, as in an archive. He cannot comment, which might be called reading the material in a cold light (of day), but this is a different coldness from that of technical refinement.

Bauman reads the ideational aspect of this history, which is the framework in which the technical aspect operates, as utopian: 'a grand vision of a better, and radically different, society' (Bauman, 1989: 91).[5] Bloch differentiates a true from a false utopia, and sees the Nazis as appropriating utopianism in a false form which does not devalue the history of millenarianism;[6] but it may be that utopianism is indelibly marked by the Nazi past, or more accurately that the Nazi rhetoric drew out of idealism a strand which was always there. This argument is compatible with Adorno's that the knowledge which arises in modern autonomy is a basis of power and has a potentially dominating and destructive force. Bauman continues: 'Modern genocide is an element of social engineering, meant to bring about a social order conforming to the design of a perfect society . . . This is a gardener's vision projected on a world-size screen' (Bauman, 1989: 91). He goes on to argue that technologically delivered genocide is consistent, if uninhibited in its expression, with a dream of perfect order. It is this which engenders totalitarianism, but such dreams are not new:

> They spawned the vast and powerful arsenal of technology and managerial skills. They gave birth to institutions which serve the sole purpose of

> instrumentalizing human behaviour to such an extent that any aim may be pursued with efficiency . . . They legitimize the rulers' monopoly on ends and the confinement of the ruled to the role of means. They define most actions as means, and means as subordination.
>
> (Bauman, 1989: 93)

He suggests such tendencies reflect deep dislocation in society, and that instrumental rationality and its mechanisms of delivery are as extant now as in the 1930s. But are institutional remembrances critical, or do they subsume this extreme history in an acceptable narrative of universal humanism?

At the risk of simplification, Holocaust memorials tend either to representation or to documentation. In the camps and sites such as the Warsaw ghetto, memorials use either the figuration of monumental sculpture or the abstraction of, for instance, shards of stone or fissures in a rock, or a wall of tombstones. In some cases the figuration resembles Socialist Realism, but not all the sculptors were from the East block and there are in any case equivalent monuments in the West, notably Zadkine's memorial for the bombing of Rotterdam; and abstraction is used in the East block, too, so that no cold war delineation can be made. At Treblinka, 17,000 pieces of granite are set in concrete around a fractured stone pillar, surrounded by dark trees.[7] On a far more modest scale, and remotely, the *Holocaust Memorial* in the Fossa de la Pedrera, Barcelona consists of several rough stones with the names of camps inscribed on polished black surfaces. In such efforts to state a history which defies representation, and of which representations become dated by the style of art, the blank surface may seem a more lasting alternative. But there are also the buildings left on site, as at Majdanek (near Lublin) where the gas chambers and crematoria are preserved; prisoners' clothing and shoes are displayed in the barracks, and several tons of ash incorporated into a mausoleum. At Auschwitz–Birkenau (internment and death camps respectively) what was not burned by the liberating Russian troops or foraged by local people as firewood and building material was preserved by state decree in 1947 as a memorial 'to the martyrdom of the Polish nation and other peoples' (in Young, 1993: 130). Each of 19 blocks was dedicated to a specific nation-state: 'By collecting a composite memory of Auschwitz, these national pavilions preserve the essential diversity of memory here. On the other hand, Jews came to see in this pluralisation of memory a splintering of Jewish suffering into so many national martyrdoms' (Young, 1993: 130).[8] The term 'pavilions' suggests a trade fair or the Venice biennale. At least two problems are evident: the construction of a narrative reflects a framing that is pre-constructed for it; and representation in any mode may not be adequate to the content. The framing of a narrative may, too, entail a rehabilitation of history for cultural, economic, or political ends. Inga Clendinnen comments on the film *Schindler's List* (1993): 'even Spielberg sweetens the horror of his concocted scenes in providing the consolatory figure of the little girl in red, herded with the victims, then making her perilous way

6.1 • Concentration camp memorial, Fossa de la Pedrera, Barcelona

back to precarious safety' (Clendinnen, 1999: 175); and she sees cinema as enforcing a narrative and so unable to respect the fragmented nature of evidence, though she concedes that drama can give insights into human situations.[9] Ernst van Alphen writes on *Shoah*, 'If the shaping of facts into a narrative . . . is unable to do justice to the facts, then the only mode of representation . . . is the archival mode: the collecting, ordering, and labelling of facts, items, pieces of evidence, testimonies' (van Alphen, 2001: 46–7). The Holocaust Memorial Museum is described as a living museum as well as a narrative museum. Perhaps *Shoah* could be called a living film in that it does not *include* but *is* the rememberings of actors in the situation; but it is not a narrative – Lanzmann is clear about that. How can the square be circled in a museum? Taking the phrase 'Creating a Living Museum' as chapter heading, Jeshajahu Weinberg, the museum's first Director, says 'The story told in the Museum describes the roles of the actors who were involved in the Holocaust: the perpetrators, the victims, the bystanders' who respectively 'wanted the world not to know', 'wanted the world to know' and 'wanted the world not to know that they knew' (Weinberg and Elieli, 1995: 17). He admits no description can evoke the experience but cites the documents buried to be found later by those who did not survive as justification that the story must be told. The question is what (whose) story for whom; but it is also whether a composite story conveys the personal memories of those who were there. Weinberg's categories already seem to be a masking of complexities.

From the outside, the Washington museum fits into its site while having its own character.[10] The hexagonal Hall of Remembrance suggests a synagogue, and the brick towers and gables of the exhibition spaces allude to the gateway to Auschwitz II–Birkenau, a visual memory emphasised inside.[11] The bare brick surfaces and metal fittings, and scale of rooms and stairways, offer what Weissberg describes as 'an overwhelming visual experience' in which the inscription of names on glass walls differs from that of the *Vietnam Veterans Memorial* in that they 'do not bear the heaviness of any gravestone' but seem 'to dance in air and, at the same time, diffuse the light and provide an oblique view of the floor below' (Weissberg, 2001: 19). Weissberg notes that the internal bridges which evoke those of the Warsaw ghetto 'are crafted in perfectly rendered steel', and that the architecture alludes not only to the models of the ghetto and concentration camp 'but also to the refraction in which we are now forced to view the historical events' in a building 'praised for its aesthetic satisfaction, even pleasure' (Weissberg, 2001: 20).

A large part of the museum's collection consists in documentary items collected from European countries, including photographs, papers, personal objects, such as shoes from the camps, and larger items, such as a railway boxcar and a Danish fishing boat, all obtained by voluntary donation (in the manner of other museums that collect worldwide). Objects enfold stories; suitcases, for instance, speak of journeys which begin and end – at Auschwitz bearing the names, birth dates and numbers of victims in 'a registration system' for the machinery of annihilation (Rogoff, 2000: 44).[12] To counter the oblivion of the suitcases, Irit Rogoff turns to pages from the diary of Charlotte Salomon, who was killed at Birkenau: 'replete with the images and dramas of everyday life, they bring us to the moment of departure and beyond with a full recognition of the abundance and breadth of the life that had been lived *in situ*' (Rogoff, 2000: 47). But the design of the museum building is itself a narrative. The process of induction into it is designed to forge an identification between visitor and victim, a relation which for those without a personal connection is remote. Seeking to individualise that relation, the designers installed a three-storey tower of photographs of the inhabitants of Ejszyszki in Lithuania, 29 of whom (of 3,500) survived. But the most pointed device of this desired relation is a personalised identification card bearing biographical details of a victim and a passport-style photograph issued to visitors on entry.

The form of the identification document (which identifies a victim and a visitor and seeks identification between them) is reminiscent of a US passport.[13] But victims are not for a day; and the passport-like document represents a permission which can be inspected or denied, and is ersatz compared to the identity cards Jews were forced to carry in Germany. Marianne Hirsch writes of the Nazi gaze that constructs these:

> It is the determining force of the identity pictures Jews had to place on the identification cards the Nazis issued and which were marked with an

> enormous J in gothic script. Those pictures had to show the full face and uncover the left ear as a telltale identity marker. In these documents, identity is identification, visibility, surveillance, not for life but for the death machine that had already condemned all those thus marked.
>
> (Hirsch, 2001: 235)

Weinberg and Elieli say the museum is unlike most historical museums because 'it is based on a narrative rather than on a collection' (Weinberg and Elieli, 1995: 49). The entry document indicates a key element in the narrative: a reconstruction for what is now a second- or third-generation public. The museum is closer, it seems, to Spielberg than Lanzmann, setting up its emotional encounters rather than cataloguing them:

> Comprehension of the narrative and its meaning is not only an intellectual but also an emotional experience. The emotional effect of the narrative in the museum exhibition is comparable to that of the narrative in novels, plays, or motion pictures. All of them are based on plot. The plot triggers identification, which envelops us mentally and forces us to relate to the meaning of the story line.
>
> (Weinberg and Elieli, 1995: 49)

The story line is spelt out by ordinary objects such as the shoes of victims as well as in images of extraordinary dehumanisation. Weissberg remarks that when the chairperson of the Museum's national campaign for funds was asked to pose for a photograph holding one of the shoes, he froze.[14] But the shoes and the grainy photographs are, perhaps, tropes of Holocaust culture: 'These are images of mourning. The camera completes, paradoxically, the work of a technology that had death as its primary imperative, turning human beings into objects in the first place. These are not images of trauma, but of belatedness' (Weissberg, 2001: 25).

Scenes of mass executions are screened behind protective walls to give visitors a choice, and children exemption. The final exhibit is a film composed of testimonies from survivors: 'pain, suffering, and anguish, but also resilience, compassion, and hope' (Weinberg and Elieli, 1995: 55). But if the story line has in common with cinema that it renders its material accessible, in a strange way familiar, there are complexities. Marianne Hirsch notes, for instance, that a small number of photographic images of the Holocaust (of the 2 million extant) are frequently used[15] to re-inscribe a specific narrative which while initially shocking is now accepted through visual replays.

Perhaps the Washington museum acts in a similar way, primarily not for survivors but for a second generation able, in a way the first might not have been, to integrate Holocaust remembering into a North American way of life. In part, and I say this not to belittle the depth of feeling involved, the museum uses the extreme history it commemorates, and to which it lends coherence as narrative, as Other to the aspirational society in which it is situated. It is a national museum in depicting the not-here, affirming the civilisation of the state which houses the museum by setting it against barbarism. The situation of the Jewish Museum in Berlin is different, in a state of both victims and perpetrators. It has a small collection of archive material hitherto on temporary display at the Martin Gropius Bau, but there was a previous Jewish Museum in Berlin, which opened in January 1933 with an exhibition of paintings by Max Liebermann and members of the Berlin Secession. Restricted to Jewish visitors by the Nuremberg Laws, and its shows to Jewish (that is, degenerate) artists, it was dismantled in 1938 and its artifacts lost. Fifty years later the Senate of the German Federal Republic approved funding for a new building to house the Jewish Museum Department of the Berlin Museum, established in 1961 when the Märkische Museum of the city's history became inaccessible from the West, on the other side of the Wall. Young cites Libeskind's description of the design based on a broken six-point star as on the verge of becoming, but sees it as on the verge of a collapse of architectural assumptions:

> Indeed, it is not the building itself that continues his architecture but the spaces inside the building, the voids and absence embodied by empty spaces: that which is constituted not by the lines of his drawings, but those spaces between the lines. By building voids into the heart of his design, Libeskind thus highlights the spaces between walls as the primary element of his architecture.
>
> (Young, 2001: 187–8)

The entrance is under the old museum; paths lead to a Holocaust tower, a garden of exile, and the star which houses the main circulation space. The building is cut through by six voids. The exit is through the garden, which contains 49 concrete columns filled with earth, in 48 from Berlin and in one from Jerusalem. They are planted with willow oaks and the plane of the ground is tilted. There is a semblance between Libeskind's statement that the Berlin museum is 'not a collage or a collision or a dialectic simply, but a new type of organization . . . around a centre which is not, the void, around what is not visible' (Libeskind, 1992: 87, in Weissberg, 2001: 21) and Adorno's that the primary colour of radical art 'is black', that the impoverishment of means in the idea of blackness informs what is written, painted and composed when avant-garde art pushes impoverishment to 'the brink of silence' (Adorno, 1997: 39–40). Leaving aside the difference between the solidity of blackness and a void, could the void used here be a trope of Holocaust culture as much as the grainy photographs, the shoes, the suitcases? Voids can be reified like anything else, and a tired joke is that people burst into tears in front of paintings by Rothko (but when they saw the prices).

Names can be reified, too; Adorno uses Auschwitz to represent the Holocaust, and is not alone in doing so.[16] But before looking to his comments on poetry (more accurately cultural criticism) after Auschwitz, I want to draw attention to the sense of ambivalence and paradox in his work. In his 1931 *Habilitationsschrift* on Kierkegaard's theology of sacrifice,[17] working with Paul Tillich, Adorno wrote that human power over nature 'remains dedicated to its annihilation in spirit rather than to reconciliation' (cited in Rabinbach, 1997: 175). This leads to *Dialectic of Enlightenment*,[18] and themes in his aesthetic theory. His refusal of solutions goes against the philosophical grain, yet follows a recognition that the worst outcome would be closure, that the question must be kept open. In *Dialectic of Enlightenment*, the world is disenchanted. Disenchantment is a freedom from spells, from a power like nature in which there is no intervention.[19] It is only after disenchantment that autonomy is possible, and through it the construction of an autonomous subject which is foundational to modernity. But autonomy engenders the fusion of knowledge and power as domination. There is no regression to innocence, nor a real freedom from the myths which rationality replaces.[20] If the power-over to which the subject was subjected becomes the subject's power, myth (the form of the old power) is reproduced within rationality: 'just as the myths realize enlightenment, so enlightenment with every step becomes more entwined in mythology' (Adorno, 1997: 11–12, cited in Rabinbach, 1997: 17).[21] Adorno sees a quest for authenticity in art as similarly caught in a double-bind, bound to be compromised, but not without exit. In *Negative Dialectics* he writes: 'Art is semblance even at its highest peaks; but its semblance, the irresistible part of it, is given to it by what is not semblance' adding that semblance is a promise of nonsemblance (Adorno, 1990: 404–5).[22]

Adorno avoids silver linings – the clouds are dark, the reality fractured – to hint at a sedimentation (to use his term) of the dark reality within forms which, in negating themselves (their forms), imply a negation of that which darkens the reality. The non-semblance is not a new art beyond art, but a seam within it which fractures semblance. This is what enables art, which is always socially produced and thus carries the terms of the oppressive conditions of its making, to interrupt the encounter with those conditions. A case of this (but not one to which Adorno refers) might be the auto-destructive art of Gustav Metzger in which acid is poured on nylon to destroy the material from which the work is made, as an attack on the art market's need for commodities but also to push the process of making art to an extreme point which almost destroys it, leaving only a record of the encounter as the presence in art's histories.[23] Outright destruction, however, is not semblance. Adorno deals with literary and musical cases in which form, or a sense of a language which becomes form, is retained to become a form of alienation. In Samuel Beckett's writing he reads 'the objective condition both of consciousness and of the reality that shapes it', while in *Godot* and *Endgame* 'Art emigrates to a standpoint that is no longer a standpoint at all because there are no longer standpoints from which the catastrophe could be named or formed' (Adorno, 1997: 250). In *Aesthetic Theory*, Adorno extends the argument of *Negative Dialectics*:

> Art can only be reconciled with its existence by exposing its own semblance, its internal emptiness. Its most binding criterion today is that in terms of its own complexion, unreconciled with all realistic deception, it no longer tolerates anything harmless. In all art that is still possible, social critique must be raised to the level of form, to the point that it wipes out all manifestly social content.
>
> (Adorno, 1997: 250)

This does not mean, as a superficial reading of several passages in *Aesthetic Theory* could be made to infer, that art's social and aesthetic dimensions are incompatible; the point is to maintain tension between the conceptual protagonists rather than resolve, hence close, the argument. When he writes that art is as abstract as social relations have become, that 'the concepts of the realistic and the symbolic are put out of service', Adorno is not saying give up but that new concepts such as sedimentation are required. On Beckett, for instance, 'This shabby, damaged world of images is the negative imprint of the administered world' (Adorno, 1997: 31); 'it is resistance in which, by virtue of inner-aesthetic development, social development is reproduced without being imitated' (Adorno, 1997: 226).[24]

The difficulty is that work *after* Auschwitz may represent it, or reduce history to kitsch. This is the danger in face of which Adorno is inferred as proscribing poetry. But he says:

> The more total society becomes, the greater the reification of the mind and the more paradoxical its effort to escape reification on its own. Even the most extreme consciousness of doom threatens to degenerate into idle chatter. Cultural criticism finds itself faced with the final stage of the dialectic of culture and barbarism. To write poetry after Auschwitz is barbaric. And this corrodes even the knowledge of why it has become impossible to write poetry today. Absolute reification . . . is now preparing to absorb the mind entirely. Critical intelligence cannot be equal to this challenge as long as it confines itself to self-satisfied contemplation.
>
> (Adorno, 1995: 34)

There are two issues: that culture cannot be redeemed from the trash condition of mass culture (to which Socialist Realism offers only an equivalence); and that the possibility for critique is corroded in total reification.[25] The problems are not identical. On the former, Adorno is seen as conservative, disliking jazz and finding in mass culture (the culture industry) a means of mass deception. But his view of high culture is not necessarily elitist. Robert Witkin argues on the contrary that the music of Schoenberg, Berg and Webern was heard by Adorno as being understood by a wide audience, found threatening because it distils anxiety, loneliness, and suffering,[26] underneath which is the abolition of the individual subject by technocratic power. This is outdated when the subject is deconstructed as a formation of identity; yet the socially produced and contingent subject may remain a witness to routine destruction. I tend towards Adorno's retention of a need for seriousness in cultural production as a means to criticise, when mass culture is a means of globalised deception now adept at subsuming all efforts to depart from its monopoly. The difficulty is that of representation, especially of suffering. In *Aesthetic Theory*, Adorno writes: 'The artwork is not only the echo of suffering, it diminishes it; form, the organon of its seriousness, is at the same time the organon of the neutralization of suffering. Art thereby falls into an unsolvable aporia' (Adorno, 1997: 39). Or, for Thomas Huhn reading Adorno, 'sublimation claims that it can pay the bill for suffering and death. This reconciliation with the status quo is thus the effacement of suffering'; the Holocaust becomes an artifact, but for Huhn it is already outside redemption because 'in genocide there is, quite literally, no death but only extermination' (Huhn, 1990: 292–3). There is, in other words, nothing left to redeem. For Huhn reading Adorno, the impossibility of poetry after Auschwitz is that of redemption when art attempts to redress suffering, but depends on suffering 'for its very existence and motive force' (Huhn, 1990: 293). Adorno writes in *Negative Dialectics*: 'After Auschwitz, our feelings resist any claim of the positivity of existence as sanctimonious, as wronging the victims; they balk at squeezing any kind of sense, however bleached, out of the victims' fate' (Adorno, 1990: 361).[27]

This is the content that Paul Celan renders in a fractured language of repetition, caesura, new words and inversion of anticipated relations within a form. There are references to an imagery of the camps and Celan (a name adapted, via Ancel, from Antschel) was a forced road-labourer in Moldavia; his father died of typhus in an internment camp in Transnistria and his mother was shot in the neck. His most reprinted poem, '*Todesfuge*' ('Death Fugue', 1945), references death as a master from Germany, and the Aryan and golden-haired Margarete, the Semitic and ashen-haired Sulamith. Lisa Saltzman reads it as a translation into language of the memory of the death camps, 'transforming trauma into lyric, history into poetics, atrocity into aesthetics' (Saltzman, 2001: 75),[28] as if a flouting of Adorno's proscription (except, as I indicate above, it is not a proscription).[29] Celan had reservations about his early work, yet Michael Hamburger writes in his Introduction to the *Collected Poems* that 'the power and pathos of the poem arises from the extreme tension between its grossly impure material and its pure form . . . an art of allusion that celebrates beauty and energy while commemorating their destruction' (Celan, 1996: 24). Celan's later work is more condensed and contorted, as in '*Engführung*' ('Straightening'), which, citing an unmistakable track, stones and grass, obliquely reminds me – but I may well be quite wrong about this – of the photograph of the entry to Birkenau.[30] Language disintegrates as the poem unfolds. Celan returned his soul to its maker in 1970.

II COUNTER-NARRATIVES

I begin this second section of the chapter with a work by Jochen Gerz, made between 1990 and 1993 in Saarbrücken. It consists of 2,146 cobble stones, each dug up, inscribed underneath with the name of one of the 2,146 Jewish cemeteries which existed in Germany before 1933 (and the date on which information on the cemetery reached Gerz), and replaced. There is no sign of the work, or notice announcing it; the marked stones are interspersed with others which remain unmarked, unexcavated. Gerz says of *2146 Steine–Mahnmal gegen Rassimus* (*2146 Stones–Monument against Racism*) that he had no commission for the work, undertaking it at first by night in transgression of the building code with a team of students and assisted by information provided by Jewish organisations. The site is the square outside the seat of the provincial parliament, Saarbrücker Schloss, used in the Nazi period as an assembly point for Jews. The work was commissioned retroactively by the city, and the square renamed *Platz des unsichtbaren Mahnmals*. Some groups, Jewish and non-Jewish, objected that the invisibility of the work, which could be called a buried or an anti-monument, was complicit with the buried status of German-Jewish history; but the point seems, to me, that the work is provocative exactly in exposing the invisibility of that history; and that this differs from the insertion of voids in Libeskind's museum. The work in Saarbrucken is one of a number of buried monuments in Germany from the late 1980s and early 1990s, a time when the generation after that implicated in Nazi history began to re-examine that past, perhaps like Gerz (and Herman Prigann whose work is discussed in Chapter 5) recalling childhood experiences of bomb sites.[31]

A difference between these buried monuments and Libeskind's voids is that the buried monuments state a history. By refusing to be visible they take back the decision at least (passive resistance, like a refusal to utter which is not Silence), while the voids represent an absence, as if a history is lost, or as if there is a prohibition on its recall from silence, the allusion in silence to universality.[32] So, I think it is legitimate to call the buried works counter-monuments or counter-narratives indicating a metaphorical sense of the underground, as the resistance was called, as a newly specific resistance against fascism. A further difference is the process of making the counter-monument in collaboration with Jewish organisations. In *Le Monument vivant* (*The Living Monument*, 1995–6) in Biron, France, Gerz similarly invited dwellers to contribute responses to an unpublished question, to be put on small plaques affixed to the stone obelisk of the town's newly refurbished war memorial. One reads:

> When you are twenty year old, it seems impossible to die. Perhaps for freedom. You thank those who died for you. Freedom always comes first, but you never know what the future will bring. There have always been war. Among us Europeans as well. I lived through the 1940s, and in spite of all the dead and the horrors, we were nonetheless very happy.
>
> (Museum for Modern Art, Bolzano, 1999: 74)

Others speak of the ordinariness of volunteering, wanting to go to one place but being sent to another, being declared unfit.

In *Mahnmal gegen Faschismus* (*Monument against Fascism*, 1986–93) at Harburg near Hamburg, Gerz and his partner Esther Shalev Gerz inserted a galvanised column into a shaft near a shopping centre, to sink at intervals until buried. Passers-by were invited to inscribe their name on it using a steel pen provided. Gerz states 'Either the monument "works" – which is to say that the initiative of the population renders it superfluous – or it remains a monument to its not having worked, as a meaningless ornament' (Museum for Modern Art, Bolzano, 1999: 54).[33] The column was soon covered in graffiti – anarchist signs, lovers' names and marks of contemporary racism. Young argues that in inviting its own violation, as he puts it, the anti-monument 'humbles itself in the eyes of beholders accustomed to maintaining a respectful, decorous distance' and in an enforced desanctification 'undermines its own authority by inviting and then incorporating the *authority* of passersby' (Young, 1993: 33). But he claims the artists were taken aback by the graffiti,[34] though it could be seen as exposure of conditions, like direct social research. Like Lanzmann's film it uses no artifice. In a recent lecture, Gerz said:

> The main thing about social life and vitality is that we cannot choose our neighbours, nor neglect the incessant flux of migration that challenges and changes us. It is not a good service to the community to argue that fear of risk and desire for security are virtues per se. If we silence issues, because they are difficult, we will become their prisoners. Art has always been a way to move into the space which has not yet been pacified, opened or socialised (as we say today) by mutual understanding. Art is not only for décor, a pleaser, an easy time out from the urgencies of life. It is itself an urgency of life. Art is not made in museums in the city centre. Art is created in difficult neighbourhoods, often in shared spaces in overcrowded suburbs.
>
> (Gerz, 2001)

These conditions (which include the less crowded and less shared spaces of universities) do not guarantee coherence. At the same time, coherence is the artificial in narratives; its lack in a counter-narrative does not mean incoherence. As the statement of an underground monument differs from that of a void, so perhaps the responses of dwellers make a non-coherence in which the conditions of dwelling are made visible. In the more rarefied space of Berkeley, in *Das Berkeley Orakel* (*The Berkeley Oracle*, 1998), Gerz used the internet to elicit questions in five languages, selecting 40 to display in an installation at the Berkeley Art Museum – for instances 'What is the taste of the sky?' 'What is beauty?' 'How far need I run to escape myself?' 'When do lies become truth?' and 'Why is my mother like this?' (Gerz, 1999). Lawrence Rinder remarks that in placing the questions throughout the museum, Gerz fractures

6.2 • Jochen Gerz, Harburg monument against fascism (1986–93).
Photograph courtesy of J. Gerz

6.3 • Jochen Gerz, Harburg monument against fascism (1986–93), detail of explanatory plaque.
Photograph courtesy of J. Gerz

'its Apollonian syntax' to tilt the balance towards the Dionysian (Rinder, 1999: 25) – too Nietzschean for me.

The spaces these works construct can, I think, be called dialogic (referencing Mikhail Bakhtin[35]); or spaces of publicity (a term used by Hannah Arendt[36]). Arendt, an observer at Eichmann's trial in Jerusalem,[37] sees deprivation of publicity (of visibility) as an oblivion which is a precondition for annihilation and itself painful. Kimberley Curtis reads Arendt as 'tone-deaf' to immediate concerns for social justice, but writes of her concern for oblivion that:

> This form of injustice is . . . prior to social injustice in the sense that the degradation of obscurity is a primary precondition of our capacity to inflict and sustain the suffering involved in the many forms social and economic inequality take . . . [which] weakens those who suffer it in ethically troublesome ways. Thus in what has been perceived as Arendt's insensitivity . . . I find the essence of a particularly compelling humanism that owes much to her phenomenology and wells out of her concern with how to intensify our awareness of reality.
>
> (Curtis, 1999: 68)

Publicity joins what Arendt sees as inner and outer realities (a background of repetition and foreground of new eruption) as site of natality, another key Arendtian concept – the movement to a conscious self which can be formed only amid the perceptions of others (in both senses), and through the performative. Oblivion thus denies the space of performance: 'Denied that movement in relation to others in a public sphere, denied the sense and pressing presence of speaking and acting beings, our own urge to appear remains unprovoked' (Curtis, 1999: 73). This denial is staged in Dan Gretton's *killing us softly*, a developing series of performances for an audience of eight – within the larger project *90% Crude* discussed in Chapter 8 – which seeks to draw provocatively helpful parallels between the operations of the Holocaust and, in a different sphere, those of the globalised oil industry and other aspects of transnational capital. Gretton accepts that the parallel is indirect, but Arendt's observation of Eichmann's inability to see the specificity of his situation, which she links to his dwelling on trivial points during his trial, indicates a possible path of enquiry – in the compartmentalisation of intelligence, or splitting-off of certain kinds of awareness from consciousness – which may enlighten the ability of some humans to take destructive decisions while being respectable and educated citizens, in some cases holders of high academic qualifications. Gretton also draws attention, as Bauman (above), to the industrialisation of killing in the Holocaust, another link to contemporary corporate mentalities.[38] In its first performances, in 2001, the work was four hours long and consisted of a narrative with visual aids and the playing of a viola, delivered to an audience sitting silently in separate black-painted cubicles. It is now a 10-hour work with more breaks and a facilitated discussion after the performance. The immediacy of the small scale and the endurance of the time the work takes are seen by Gretton as central to the engagement he requires from those who participate – by invitation and warned beforehand that the work is testing. That engagement centres on an argument within the listening consciousness.[39] I have some reservations as to the mode of narration and reception,[40] but the work coincidentally examines some of the implications of Bauman's *Modernity and the Holocaust* (not Gretton's source).

I want to move now to a broader focus on critical reflection in contemporary art through John Goto's *Tales of the Twentieth Century*, though the first case from it concerns the Holocaust. Dietlef Mühlberger writes in an introductory essay to a catalogue for John Goto's *Terezin* (1988)[41] – a sequence of works using oil paint, text and adapted photographic and documentary images – that it is essential to confront the Holocaust, but 'one feels helplessly inadequate . . . Neither the suffering and trauma of the victims of the persecution, nor the cruelty, inhumanity and sheer barbarism of the perpetrators of the crimes, can be adequately conveyed in words' (Goto, 1988: 14). In pictures, then? Does the fracturing of the images, or their added toning and layering, sometimes daubing over them so that it is not certain what is in front and what behind, figure or ground, allow something to be said which cannot be put into the words of ordinary speech? Does it open a fissure of meaning in what has meaning only as total negation? As in Celan's poetry, the references are oblique; some need the artist's notes to elucidate what transpires as a drama unfolding in which the camp, for instance, is both a holding base and a centre of artistic production: 'In the *sonderewerkstatte* (special workshop) at Terezin, copies of paintings by Rembrandt and Rubens were made for the Germans' (Goto, 1988: 50). The image *Rembrandt in Terezin* shows Jews assembled at the fairground in Holesovice, on the outskirts of Prague, for deportation; it is combined with a transcription of a drawing by Karel Fleischmann, 'First Night of New Arrivals', made at Terezin, a detail of a Rembrandt crucifixion with impasto, and vertically dividing the composition a black shadow in the shape of an artist's easel. Others trace, fragmentarily, the presence of Friedl Dicker-Brandeis (Freidlika Brandesova) who taught art to children in Terezin, and was herself a student of Johannes Itten in Vienna and at the Bauhaus (closed by the Nazis in 1933).[42] Dicker-Brandeis was killed. Among others re-deported from Terezin was Kafka's sister Otla Davidova, a voluntary children's nurse.

While much Holocaust art deals with the unspeakable, as it were, Goto draws out traces of biography that can be visually hinted at, through fragments of texts and a street plan showing the normalizing street names given to the alleys of Terezin in a beautification programme prior to a visit by the Red Cross in 1944, the semblance of normality that integrated the camp in the Nazi machine, and in the maintenance of the fake idea that the Jews were deported to new settlements where work made people free. Of course, some were held in labour camps, or like Primo Levi had skills that could be used in the war effort (Levi was a chemist, but others were pastry cooks or could work in arms factories) and survived; and not all were Jews – Goto uses a red triangle (the sign for political prisoners) in one work, a yellow triangle (Jews) in another. But perhaps the care of making the work and adding paint, put on by hand, like the individuality of the images of persons since rendered anonymous (rendered, like an industrial product also) is a kind of antidote to the machine's totally mechanical processing. There are impurities in the works, juxtapositions that slip, things not found in the perfect garden.

Terezin was not typical – a holding camp not a death camp – but it seems as important in its way that the breadth of the apparatus for purification of the Nazi state, and its normalization in Germany in the 1940s through this model ghetto, is brought home[43] as it is to not forget the most extreme acts. The history of Terezin is less adrenalin-producing than that of Auschwitz and Birkenau, or . . . but perhaps because of that is more open to analysis in the way suggested by Bauman, but not only in that way. Goto draws attention to the presence of artists in Terezin, its links to the Bauhaus (which shaped the kind of art education Goto and I had in the late 1960s), and to the importance of culture (*kultur*) to many of its inmates.[44] In *highway of fate* paint is applied thickly, like Kiefer, making a mess of the image of the road to the rail tracks. In the opposite direction the prisoners called the road the highway of memory.

Goto visited Terezin in 1983: 'I went, I saw, I photographed, I knew I had no knowledge of what happened' (conversation, February 20th, 2003). Through archive material and historical and autobiographical accounts, he sought to gain some impression of the cultural as well as everyday realities of the camp, its links to the Bauhaus, and to find a device to hold together a

6.4 • John Goto, *Rembrandt in Terezin* (from *Terezin*) (1998). Photograph courtesy of John Goto

collection of images resulting from the encounter. In this way, the character of Friedl Dicker-Brandeis acts as protagonist, seen in the opening image and again later with Itten. Goto is clear about his identification with the prisoners, but accepts the moral difficulty of the subject, and in particular of the question of voice – of whose voices, whose histories are subsumed in the artist's voice when it is only the artist who, at a remove of decades, has the opportunity to speak. He admires the work of John Heartfield, but does not try to copy the approach any more than to contrive a silver lining. In the catalogue for *Terezin*, Craigie Horsfield writes 'Art's claim to healing or to redemption fails utterly, frail in the face of the exorbitance of evil'; and that (in keeping with Bauman, though Bauman's book was published a year later) 'the terrible fact of the Holocaust was not in its singularity but in the certainty of its constancy' (Horsfield, 1988: 54). From this, he argues, follows a responsibility to recognise this history:

> These pictures come back . . . to this, a dark and piteous thing that must mark our lives is illumined with the passion and faith of men and women whose inheritance we reject at risk to ourselves and our children. It is never the recurring evil that is unique but the always singular and distinct voice of another that speaks. Its true epitaph is in our recognition, our fellowship, and our resistance.
>
> (Horsfield, 1988: 55)

For Goto, 'to sit in a library is not enough; I'm glad I went there and photographed it' (conversation, February 20th, 2003); the work is a form of witness but necessarily mediated through the material (in the sense of means, photographic and painterly) and process of making the work as well as his own response to the material. Today, in a de-politicising world, he sees resistance as difficult after the collapse of socialism, and has doubts about much work which claims to be documentary or engaged.[45]

His own work has moved in two directions since *Terezin*. One is in the use of digitally manipulated photographic images to renegotiate the possibility

6.5 • John Goto, *Monument* (from *The Commissar of Space*). Photograph courtesy of John Goto

of history painting in large-scale and complex figure compositions. In these, there is something out of place, which interrupts the semblance of familiarity with the scene, and works against the immaculately smooth surfaces. Goto made another journey in 1994 to investigate the spaces and circumstances of the later years of Kasimir Malevich, from which he produced *The Commissar of Space*, a digital work exhibited as a sequence of large full-colour canvases.[46] Malevich died from cancer in 1935 and his ashes were buried beneath a cube designed by Nikolai Suetin at Nemchinovka. His funeral was paid for by the Leningrad City Council but his paintings were no longer required. After 1928, when Stalinism began to restrict cultural activity, Malevich back-dated new works to the pre-revolutionary period. They depicted peasants in highly coloured, icon-like style and were exhibited in 1929 in Moscow and Kiev. The director of the gallery in Kiev was arrested, and it took Malevich more than two years to retrieve the work. The context for this rejection of a subject-matter that is also handled by Socialist Realism is the deportation of kulaks (the peasant-farmers of the Ukraine) and collectivisation of agriculture to achieve a 55 per cent increase in production. Malevich's paintings of the late 1920s and early 1930s resembled his pre-Suprematist work – which has an icon-like quality in another way, referencing the desert of the life of monastic contemplation – in reasserting the autonomy of the picture-plane, and are hence at odds with the representational requirements of Socialist Realism.[47] Brandon Taylor notes a remark by Malevich to a former student, Anna Leporskaya, that he found himself 'hidden in clouds and dots of colour' and declared himself 'the Chairman of Space' (Taylor, 1998: 18). In *Monument*, from the Malevich sequence, Goto collects people who had some presence in the circumstances of his life, while a white figure representing Malevich himself floats above a podium. Included are Gorki, Gerasimov, Stalin, Kirov, Suetin, Mandelstam, Una and Natalia Malevich, Pasternak and Tatlin, among others, and a dog. Behind are trees. The figures are evidently transposed from elsewhere, adapted to scale and fitted into what becomes a large figure composition in the academic manner. The digital adaptation is impeccable and sets up a tension between the naturalness of the image as a whole and the unnaturalness of the circumstance in which so many figures in history and in Malevich's life should be together, as if at a funeral or the unveiling of a statue which does not exist. If history painting, the highest genre in the academic hierarchy, is defined by 'its public and ethical form; its principles of historicity and narrativity; and its didactic intent' (Seddon, 2000: 82), Goto plays with it for post-modern times in which grand narratives are no longer viable but in which the requirement to say something is not annulled. The narratives are broken, and logically incompatible elements put side by side in the way that ideologically compatible but temporally or spatially incompatible elements are combined in historical portraiture and history painting in the academies. There is no plan for history now but the circumstances and the seemingly inconsequential moments in those circumstances can be recomposed to draw attention to

6.6 • John Goto, *Marks & Spencer* (from *Capital Arcades*). Photograph courtesy of John Goto

survivals or cancellations of human values such as love and the right to life, and the fading of it all.

The other direction in Goto's work through the 1990s is found in a sequence of images that present acts of human symbolic violence in enlarged, monochrome but otherwise unmediated images of damaged rood-screens photographed in churches in Norfolk, Suffolk and Devon.[48] The faces painted on wooden screens as part of a Catholic tradition have been attacked, using swords, knives, hammers put through the wood, almost as if (or as if) they were real not painted. Some have been broadly cut across the face; others have an eye removed or are transfigured by a grid of cuts as if to visit them with complete abolition as the visible presence of an angel, saint, holy martyr, king or donor – like the statues that fall in revolutions (see Chapter 1). It could be argued, perhaps was, that the figures represented privilege, and the superstition of the old religion, but what transpires is the violence of suppression. It is as if Adorno's idea that myth is reproduced in the terms of rationality has taken the form of a reproduction of hate in the terms of abolition of a religion of divine love, that release from the bonds of a hierarchic religion has unleashed a wilder and no longer mediated energy, which is expressed in raw destructiveness.

The universal happy ending is now globalised through fast food, fizzy sugar-drinks, and the culture industry. But if the Nazi history represented in museums is remote, or is inextricable from television history programmes, Hans Haacke's

And You Were Victorious After All (1988), a hollow obelisk placed over the *Mariensäule* in Graz to renew memories of a similar covering put on it for the *Anschluss* in 1938, was fire-bombed by a neo-Nazi. Michael North interprets the title as an intricate irony 'that the Nazis were wrong, that they were not to be victorious, but simultaneously that they were right, that there remains a layer of Nazism just under the surface' (North, 1992: 26). There may be, as Gretton argues (above), parallel destructive scenarios to critique. The difficulty remains that representation is not adequate while art reaches only a minority of the public. Gretton works with that by adopting a micro-scale for the work while seeing the shift in individual consciousness it may evoke as a shift in the matrix. Goto sees irony and satire as viable means to criticise, observing 'we could end up in perpetual mourning for socialism, which would be comforting but not useful' (conversation, February 20th, 2003). In a very different way, Gustave Metzger revels in an excess he has taken back from history, reaching out to black holes and galaxies, which are nature.

> From the burning core of our earth to those expanses where we could never go, and if we did, never survive. Do we share the burning to the point where everything disintegrates? What is the fever pitch of a person in love; or the maniacal excess of a creative act; or the blind onslaught of a killer? What has it to do with the torrents within a galaxy?
>
> It is the extreme and the excess. It is the burning to the point threatened by dissolution. Dissolution does not take place – any more than the galaxy dissolves – despite its white heat.
>
> (Metzger, 1999: 47)

I let him have the last word here.

NOTES

1 Young cites the justification for the project by President Carter in 1979: 'it was American troops who liberated many of the death camps, and who helped to expose the horrible truth . . . Secondly . . . we must share the responsibility for not being willing to acknowledge forty years ago that this horrible event was occurring. Finally, because we are humane people, concerned with the right of all peoples, we feel compelled to study the systematic destruction of the Jews so that we may seek to learn how to prevent such enormities from occurring in the future' (Young, 1993: 336). Finkelstein sees Carter as placating the Jewish vote, and notes failed attempts to establish an African-American museum (Finkelstein, 2000: 72–3).

2 Young notes a 10-minute stoppage of work by 500,000 Jews in New York on December 2nd, 1942; and that projects for Holocaust memorials began in the US in 1948 (Young, 1993: 286, 290).

3 Bauman cites Sabini and Silver (1980) as giving a notional figure for deaths in pogroms of 100 a day (equivalent to 146,000 in four years). The point is also that 'Contemporary mass murder is distinguished by a virtual absence of spontaneity . . . and the prominence of rational, carefully calculated design . . . an almost complete elimination of contingency and chance, and independence from group emotions and personal motives' (Bauman, 1989: 90).

4 The text is from material used by Dan Gretton in *killing us softly*, a performance work in which the Holocaust is compared, in a way Gretton sees as 'provocatively helpful' (conversation, June 23rd, 2003), with the destructive operations of the oil industry. The work is discussed in section II of this chapter.

5 The *Faber Book of Utopias* includes an extract from *Hitler's Table-Talk, 1941–44*, recorded by Borman, describing Hitler's plan to turn the Ukraine into an farm for a healthy Aryan peasantry (Carey, 1999: 423–5). Also included is H. G. Wells' *Anticipations* (1901); Wells writes of whole masses of the human population as 'inferior in their claim upon the future' and see their flaws as 'contagious and detrimental in the civilizing fabric' (in Carey, 1999: 369). Bauman cites R. W. Darré, who uses the model of a garden in which the weeds must be eliminated, as metaphor for population control (Bauman, 1989: 113–14).

6 'alongside this crudeness there is also an undercurrent of very old dreams. The strongest is that of the "Third Reich" ... Music on the square piano, bands in beer gardens sang out to him, when there had already long been a Kaiser: "A crown lies in the deep, deep Rhine"' (Bloch, 1991: 57). Bloch contrasts the Nazis' false millenarianism with the Kingdom of the Third Gospel of Joachim of Fiore (see Bloch, 1986: 509–15). He also notes the second-hand character of Nazi forms: 'The Nazi was creative, so to speak, only in the embezzlement at all prices with which he employed revolutionary slogans to the opposite effect' (Bloch, [1937] 1991: 117).

7 See Young, 1993: 113–208 on monuments in Germany and Poland.

8 Young is critical of the nationalization of Holocaust memory, noting that at Majdanek Jews were 80 per cent of those killed, and that an inscription in the camp links Hitler's rise to the help of German industrialists: 'Hitler merely did the dirty work of big business, and fascism was only a form of monopoly capitalism run amok ... Subsumed once in an economic critique of the camp, the murder of Jews is submerged yet again in the national identities of victims' (Young, 1993: 122). Cf. Bauman's argument above.

9 Hansen notes the divergent receptions of *Schindler's List* by a mass public and intellectuals in the US and Germany; citing Hoberman (1993) 'Is it possible to make a feel-good entertainment about the ultimate feel-bad experience of the twentieth century?' (in Hansen, 2001: 130), she questions the adequacy of a fictional narrative framed by hierarchies of gender and race, and which unifies individual stories. She argues, too, against Lanzmann's criticism of the film as reducing the dialectics of the problem 'to a binary opposition of showing or not showing' rather than casting it as competing representations and modes of representation (Hansen, 2001: 134). See also Hirsch (2001) on Spiegelman's cartoon *Maus*; and Kugelmass (1992) on US Jewish tourism in Poland.

10 See Griswold, 1992: 82–6 on the architecture of the Mall; Young, 1993: 337–47 on the design process; and Weinberg and Elieli, 1995: 25–31 for appreciation of the design as effected. An adjacent Federal building was added to the museum rather than demolished as Freed wanted, to house a café and administrative offices (Weinberg and Elieli, 1995: 27). The public spaces of the site contain art commissioned through a Percent for Art scheme.

11 'Looking around [the Hall of Witness], visitors realize that they are surrounded by red brick walls and dark-gray steel structures, reminiscent of a prison building. The hall's roof is an enormous skylight, whose glass-and-steel structures are skewed, a hint at the state of the world in Holocaust times. A wide stairway leads up to the second floor, ending in a brick gate whose arch is shaped exactly like the gate to the death camp of Auschwitz–Birkenau ... The pattern of the bricks in the walls is reminiscent of the barracks at Auschwitz I. The northern side of the building is made up of a row of four interlinked brick structures, shaped like outsized Auschwitz watchtowers' (Weinberg and Elieli, 1995: 25).

12 Rogoff compares them to the suitcases of immigrants at Ellis Island, piled up high 'as a display of a mass, defying their very individual markings' in a 'display strategy that wants to insist on driving home both its quite natural disapproval of what took place but also its hopes that this act of museumification serves as a kind of amends' (Rogoff, 2000: 44). Rogoff then quotes Felman's 'In an Era of Testimony – Claude Lanzmann's SHOA', *Yale French Studies*, 79, pp. 39–82 (see also Felman

and Laub, 1992) that the aim of the Nazi persecution was to make Jews invisible, being reduced to the term *figuren* as 'that which all at once cannot be seen and cannot be seen through' (in Rogoff, 2000: 45).

13 Weinberg and Elieli write 'staff had to wrestle with the question of how to project the human face of the victim', that it was essential to see victims as individuals, and that the card means the visitor 'goes through the exhibition of the Holocaust with a companion' (Weinberg and Elieli, 1995: 72). Young cites Rosen to the effect that many Jews scrambled to acquire false papers in order to survive, and that 'There is a reverse principle at work here, as if everyone were expected to enter the museum an American and leave, in some fashion, a Jew' (Rosen, 1991, quoted in Young, 1993: 344).

14 'Let me tell you, when this little shoe was handed to me, I froze. Bear in mind that I am a former partisan. I was hardened in battle and I deal with this Holocaust story almost on a daily basis. But when I held in my hand that shoe – the shoe of a little girl who could have been my own granddaughter – it just devastated me' (Weissberg, 2001: 21–2, citing an undated letter from Miles Lerman to potential donors, 1993). Weissberg comments: 'They become useless objects; and their owner's real sufferings take place after they were left behind. Faded to a uniform colour that masks their individual shapes, they are unable to tell any stories of their bearer . . . Homogenized as one group, these shoes speak as a mass and exemplify mass murder' (Weissberg, 2001: 23).

15 Hirsch cites Milton (1991) on the quantity of images, and specifies those of the gate to Auschwitz I with the sign *Arbeit macht frei*, and of the rail tracks leading to the long building with gable and gateway of Auschwitz II–Birkenau, to which trucks took prisoners directly: 'The specific context of these images has certainly been lost in their incessant reproduction . . . these images have come to function as trope for Holocaust memory itself' (Hirsch, 2001: 226). But she takes this as construction of a post-memory (the response of the second generation to the trauma of the first), a product rather than a screening of the Holocaust which connects the second generation to the first.

16 Rabinbach cites Jaspers as identifying the centrality of Auschwitz, placing 'the caesura at the centre of philosophical reflection'; he quotes Habermas: 'Auschwitz has become the signature of an entire epoch – and it concerns all of us. Something happened there that no-one could previously have thought even possible.' (in Rabinbach, 1997: 163–4, citing Habermas from 'Historical Consciousness and Post-traditional Identity', Habermas, 1989: 251–2).

17 See Jarvis, 1998: 9–10 on Adorno's thesis; and 11–15 on his flight to England (registering as a doctoral student at Merton College, Oxford), invitation to join the Frankfurt Institute in New York (on a programme of social research on radio listeners), and development with Horkheimer of *Dialectic of Enlightenment*.

18 Rabinbach refers to the influence of Benjamin's *Trauerspiel* (Benjamin, 1985), and notes that Benjamin's death occurred while Adorno and Horkheimer were beginning to prepare the work which became *Dialectic of Enlightenment* (Rabinbach, 1997: 174–81).

19 'the fully enlightened earth radiates disaster triumphant. The program of Enlightenment was the disenchantment of the world; the dissolution of myths and the substitution of knowledge for fancy' (Adorno and Horkheimer, 1997: 3).

20 'redemption can only be salvaged by a thinking that radically refuses any compromise with magical practices, myth, or the transposition of worldly events into symbols . . . only disenchantment . . . can bring about salvation' (Rabinbach, 1997: 179).

21 Rabinbach uses the 1972 edition (New York, Seabury Press), in which the passage is on p. 16. On the development of the book from a presentation volume for Friedrich Pollock in 1944 to the 1947 published edition (Amsterdam, Querido Verlag), see Rabinbach, 1997: 166–73. See also Jay, 1984: 36–40; Jarvis, 1998: 20–43.

22 See Jay, 1984: 53–4, who concludes that 'in its very refusal to subordinate nature to thought, matter to spirit, de-aestheticized art provided a flickering utopian model of what mankind [*sic*], despite everything, might become' (Jay, 1984: 54).

23 'Auto-destructive art is primarily a form of public art for industrial societies' (Metzger, 1959, 'Auto-destructive Art', in Hoffmann, 1999: 26). In 'Manifesto World' (1962), Metzger says: 'The artist must destroy galaxies. Capitalist institutions. Boxes of deceit. You stinking fucking cigar smoking bastards and you scented fashionable cows who deal in works of art' (Hoffman, op. cit.). In 1974, Metzger (who left Germany as a child in 1939) was included with Beuys and Haacke in the show 'Art into Society – Society into Art: Seven German Artists' at the Institute of Contemporary Arts, London, but, of Polish Jewish family, refused to be called German and agreed to participate by not exhibiting a work but writing for the catalogue an essay proposing an art strike as antidote to an art of social habilitation.

24 Lawson, in *Closure*, writes 'Art in this sense is the pursuit of openness and the avoidance of closure. As a result the artist is one who is not engaged in an attempt to provide closure but seeks instead to point towards the residue that lies outside of closure' (Lawson, 2001: 206). See Harding, 1997: 53–64 on Adorno and Beckett.

25 Jarvis notes a link to Benjamin. In *Aesthetic Theory*, in a passage on *Sachlichkeit* (objectivity), Adorno writes 'Even the highly cultivated aesthetic allergy to kitsch, ornament, the superfluous, and everything reminiscent of luxury has an aspect of barbarism, an aspect – according to Freud – of the destructive discontent with culture . . . progress and regression are entwined. The literal is barbaric. Totally objectified, by virtue of its rigorous legality, the artwork becomes a mere fact and is annulled as art' (Adorno, 1997: 61). Jarvis translates this as 'the barbaric is the literal' (Jarvis, 1998: 145) and describes it as 'an oblique comment upon Benjamin's more famous aphorism that "there is no document of civilization which is not at the same time a document of barbarism"' (from Benjamin, 1973: 248).

26 'These composers recorded the catastrophe of the overwhelming power of the rational-technical machinery of modern collective institutions . . . through the resistant spirit and suffering of the subjectivity mutilated by it' (Witkin, 2003: 61). Witkin quotes from 'On the Fetish Character of Music and the Regression of Listening': 'The whole cannot be put together by adding the separate halves, but in both there appear . . . the changes of the whole, which only moves in contradiction . . . Between incomprehensibility and inescapability, there is no third way; the situation has polarised itself into extremes that actually meet' (Adorno, 1991: 30–1, quoted in Witkin, 2003: 61–2).

27 Marcuse writes in *The Aesthetic Dimension*: 'Auschwitz and My Lai, the torture, starvation, and dying – is this entire world supposed to be "mere illusion" and "bitterer deception"? . . . Art draws away from this reality, because it cannot represent this suffering without subjecting it to aesthetic form, and thereby to the mitigating catharsis, to enjoyment. Art is inexorably infested with this guilt' (Marcuse, 1978: 55). He continues that art is not released from remembering but must not let go either of the idea that 'The revolution is for the sake of life, not death. Here is the perhaps most profound kinship between art and revolution' (Marcuse, 1978: 56).

28 Celan, 1996: 62–5. See Saltzman, 2001, figs 1 and 2 for paintings by Kiefer (1981–3) that evoke Celan's poem, the latter an image of a dark, brick-vaulted chamber titled *Sulamith*. Saltzman writes 'With the conferral of a name, Sulamith, and with the evocation of both Hebrew Bible and Celan's poem, it is presumed that the darkened chamber is transformed, transfigured, translated, into a site of Jewish memory, a Holocaust memorial in painterly form' (p. 81); she is uncertain what it represents – a piece of Nazi architecture, a stasis, an absence except in a name, and notes that 'Death Fugue' was translated by Greenberg in 1955. Levi includes the poem in an anthology *The Search for Roots: A Personal Anthology* (2001a, Harmondsworth, Penguin, pp. 198–200), saying 'I wear it inside me like a graft' (p. 198). See also Levi, 2001b: 42–4; Felstiner, 1995: 22–41, 118.

29 Adorno makes two references to Celan in *Aesthetic Theory*, one in passing bracketing him with Beckett (1997: 219), the other at more length on the hermetic aspect of poetry after Mallarmé; he reads Celan as inverting the experiential content of hermetic poetry: 'His poetry is permeated by the shame of art in the face of suffering that escapes both experience and sublimation. Celan's poems want to speak of the most extreme horror through silence. Their truth content itself becomes negative. They imitate a language

beneath the helpless language of human beings, indeed beneath all organic language: It is that of the dead speaking of stones and stars . . . The infinite discretion with which his radicalism proceeds compounds his force. The language of the lifeless becomes the last possible comfort for a death that is deprived of all meaning' (1997: 322).

30 The German title means a device in the composition of fugues (the Italian *stretto*), or reduction; Hamburger reads ambiguity throughout the poem: 'since Celan took every word as literally as possible, often breaking it down etymologically in the manner of Heidegger, one of his reasons for choosing the German word must have been that it characterized not only the structure of his poem but its theme as well' (Celan, 1996: 26). Felstiner gives an account of the poem's circumstances, which include the use of 'Death-fugue' in schools, the banning of Alain Resnais' film *Night and Fog* – a black-and-white documentary of the death camps which opens with colour footage of grass growing where the prisoners were, and ends with the camera panning again in colour over the remains of Auschwitz; Celan wrote a German translation of the narrative ending 'we do not hear that the scream never falls silent' (Felstiner, 1995: 93) – and Celan's perception that, if his earlier poem was becoming accepted he must write something resistant. Felstiner translates *untrüglichen* as unerring (not unmistakable as in Hamburger's translation), and the poem's first term (*Verbracht*) as taken (not driven), not quite deportation but with something of the resonance and the unstated destination implied by deportation. The reference to grass is to Resnais' film, and a biblical use in Isaiah (40: 7). Felstiner notes the asterisks that punctuate the poem, linking end to beginning: 'Repeat this to the six millionth degree' (Felstiner, 1995: 125).

31 Michalski suggests a link to the Jewish prohibition on images, and quotes a saying, '*der Boden sollte unter den Füssen brennen*' (the ground should burn under the feet), as reference to catharsis. Among other buried monuments are Horst Hoheisel's *Aschrott Fountain* (1988) in Kassel (see Young, 1992: 70–6, and Young, 1993: 43–8), an inverted fountain on the site of one destroyed by the Nazis in 1939; and Micha Ullman's *Empty Library* (1995), Berlin, an empty underground room on the site of a Nazi book-burning in 1933. For discussion of the anti-monument see Young, 1992: 54–5.

32 Gerz proposed a *Monument for the Murdered Jews of Europe* in response to a competition in Berlin in 1997. It consists in a building of three rooms (designed by Iranian architect Nasrine Seraji) of memory, replies and silence, with a plaza in which 39 posts carry the word Why? written vertically in digital lighting in the languages of Europe's Jews. In the three rooms, respectively, members of the Shoah Foundation (set up by Spielberg) would collect memories from survivors, visitors write answers to the question 'why did it happen?' in books lining the glass wall, and listen to 'the eternal e' of US composer La Monte Young. The answers were also to be engraved on the plaza (Museum for Modern Art, Bolzano, 1999: 80–5).

33 See Young, 1992: 56–66; 1993: 28–37; Michalski, 1998: 182–4. Cf. 'we will one day reach the point where anti-fascist memorials will no longer be necessary, when vigilance will be kept alive by the invisible pictures of remembrance' (Jochen Gerz, quoted by Doris von Dratein, cited in Young, 1992: 60, n. 12).

34 'illegible scribble of name scratched over names, all covered over in a spaghetti scrawl . . . People had come at night to scrape over all the names, even try to pry the lead plating off the base. There were hearts with "Jurgen liebt Kirsten" written inside, stars of David, and funny faces . . . swastikas also began to appear: how better to remember what happened than by the Nazi's own sign? . . . when city authorities warned of the possibility of vandalism, the Gerzes had replied, "Why not give that phenomenon free rein and allow the monument to document the social temperament in that way?"' (Young, 1993: 35).

35 See Brandist, 2002; Bell and Gardiner, 1998; Mandelker, 1995. In the first, see particularly chapters 3 and 4 on dialogue as a discursive enactment of inter-subjectivity; in the second, see Bell, 1998: 49–62 on dialogue and cultural change; in the last, see Aronowitz, 1995: 119–36 on the idea of literature as social knowledge.

36 See Arendt, 1958: 22–78 on the duality of private–public. Arendt's position is derived from her reading of the *polis*; for

instance, on the term public: 'It means, first, that everything that appears in public can be seen and heard by everybody and has the widest possible publicity. For us, appearance . . . constitutes reality. Compared with the reality which comes from being seen and heard, even the greatest forces of intimate life . . . lead an uncertain, shadowy kind of existence unless and until they are transformed, deprivatized and deindividualized, as it were, into a shape to fit them for public appearance' (p. 50). See Sennett, 1995: 31–86 on the gendered naked and clothed, open and enclosed, in Greek society in the time of Perikles. In a lecture at the London School of Economics (April 12th, 2003, in the programme 'CivicCentre' coordinated by Alan Read of the University of Surrey, Roehampton), Sennett spoke of the visual and social arrangements of the *pnyx* (assembly) in Athens, where the speaker stood silhouetted against the sky while the listeners sat among those who knew and could vouch for them in an attitude not only passive but, in the body language of that society, vulnerable.

37 Arendt perceived Eichmann as unable to understand the reality he helped produce. Curtis summarises: 'Eichmann, it seemed, was incapable of countenancing the particularity of the world, neither his specific particularity . . . nor the particularity of others.' (Curtis, 1999: 47).

38 Gretton writes: '*killing us softly* is an attempt to go beyond our *knowledge* of acts of inhumanity into a different state of *experiential* understanding. Many of us have made our own journeys through the territory of genocide and the Holocaust . . . At different times of our lives we have chosen to read or not to read, to watch or not to watch, to engage or not to engage. However, as the Swedish writer Sven Lindqvist suggests, perhaps we have been approaching this territory with the wrong kind of map . . . By shining light into the world of the bureaucrats, planners and businessmen who contributed to Nazism and the Holocaust, *killing us softly* raises a critical question as to whether such an event can be viewed as a finished, historical episode or whether the psychology and behaviour that enabled genocide to occur then is not only still present today, but exists quite specifically in the mindset and activity of individuals working for transnational corporations' (publicity material supplied by PLATFORM, 2003).

39 I have worked from documentary material supplied by Gretton, various conversations with members of PLATFORM (see Chapter 8), and participation in an early performance. Gretton is currently (2003) developing the material as a book, in discussion with John Berger. One aim, of *killing us softly* and *90% Crude* as a whole, is to 'contribute to a change in public perception of the oil industry so that, over a period of five to ten years, talented graduates will feel that the oil industry is so dubious ethically that, effectively, it becomes a no-go area' for employment, as happened to an extent already in the nuclear industry (letter from Gretton, July 15th, 2003).

40 I have two areas of concern: the centrality of Gretton's role as narrator; and the panopticon effect which puts the participants in passive isolation during the narration, which I discuss in Chapter 8. Two letters from previous participants, among a large body of feedback material collated by PLATFORM, state: 'The experience of being in a small, dark space with no easy way of escape stays with me; having to face words and images, a familiar part of my middle-aged mental landscape, yet so terrifying they've been shut away in a mind box labelled "the incomprehensible, the unbearable"'; and: 'It was gripping, time became irrelevant, huddled in our little black cocoons, complete with comfort blankets, we were immersed into a story that none of us wanted to hear but somehow we found impossible to resist' (supplied by Gretton, letter July 15th, 2003).

41 The fortress of Terezin (Theresienstadt), built in the reign of Joseph II 60 km north of Prague, is in the form of a 12-point star. The local population (3,700) was evicted and the town used as a ghetto and transit camp, easily sealed because entrance was only by six gates in the old walls. Around 155,000 people, mainly Jews from Czechoslovakia (including most of the Jews of Prague), Austria and Germany, some from Holland, Denmark and Poland, and some political prisoners, were sent to Terezin between 1941 and 1945. 88,000 were transported to the East, the leaders of the ghetto (the Council of Jewish Elders) often being involved in the composition of the transports. Although initially the rules were draconian – separating genders, banning smoking and movement between barracks – and

6.7 • PLATFORM, *killing us softly*, feedback day after the first cycle of performance, July 2000.
Photograph courtesy of PLATFORM

rigidly enforced (16 were hanged for petty transgressions), in 1942 some relaxation occurred; a limited postal service was permitted, and the camp became a centre for elderly and in some cases well-known German Jews, and those holding military decorations from 1914–18. In 1943 improvements were made to the water, toilet and cooking facilities by the prisoners. There were lectures, performances by orchestras and choirs, a cabaret group. Art was part of normalisation. A local currency was introduced (the ghetto crown) for use in shops stocked with the belongings of new arrivals. As Mühlberger says (from whose account the above is derived), 'the German authorities organizing the systematic murder of European Jewry hit on the idea . . . of transforming Terezin into a "model ghetto" which could be used in Nazi propaganda to counter the growing rumours of atrocities' (in Goto, 1988: 17). In 1944, as a result of Danish pressure, permission was given for a visit by the Red Cross; houses were painted and 1,200 roses planted. Few of the artists survived. Of the 15,000 children who went through the camp, 100 survived.

42 Around 4,000 drawings by children at Terezin are archived in the State Jewish Museum, Prague.

43 Young records that one survivor, asked whether she had been in a camp, replied 'Yes, but forgive me – I was only in Terezin' (Young, 1993: 244).

44 'the artists who lived and died at Terezin . . . are seen with an immense compassion. They came out of a world in which culture and Art were of the greatest importance. They lived in a vile time and carried with them . . . a faith in Art and its power. To us such faith is delusion but we are wrong and how vulnerable is our time' (Horsfield, 1988: 55).

45 Goto sees Lanzmann's *Shoah* as an authentic documentary. When we both taught at the same University in 1998, I assisted him in a course for second-year undergraduates in art dealing with the making of work from everyday and extreme experiences, in which *Shoah* was screened (in more than one session).

I introduced poems by Celan and settings of them by Georgian composer Giya Kancheli. We were told that such material was inappropriate for an art degree course; a meeting of the year group was called at which objections were solicited. We both resigned soon after.

46 See Goto, 1998, which includes an essay on Malevich by Brandon Taylor (points from which are summarised below). Goto's work and Taylor's essay are available at http://www.johngoto.org.uk.htm and a book, *Ukadia* was published in 2003 by the Djanogly Gallery, Nottingham. More recent digital series are *Capital Arcade*, in which the compositions of paintings from the European tradition have been adapted for the consumption-scape of a mall, and *High Summer* in which landscapes in the eighteenth-century picturesque style are manipulated to draw attention to present actualities. A few images from *Capital Arcade* are reproduced in the 2nd edition of *The City Cultures Reader* (Miles, Hall and Borden, 2003).

47 See Taylor, 2000; Clark, 1997; Bown and Taylor, 1993.

48 *Loss of Face* was a selection of these (of which there are more than 100) exhibited at Tate Britain in 2002. The catalogue states: 'The Reformation created a dramatic rupture in the development of the visual arts in Britain, and many have argued that the dominance of literature in British culture is directly attributable to the events of this time' (in Goto, 2002, n.p.)

7

1993 (II)

PARTICIPATION AND PROVOCATION

•

In this chapter I examine possibilities for art and architecture as critical interventions in urban societies. Looking selectively at visual practices from the 1970s to the present, I draw out two overlapping tendencies: work that involves the participation of others in its making; and work that, while not involving co-production, seeks to provoke active reception in exposure of social, cultural, and economic conditions. Permeating the chapter is Walter Benjamin's insight (1934) that intervention can take place, and must, in the means of production. But I ask what that means, and whether the means of production includes the conceptual and linguistic categories through which we describe and prescribe a world. Looking to more recent critiques, I introduce texts by Peter Bürger, John Tagg and Rosalind Krauss before discussing Benjamin's essay on 'The Author as Producer'. In the second section of the chapter, I link Benjamin's thesis to participatory and provocative art from the 1970s to 1990s, and a parallel history of radical planning, looking in particular at the work of Mierle Ukeles. In the third section I turn to provocation in the work of London-based artists Cornford and Cross, and in a project by the Lisbon-based group Extra]muros[. The emphasis in this chapter on cultural agency in social formations is complemented by that of Chapter 8 on environmentalisms. I begin here with an account by Chinese-American artist Mel Chin of his performance at the Dia Art Center, New York at 2 p.m. on April 25th, 1993, in which he draws an analogy between the actions of a sniper and those of a viral particle entering a bloodstream.

1. After introduction you have 30 seconds to . . .
   Pick up Remington M700.30–06 bolt-action sniper rifle, altered (with wireless microphone) Raytheon Night Vision telescopic sight, from hidden place under the panel table. Load rifle with empty shell. Walk up to the podium and get into position. Aim at the audience slightly above the heads and toward the far NE corner.

[. . .]
3. Pull rifle trigger. Audible: 'click'. Sound man: 'HIT'
4. Eject the shell.
[. . .]

(Chin, 1999: 69–72)

I THE MEANS OF PRODUCTION

After firing the rifle Chin read his text:

> I begin with the constructed voices of two who abnormally mark the conclusion of life with unerring precision – a covert peacetime Marine sniper (whose accuracy is 98% at 1,000 meters), and a virion's pathological trek within a human host.
>
> (Chin, 1999: 74)

The sniper and the virion are specialised entities denoting the power of the military–industrial complex and of evolutionary processes. Chin sees the analogous modes of operation they represent as elective appropriations for an art of 'reconnaissance and reflection for a method to provoke' (Chin, 1999: 74). Observing a fusion of the two in the pre-emptive use of post-immunological tools by US army doctors, he asks 'If the antidote is first, can a more predictable and virulent one-shot-one-kill genetic poison packed in a gp160 envelope be far behind? Will its delivery be through deceptive social contact? In an embrace at the next summit?' (Chin, 1999: 75).[1] The virion infiltrates its host, a tactic Chin sees as viable for art in a situation in which outright resistance is unwise – 'The benefits of protracted war are little' (Chin, 1999: 76). He cautions that those who de-centre power need to confront 'the fascism that causes us to love power, to desire the very thing that dominates and exploits us' (Chin, 1999: 77); and, dissatisfied with the identity of artist, argues that those who intervene in 'the multitude of systems that comprise our culture . . . may wish to pack a sniper/viral mind set' (Chin, 1999: 75, 77). To the insecurity bred by surveillance he poses the unadvertised and undetectable:

> The frightening conditions that are imposed by a sniper are no longer lurking in the historic journals of war. They are in your face though out of sight. Such mechanics need not be taken as negative models but as successful working models that are worth taking seriously. Such expeditions into these non-traditional venues are especially fat targets or assignments for art.
>
> (Chin, 1999: 77)[2]

The viral metaphor is used, too, by Jane Trowell, who writes of 'a viral quality, slipping a proposition into the bloodstream under the guise of a safe publication' (Trowell, 2000: 107). Her reference is to *Ignite*, a spoof newspaper (1996–7) produced by PLATFORM and distributed to commuters at London rail termini. *Ignite* looked like a regular evening paper but carried stories not found in the mainstream media exposing the environmental impact of Londoners' oil consumption and the effects of the global oil industry on the natural and social environments of the countries of extraction.[3]

While the public for *Ignite* was identified only as those who use London's rail termini, Chin's performance at DIA addressed a more specific public of cultural consumption. Similarly, the Guerrilla Girls[4] take the artworld as their public, in 1987 attaching peel-off stickers to the doors and windows of New York galleries:

> WE SELL WHITE BREAD
>
> ingredients: white men, artificial flavorings, preservatives
> * contains less than the minimum daily requirement of white women, and non-whites
> a public service message from Guerrilla Girls
> conscience of the artworld.
>
> (Guerrilla Girls, 1995: 51)

Elsewhere, they point out that US bus companies employ 49.2 per cent women drivers while in 33 New York galleries the extent of women's representation is 16 per cent;[5] and explain the advantages of being a woman artist:

> Working without the pressure of success
> Not having to be in shows with men
> [. . .]
> Knowing your career might pick up after you're eighty
> Being reassured that whatever kind of art you make it will be labelled feminine
> Not being stuck in a tenured teaching position
> [. . .]
> Not having to choke on those big cigars or paint in Italian suits
> Having more time to work after your mate dumps you for someone younger
> [. . .]
>
> (Guerrilla Girls, 1995: 53)

These projects convey outrage in representations of reality; but publics participate in them, as in *Ignite*, post-production. In the 1990s, with roots in the radical art of the 1960s, a new emphasis began to be placed by some artists on participation in the making of the work. This, too, has several strands, from the provocation of uncomfortable awareness of the complicity of normalised assumptions in a culture of dominance, to an empathy in which stories other than those of the dominant regime can be voiced.

As artist in residence at the Anderson Ranch Arts Centre at Snowmass, Colorado in 1994, Chin placed spoof parking permits on the windscreens of cars in the Centre's car park, asking drivers to write in their car make and registration, then tick boxes for race, income level and sexual preference. For race he listed Black, Asian, Hispanic, Native American and Other; for income, below $6,000 and over $100,000; and for sexuality only bisexuality, homosexuality, trans-sexuality, and asexuality (or other). The project draws its public into an investment of time, through which to deconstruct categories of producer and consumer as separate entities – a modest claim, not world revolution, but more than that the reader completes the work (which is never completed). If such interventions are incidental this is not to say they are accidental. While Chin sets up a situation in which the simulation of bureaucracy conceals its subversion, other kinds of project use evoke cultures denied visibility and audience in an affluent society.

For a second case, I look to Jackie Brookner's *Of Earth and Cotton* (1998). Brookner visited the homes of ex-sharecroppers (who picked cotton in the depression years) to model their feet in clay, sitting on the floor listening to them talk (or not as they wished), exhibiting the feet on a ground of clay as a gallery installation. Unlike Atherton's life-casts (Chapter 5), the feet are made from observation. Clay has a specific meaning for sharecroppers, and the work does not seek monumental status through a transposition into a material of high or official culture. Brookner speaks of not wanting to have people 'get their feet stuck in plaster' and of a space of intimacy which opened as they watched 'this bunch of mud turn into their own feet' (de Boer, 2000: 24). To make something in clay takes time, and may distract from the self-consciousness of the encounter to allow stories to be told which would otherwise remain unheard. There were no stereotypical responses; Brookner's question as to what they felt about 'the land' produced blank looks: 'I had to learn to ask more specific questions . . . they knew what soil cotton would grow well in and what the rain does' (ibid.).

In both of these otherwise differing projects, the co-production of the work (which retains the expertise of the professional but equates it with the cognition or expertise of non-professionals) introduces a collapse of the separation of production and reception. Since the late 1960s there have been many refusals in art (and to an extent in architecture) – of individual practice in favour of group working, of the object in favour of documentation and process, and of institutional spaces of dissemination:

> For the past three or so decades visual artists of varying backgrounds and perspectives have been working in a manner that resembles political and social activity but is distinguished by its aesthetic sensibility. Dealing with some of the most profound issues of our time – toxic waste, race

> relations, homelessness, ageing, gang warfare, and cultural identity – a group of visual artists has developed distinct models for an art whose public strategies of engagement are an important part of its aesthetic language.
>
> (Lacy, 1995: 19)

Lacy's terms reflect dominant usages like homelessness (rather than eviction),[6] but in work by Alan Kaprow,[7] Martha Rosler,[8] and Hans Haacke,[9] among others, interventions have been made not only in the production of visual culture but also in the production of meaning, and a recognition that the meanings of urban spaces and spatial practices are – as Lefebvre argues (Chapter 4) – produced not given or universal.

I move now to texts by Peter Bürger, John Tagg, and Rosalind Krauss before reconsidering Benjamin's of 1934, in which he says:

> Instead of asking: what is the position of a work *vis-à-vis* the productive relations of its time . . . instead of this question, or at any rate before this question, I should like to pose a different one . . . I should like to ask: what is its position *within* them?
>
> (Benjamin, 1983: 87)

Bürger addresses this, indirectly in response to Benjamin,[10] by seeing the avant-garde of early Modernism – which I call a second avant-garde but which he sees as *the* avant-garde[11] – as being in critical relation to art's institutions (its means of production, mediation, validation, and dissemination):[12]

> The avant-garde not only negates the category of individual production but also that of individual *reception*. The reactions of the public during a dada manifestation where it has been mobilized by provocation, and which can range from shouting to fisticuffs, are certainly collective in nature. True, these remain reactions . . . Given the avant-gardiste intention to do away with art as a sphere that is separate from the praxis of life, it is logical to eliminate the antithesis between producer and recipient.
>
> (Bürger, 1984: 53)

Bürger continues that Tristan Tzara's instructions for a Dadaist poem and André Breton's for automatic writing are like recipes; taken literally they act as guidenotes for anyone's making of a work.[13] But for Bürger the division of art from life cannot be overcome within a bourgeois regime, while a false sublation occurs in pulp fiction and market aesthetics.[14] Perhaps a similar critique could apply to the claim made for the New York artists of the 1950s and 1960s as Promethean adventurers or gatekeepers to primordial or depth-psychological domains.[15] Bürger is more pessimistic than Marcuse – who writes in *An Essay on Liberation* of art as rupturing the codes of perception of the dominant society, recognising that 'The radical character, the "violence" of this reconstruction in contemporary art seems to indicate that it does not rebel against any one style or another but against "style" itself' (Marcuse, 1969: 47)[16] – but accepts that the Modernist avant-garde exposed art's institutional structures. In nonorganic art – which no longer claims a unity in form – Bürger sees the fracturing of a congealing system, citing Brecht's use of estrangement and discontinuity: 'Although the total return of art to the praxis of life may have failed, the work of art entered into a new relationship to reality' (Bürger, 1984: 91).

John Tagg rejects dichotomies such as art-and-society entirely, arguing that certain kinds of conservatism and certain kinds of Marxism meet when art is seen as either autonomous or determined:

> To function as a means of communication and exchange, systems of meaning must already contain social relations. Without shared conventions, patterns of usage and a community of speakers, we would not say we were dealing with a language at all. Languages, moreover, not only presuppose and contain social relations, but have a constitutive role in relation to their speakers, as their systems of differentiation and structures of address cut out the spaces of social identities that are produced, defined and fixed in relations of difference, which are also relations of power.
>
> (Tagg, 1992: 177)[17]

For Tagg there are three levels of the process: the technological and organisational capacities of production, transformation, use and consumption; the means of communication which produce relations of knowledge and power; and the techniques and modalities of power which shape the field of social relations and condition the place of subjects and bodies within a field of knowledge. The utility of this model is that it avoids a dualism of economic base and cultural superstructure, its levels intersecting both. Tagg continues:

> What we have therefore is not a cusp . . . We have specific historical sites, fields of knowledge, arenas of action, spaces marked out by incitements and constraints of production, meaning and power. It is these we must analyze rather than just inhabit, if we would calculate the possibilities of intervention and change.
>
> (Tagg, 1992: 178)

Tagg writes that, as concepts, Art and Society have parallel histories, and what is to be confronted is not the idea of a human condition but the conditions in which a specific idea of a human condition is produced.

In a passage coincidentally close to Chin's reflection on tactics, he argues 'The universalisation of the subject . . . [like that of Art] is caught in the strategy of power it claims to transcend'; that reverence for the artist as authentic voice or survivor must also go, while the moving-target-survivor subscriber – 'the grunt who susses out the war machine, keeps on the move, and gets back home; the one who knows the ground and the rules of the game and whose resistance cannot be appropriated' (Tagg, 1992: 179) – is a model for practice (like de Certeau's idea of tactics in walking.[18]) By comparison, Rosalind Krauss approaches a problem of meaning within art criticism but perhaps less meaningful outside it. In 'Sculpture in the Expanded Field' (1979), she argues that when sculpture lost its function as monument, as in Rodin's *Gates of Hell* and *Balzac*, sitelessness became its condition: 'In being the negative condition of the monument, modernist sculpture had a kind of idealist space to explore, a domain cut off from the project of temporal and spatial representation, a vein that was rich and new.' (Krauss, 1983: 36),[19] but exhausted by the 1950s, experienced as 'pure negativity' (ibid.). In this no-man's land sculpture became not-architecture and not-landscape. It is an inverse definition which preserves a negativity and allows Krauss to critique art when it becomes a hole in the ground – *Perimeters/Pavilions/Decoys* (1978) by Mary Miss is the main object of her scrutiny. The essay is eloquent, and faultless in construction; but leaves me wondering whether it says anything about meaning in other than the siteless terrain of Modernism. Yet I should not rush to a conclusion. In 'The Originality of the Avant-Garde', Krauss draws out aspects of the production, in the literal sense of casting, of some of Rodin's works. *The Gates of Hell* was not cast until 1921, after his death – so that *all* casts of it are equally (non-)original in that there is no Rodin-approved cast to which to compare them.

Krauss uses this circumstance to recall Benjamin's sense of an emptying out of aura in lens-based work. She goes on to say:

> The avant-garde artist has worn many guises over the first hundred years of his existence: revolutionary, dandy, anarchist, aesthete, technologist, mystic. He has also preached a variety of creeds. One thing only seems to hold fairly constant in the vanguardist discourse and that is the theme of originality. By originality, here, I mean more than just the kind of revolt against tradition that echoes in Ezra Pound's 'Make it new!' . . . avant-garde originality is conceived as a literal origin, a beginning from ground zero, a birth.
>
> (Krauss, 1986: 157)

She argues that originality adopts repetitive means such as the grid, as sign of pure disinterest or to figure the ground of the pictorial object, so that originality and repetition are 'bound together in a kind of aesthetic economy, interdependent and mutually sustaining, although the one – originality – is the valorized term' (Krauss, 1986: 160). This suggests Adorno's unreconciled tension between polarities. Aligning the dualism of originality–repetition to those of reproducible–unique and fraudulent–authentic, Krauss begins to address the means of production of, not so much the casts after Rodin which are her immediate focus, but the category modern art. The arena is still that of art writing, but broader implications begin to emerge. Her final case is the pirated photographic prints of Sherrie Levine – reproductions of originals from a photographic history, not as document but as work.

I have reservations as to Krauss' tendency to a self-referential criticism, and find her habitual normative masculine odd;[20] but it is a neat accomplishment to resite reproduction to the core of a genre of the unique. Yet in the self-referential aspect of Levine's effort to problematise the copy as well as the original I find a rerun of late Modernism not an alternative to it, and I wonder if that is where Krauss, too, feels comfortable. I still wonder how art might engender a critical consciousness of the means of production. Perhaps the following two cases will contribute to an understanding of this.

Rosemary Trockel's project for *TSWA Four Cities* (1990) in Glasgow used cylindrical columns, reproducing the form of functional elements in the shopping mall that was the site, flyposted with image-text sheets in colour and black-and-white to produce seemingly arbitrary but nuanced juxtapositions. A woman smiles, her hair wet and her hands on her breasts as she stands in a sea amid spray – an image of *bonheur* above a caption, 'READ WHAT THE EXPERTS HAVE WRITTEN . . . THEN *YOU* DECIDE', above an image of a typewriter – perhaps they were still in use then – from which a page emerges: 'I take part / You take part / They take the profit / I take part . . .' (Lingwood, 1990: 67). Advertising and mass media bombard consumers with ersatz coherences, but it is the slippage here which exposes the irrationality of what such media normalise and naturalise.

The second case is Alba d'Urbano's *Il Sarto Immortale* (1997).[21] Using the standard means of couture (the cat-walk show, the garment wrapped in tissue

7.1 • Alba d'Urbano, *Il Sarto Immortale*, Aachen

in an expensively produced box, the media photographs) d'Urbano designed garments in a fabric printed with a digital image of her body. The tension in fashion between concealment and the masculine gaze is interrupted; the second skin which covers a body to partly reveal it now exposes that which it covers. This short-circuiting of the gaze was exhibited, too, in large posters showing the garment worn by a model whose face, of course, was her own and not that of the person whose body-image she wears.

Both Trockel and d'Urbano appropriate the means of cultural production in the dominant society to invert it, so that the power relations which for the most part are implicit in the fields of cultural production referenced are opened to critical awareness. A reproduction of an image might do that, too, but in these cases the referents are more extensive.

At this point I return to Benjamin, who in 1934, addressing the Institute for the Study of Fascism (a group of anti-fascist writers in Paris),[22] argued that

> the rigid, isolated object (work, novel, book) is of no use whatsoever. It must be inserted into the context of living social relations. . . . Social relations . . . are determined by production relations. And when materialist criticism approached a work, it used to ask what was the position of that work *vis-à-vis* the social production relations of its time.
>
> (Benjamin, 1983: 87)

From which he moves to a restatement of the problem in terms of a position within those relations. Benjamin applies Marx's critique of Feuerbach's materialism to cultural production. Instead of recording or commenting on conditions art can produce them as part of an apparatus that shapes life and is at the same time shaped in life. Taking the example of the Soviet press in which readers' letters become editorial content, Benjamin plays with terms for writer (*Schreibender*): 'The reader is always prepared to become a writer, in the sense of being one who describes [*Beschreibender*] or prescribes [*Vorschreibender*]' (Benjamin, 1983: 90). He is critical of the German left intelligentsia, and reiterates the need for intellectuals to insert themselves in the production process. But what is it? In revolution it entails taking over the factories; or taking over the railways so that revolutionaries can be mobilised and the radio stations so that revolt can be communicated. The tactics remain valid in certain situations.[23] But Benjamin cites Brecht:

> In the interests of this dialogue Brecht went back to the most fundamental and original elements of theatre. He confined himself, as it were, to a podium, a platform . . . Thus he succeeded in altering the functional relationship between stage and audience, text and production, producer and actor. Epic theatre, he declared, must not develop actions but represent conditions . . . by allowing the actions to be interrupted.
>
> (Benjamin, 1983: 99)

Estrangement is a take-over of a theatrical means of production, but Benjamin sees photography lapsing into transcendental ways. He compares Dadaism and photographic reportage:

> The revolutionary strength of Dadaism lay in testing art for its authenticity. You made still-lifes out of tickets, spools of cotton, cigarette stubs, and mixed them with pictorial elements. You put a frame round the whole thing. And in this way you said to the public: look, your picture frame destroys time; the smallest authentic fragment of everyday life says more than painting . . . Much of this revolutionary attitude passed into photomontage. You need only think of the works of John Heartfield, whose technique made the book into a political instrument. But now let us follow the subsequent development of photography . . . It has become more and more subtle, more and more modern, and the result is that it is now incapable of photographing a tenement or a rubbish-heap without transfiguring it.
>
> (Benjamin, 1983: 94)[24]

This complements the frequently cited essay 'On the Artwork in a period of technical reproducibility' (*Das Kunstwerk im Zeitalter seiner technischen Reproduzierbarkeit*),[25] and before moving to more recent cultural practices, I emphasise two aspects of Benjamin's texts: his interest in film; and the scope to read the means of production as including language.

On film, Adorno and Horkheimer write in *Dialectic of Enlightenment*:

> Real life is becoming indistinguishable from the movies. The sound film, far surpassing the theatre of illusion, leaves no room for imagination or reflection on the part of the audience . . . The stunting of the mass-media consumer's powers of imagination and spontaneity does not have to be traced back to any psychological mechanisms; he must ascribe the loss of those attributes to the objective nature of the products themselves, especially to the most characteristic of them, the sound film. They are designed so that quickness, powers of observation, and experience are undeniably needed to apprehend that at all; yet sustained thought is out of the question if the spectator is not to miss the relentless rush of facts.
>
> (Adorno and Horkheimer, 1997: 126–7)

But Benjamin writes that what matters most is that the actor or actress represent herself or himself to the public before the camera, rather than someone else or a fictional character (Benjamin, 1970: 231):

> for the first time – and this is the effect of the film – man has to operate with his whole living person, yet forgoing its aura. For aura is tied to his presence; there can be no replica of it. . . . the singularity of the shot in the studio is that the camera is substituted for the public. Consequently,

> the aura that envelops the actor vanishes, and with it the aura of the figure he portrays.
>
> (Benjamin, 1970: 231)

The outcome is a departure from semblance. Benjamin writes, counter to Adorno's pessimism (from Hollywood, while Benjamin's sources are in Soviet and German film, though not exclusively as the following demonstrates):

> Mechanical reproduction of art changes the reaction of the masses towards art. The reactionary attitude toward a Picasso painting changes into the progressive reaction toward a Chaplin movie. The progressive reaction is characterized by the direct, intimate fusion of visual and emotional enjoyment with the orientation of the expert. Such fusion is of great social significance . . . With regard to the screen, the critical and receptive attitudes of the public coincide. The decisive reason for this is that individual reactions are predetermined by the mass audience response they are about to produce . . . The moment these responses become manifest they control each other.
>
> (Benjamin, 1970: 236)

Esther Leslie argues that Benjamin sees audiences as gaining expertise by measuring what they see on screen against the realities of life, 'and because they learn to assimilate new scenarios of potential social and physical ordering' (Leslie, 2000: 149). 'Film imprints on celluloid the alienated existence of humans' (Leslie, 2000: 152), and is made (audiences know) in takes that mimic production labour. Recognition of this is a recognition of contingency, a step towards imagination of alternative orderings of both the plot and the construction of a social order. As Leslie indicates, Benjamin's more radical insight is that perception is conditioned by technologies. Film is an industrialisation of perception, but liberates as the eye scans the screen.

I return now to the question as to what constitutes the means of production. Theatrical staging is one instance; I suggest above that gaze is another. Do they include languages and categories of thought? An instance of the power of verbal language is given (from French) by Luce Irigaray:

> (a) The plural of two genders together is always masculine . . . (b) the most valorous realities are usually masculine in our patriarchal cultures . . . (c) the neuter, which often takes the place of a sexual difference that has been erased, is expressed in the same form as the masculine; this is true of natural phenomena . . . or realities involving an obligation or a right . . . These forms of language and speech that seem to us to be universal, true, intangible, are in fact determined and modifiable historical phenomena. They entail consequences for the content of discourse which are different for each sex.
>
> (Irigaray, 1994: 27)

Earlier in her paper 'A Chance to Live',[26] Irigaray calls for a display in public places of posters depicting the mother–daughter relationship erased in patriarchy. The proposal is a provocation to the recognition of the absence of such a category of imagery, which portends an equivalent lack of an imaginary in which they would have meaning.

II VISIBILITIES

Irigaray's proposal hints that the syntax and vocabulary of a visual language may structure thought as well as being structured by it. This accords with Massey's comments on linear perspective and a privileging of visuality.[27] Perhaps there are other ways to represent a city than in the conventional plan from the god's-eye viewpoint – from engagement, street-life, or from touch, smell and hearing, or through stories.[28] In a similar change of convention which is a change in power relations, to write in the first person interrupts the claim to universality of the third, and in arenas where the boundaries of architecture, art and film collapse in new media, narrative is fractured by voices which portend new audibilities:

> Just as conspiracy theory explodes the myth of a single linear history, architecture can be seen as not the fastidious refinement of an abstract language but the site of collisions between competing ideologies. Architecture becomes a trip across the wavebands, samples of disassociated, but recognizable, story lines. A single narrative thread is lost, thereby making possible multiple readings.
>
> (Fat, 2001: 345)

Fat is a collective. Its projects seek to enable proactive participation in a terrain between fashion, architecture, and taste, through appropriations of familiar sites such as the bus shelter and the billboard:

> Members of Fat are not cultural terrorists. To explode myths and address core issues Fat works . . . from the inside out. Utilizing the tactic of leaching – intervention and recoding within existing structures such as the media, advertising spaces, prestigious/exclusive art events, urban transport systems – we aim to explore, challenge, and possibly explode current notions of what is perceived to be art and to oppose traditional conceptions of authenticity.
>
> (Fat, 2001: 341)

When public art is annexed to urban development and denotes competition among cities for places on a global culture map, it is more likely that criticality will be found in alternative art spaces (and reactions to the mainstream on its doorstep),[29] and in the projects of groups such as Fat, Cornford and Cross and PLATFORM in the UK, or Ant Farm, Taller de Arte Fronterizo (Border Art Workshop) and Group Material in the US.

Group working departs from the individualism of modern art, and has, at least in the cases noted above, tended to a politicized approach. For Benjamin, the communication of the political is achieved through the technique of the work as well as in its implicit or explicit content, but the term has more nuances in German than in English: *Techniker* means engineer (as in engineering a new society and not as in English refining a technique), and can be critical.[30] In the former sense, the group Welfare State International call themselves engineers of the imagination.[31] If technologies of production shape perception, perhaps participatory modes of working – new genre public art is the term adopted for this by Suzanne Lacy[32] – equate to *Technik*. The frequently encountered question in response to such work, 'Where's the Art?' is hence apt.[33]

I look now at three projects by Mierle Laderman Ukeles to ask how they impact power (not as domination over nature and others as intended by Adorno and Horkheimer in *Dialectic of Enlightenment* but, after Foucault, as a process of implicit coercion).[34] *Touch Sanitation* began in 1976 when Ukeles became unfunded artist in residence in the New York City Department of Sanitation:

> On July 24, 1979, I started shaking hands with the first of all New York City's 8,500 sanitationmen and officers, 'sanmen', the housekeepers of the whole City, workers in the largest of maintenance systems . . .
>
> As an artist, I tried to burn an image into the public eye, by shaking, shaking, shaking hands, that this is a human system that keeps New York City alive, that when you throw something out, there's no 'out'. Rather there's a human being who has to lift it, haul it, get injured because of it (highest injury rate of any US occupation), dispose of it, 20,000 tons every day. Our garbage, not theirs.
>
> (Ukeles, 2001: 106)

Touching the hands, which are perceived as touching filth, Ukeles subverts the invisibility of garbage collectors who 'feel so isolated they could be working on the moon' (Ukeles, 2001: 107). Over eleven months she walked the five boroughs to personally greet each sanitation worker, timing her walks to fit an eight-hour shift. This can be read as identification with the working class in a voluntary sharing of conditions (in which the artist is free to leave), or more accurately as a coinciding with the interests of a section of that class:

> As woman artist injecting myself into a 'man's world', I represented the possibility of a healing vision: not a pretend sanman, not an official investigator, not a media voyeur, not a social scientist, rather a 'sharer' in an ecological vision of the operating wholeness of urban society.
>
> (Ukeles, 2001: 106)

Patricia Phillips views the repeated handshaking as avoiding the difficulties of being overwhelmed or authoritarian when an artist performs as if the public realm were a blank space: '*Handshake Ritual* required the artist to adopt and accept the rhythms and routines of an established workplace, a site intrinsic to the public domain . . . [and] to embrace a prevailing public language' while dealing with unpredictable responses by individuals deciding their own participation. She adds: 'Since the unplanned inflects even the most orthodox municipal system, it is precisely this potent mixture of the unexpected and the nonnegotiable that illuminates possibilities' (Phillips, 1995a: 181).

Ukeles entered her project with the NYCDS as a woman artist. She refers to her roles as woman and mother, and fuses the work of both with that of an artist, in an early work (*Wash*, 1973), cleaning the pavement in front of A I R gallery, New York. In Manifesto for Maintenance Art (1969), she writes of repetitive tasks as challenging linear history, proposing a cooperative and renewable aesthetic in place of a pure and autonomous Modernism:

> Avant-garde art . . . is infected by strains of maintenance ideas, maintenance activities, and maintenance materials . . . I am an artist. I am a woman. I am a wife. I am a mother. (Random order). I do a hell of a lot of washing, cleaning, cooking, renewing, supporting, preserving, etc. Also, (up to now separately) I 'do' Art. now I will simply do thee maintenance everyday things, and flush them up to consciousness, exhibit them, as Art.
>
> (unpublished document cited in Phillips, 1995: 171)

One outcome is a collapse of the detachment of audience from subject when the audience acts on the direction and duration of a work. The handshaking accords no more significance beyond mobility to the artist's hand – it is not a zen puzzle of the touch of one hand shaking – than to the sanman's; at the same time, the frame of art is unavoidable and it is Ukeles who is known by name rather than her 8,500 collaborators. The project still draws attention to the roles of professionals outside the elite structures and conventions of expertise. For Ukeles it is a matter of lending visibility: 'Our culture thinks of itself as being so advanced and is completely dependent on these people, but has no way of seeing them' (lecture, Staten Island College, February 5th, 2002). The postmodern artist's freedom to call anything art can, it appears, be a utile intervention in the means of production, a counter to oblivion in Arendt's terms (Chapter 6). But if the sanmen were recognised and possibly felt empowered through the project, this does not mean that a similar quality can be found in all projects that involve identified recipients;[35] nor that projects that appeal to non-specific publics are unempowering. Power relations which are taken as normal can also be confronted more directly.

In *Cleaning of the Mummy Case* (1973), as artist in residence at the Wadsworth Athenaeum in Hartford, Connecticut, Ukeles played with a system of power relations in which curators were allowed to handle exhibits while

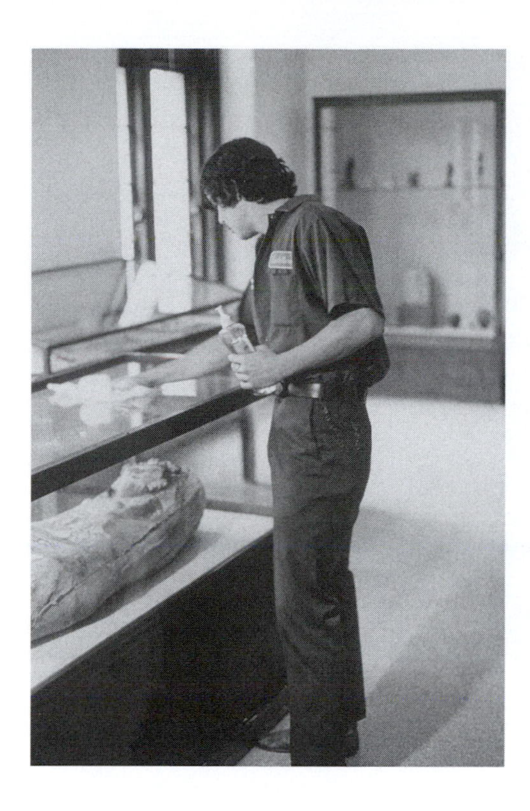

7.2 • Mierle Ukeles, *The Cleaning of the Mummy Case*, Hartford. Photograph courtesy of Ronald Feldman Fine Arts Inc., New York

cleaners maintained the cases in which exhibits were housed. Ukeles recoded a glass case containing an Egyptian mummy as an artwork:

> It remained the same case, serving the same function in an unaltered setting, but its recategorization now required an abrupt revision of museum procedures for its care and conservation . . . It is not difficult to imagine the far-reaching and bizarre circumstances that such a shift in nomenclature could produce. If sinks were declared art objects, for instance, would conservators become the guardians of public washrooms?
>
> (Phillips, 1995a: 175)

In a final gesture, she held the museum keys for a day to lock and unlock doors at whim, greeted with a frenzy when she tried to lock the curators in their room (they left rather than be locked in, to her surprise).

The emphasis of Ukeles' work from the 1980s onwards has been on a material environment, at Fresh Kills landfill site, Staten Island, and the flow of garbage from the city to it. In a gallery show, *Maintenance City / Sanman's Place* (1984), items of garbage were sorted and recoded as recyclable, perhaps valorised by gallery exhibition. More recently, Ukeles has worked on the rehabilitation of the landfill site in preparation for (and after) its closure in 2001, producing a multi-screen installation derived from interviews with dwellers,

7.3 • Robert Guerra, Mierle Ukeles and Dr Stephen Handel at Fresh Kills landfill, Staten Island.
Photograph courtesy of Ronald Feldman Fine Arts Inc., New York

ecologists, and other experts.[36] The garbage is now covered, serviced by an extensive and intricate web of pipes and pumps to manage fluid and gas seepage; large parts of the site are almost a wilderness in sight of the city's towers (not all of them now). Ukeles sees the project as 'a model for the entire world of a power to take something that was rejected and return it' (lecture, Staten Island College, February 5th, 2002). But the plans were put on hold when the site was used to take the debris from the Twin Towers.

Ukeles' office in the Department of Sanitation was closed for four months after the attack, but her greatest shock was at the dumping of human ash with garbage: 'Garbage means things stripped of identity. It all becomes the same thing. We cannot treat human remains like that – we will have to use the unlimited power of human ingenuity to make a place of honor which restores the value of human life' (conversation, February 5th, 2002). When I visited Ukeles she was still dealing with 9–11: 'I feel the need to create a space for listening to each other, while surrounded by flowing images of the unvoiced site. I want to make this place, for now, to make room for each other trying . . . to figure out how to understand this site all over again' (artist's statement). After her lecture at Staten Island College a member of the audience proposed that, since the ash of Chief Executive Officers was mixed with that of the invisible people (the cleaners and the maintenance staff) the memorial should be for surviving CEOs to begin to treat the 'invisibles' with kindness.[37]

Since the invisibility of employees categorised as menial is enforced through design and a hierarchic organisation of spaces, I wonder how architecture might address the issue. The contingency of design on engineering, and now a preponderance of design-and-build schemes in which project managers drive the process, inhibit this aim. There are exceptions – notably the NMB Bank in Amsterdam designed by Ton Alberts with participation by users[38] – while too often in both architecture and planning consultation means only imposition. But power which is donated is in any case a retention of power in the hands of the donor; empowerment requires a radical renegotiation of the terms of engagement – something equivalent to a collapse of the separation of producer and consumer. Participatory design goes some way towards this; the self-build housing piloted by Walter Segal, like Hasan Fathy's mud-brick solution for Egypt's housing problem in the 1940s, and the construction of huts in housing project margins, take it to a point at which the dweller is both.[39] As yet only a tiny proportion of houses are made this way, yet the existence of such schemes enlarges the map so that its mid-point is no longer where it was.

It is important to emphasise, countering some areas of community architecture, that participatory architecture produces new, not ersatz vernacular, forms. A striking case is the work of the Rural Studio of Samuel Mockbee.[40] Although the locations of Mockbee's projects are in the rural Black Belt (from a dark loam) of Alabama and Mississippi, not in cities, the houses and civic amenities his studio designed are distinct both in using low-impact materials and in looking decidedly new. Mockbee's values included that 'the architectural profession has an ethical responsibility to help improve living conditions for the poor' and that architectural education should go beyond design – 'paper architecture' – to building, thereby enacting 'a moral sense of service to the community' (Dean and Hursley, 2002: 1). Working as a team, Mockbee, his students and dwellers whose existing houses were in disrepair or dereliction produced new houses, which were largely free to the dweller (or client). Low budgets (mainly from funding partners) dictate the use of low-cost materials, which tend to be local and are in some cases recycled. This creates a form of building which is in keeping with a vernacular tradition but does not copy it. As Dean and Hursley summarise:

> The studio's characteristic modern esthetic was from the start nudged by typically southern rural forms and idioms: sheds, barns, and trailer. The Bryant House, for example [1994, built for Shepard Bryant, his wife, and grandchildren, using hay-bale construction], is all porch and roof, a steeply raked acrylic structure supported by slender yellow columns . . . But even the most futuristic constructions look anchored in their neighbourhood.
>
> (Dean and Hursley, 2002: 9–10)

Mockbee's studio also produced student housing and amenities such as a baseball field, and a chapel. At a highly localised level, life indoors and outdoors (private and public in an Arendtian sense, from a classical model) may merge; but in a large city the problem is more complex. In that situation, the question may be not the housing of sociation, which takes place anyway, but the invitation to democratic exchange.

The division between dwelling and what constitutes a public realm is problematised by the Portable Democracy Tent by Wendy Gunn and Gavin Renwick, erected in Istanbul in 1989–90, and then in Athens, Belgrade, Budapest, Prague, Berlin and Glasgow: 'Conventional design briefs seem to encourage the abstract definition of space and isolation of function, over a design that evolves from an investigation of the social processes of surrounding context, the realities (the potentials) of climate, the process of habitation' (Gunn and Renwick, 1998: 95). The interior housed conversations on democracy, in a modern adaptation of a pre-modern nomadic architectural form. This, too, appears to intervene in the means of production, in this case the production of society. Just as the *agora* and the *pnyx* in Athens condition the performativity of speaker and audience,[41] the tent deconstructs relations between publics, and between public and private utterances. Inside, it is all public, just as the personal (utterance) is political.

In contrast, Jeremy Hill sees community architecture[42] as mirroring mainstream architecture's utopianism while reproducing its erasure of political content:

> community architecture avoids a direct discussion of style through its focus on the process of collaborative design . . . despite this disavowal of style it, slips into the argument anyway. There is an underlying assumption that a certain vernacular will emerge effortlessly from the process of collaboration because that is what people most naturally relate to.
>
> (Till, 1998: 68–9)

This prescribed vernacularism which is a reaction not a handing-over of means has no tradition in which to be rooted, and the communities assumed as constituencies for it no longer exist in major cities, if they ever did.

Martin Albrow argues, adapting Arjun Appadurai's (1990) concept of an ethnoscape as a socioscape, that if conventionally migration was met with a requirement of assimilation into a local culture linked to place, this is impossible in a multi-ethnic urban environment; and that there can be 'localities without community and cultures without locality' (Albrow, 1997: 41–2). Albrow sees localities as integrated in global patterns of communication while dwellers have uneven patterns of participation in social networks: 'For each person their place in the locality represents a point where their sociosphere literally touches the earth. But for each person who is viewing other people there can only be a very partial idea of the relevance of locality for others' sociospheres' (Albrow, 1997: 52).

Albrow uses a new vocabulary for a new situation. He asks: 'So where is community here?' and answers 'We know precious little about the ways in which the different sociospheres relate to each other except in stereotypes formed in the stage of nation-state sociologies' (Albrow, 1997: 52–3). A world of migrations and flows spawns forms of sociation which seem from previous assumptions of stability to be fractured, though they may cohere just as children who skip through web sites or television channels build a multi-focus attention span. The implications for planning in multi-ethnic cities are investigated by Leonie Sandercock, who cites Iris Marion Young on identity politics and justice (see Chapter 9) and concurs that policy needs to relate to group-consciousness:

> A politics of difference . . . must be able to take on board some (redefined) notion of the good of/in that society. This does not necessarily mean the return of the outmoded concept of 'the public interest', but it does demand the creation of a civic culture from among the interactions of multiple publics.
>
> (Sandercock, 1998a: 186–7)

Like the viral artist, the radical planner operates in crevices: 'the identity of the radical planner . . . is that of a person who has, essentially, gone AWOL from the profession . . . to work in opposition to the state and corporate economy' (Sandercock, 1998a: 99–100). Imbalances are addressed in a voice that has the authority of governance:

> Feminist theories of language often start by showing how language forms one's sense of reality, order, and place in the community. As such, language can be limiting as well as empowering . . . Empowering language and dominant forms of communication are frequently acquired through formal education. When education is unequally distributed, inequalities in communication will be accentuated.
>
> (Sandercock and Forsyth, 2000: 449–50)[43]

Planning may not now be, in the Chicago School mould, a rational linkage of technical solutions to conditions as if technical expertise had the status of a (Kantian) disinterested judgement, but a process of listening and handing-over, which will require new vocabularies for local knowledges.

In 2001, planning workshops in St Adrià de Besòs, a municipality adjacent to Barcelona, were initiated by researchers at the University of Barcelona in response to plans for a World Forum of Cultures to be held in 2004. The background to the Forum is Barcelona's claim to be a world city, attested by the opening in 2001 of a World Trade Centre designed by I. M. Pei, at Port Vell. The working-class district of Poble Nou has been recoded as a knowledge quarter, and the remaining section of the waterfront is under development. Rail yards, sewage and power plants will become a solar energy and water treatment park by a marina, and residual publics will be peripheralised. Working with local groups in St Adrià de Besòs, where political allegiances range from communist to fascist, researchers began an analysis of needs, producing material for a public exhibition and workshops in which dwellers contributed ideas for future redevelopment.[44]

III PROVOCATION

Perhaps it is not so much a matter of mending fences as putting them up. In Albion Square in Stoke-on-Trent in 1996, two artists erected a steel security fence around three tracts of grass between the bus station and the shopping centre. As David Cross and Matthew Cornford assembled the work, they were asked by many passers-by what they were doing. Once it was known this was art, and paid for from the public purse, there was predictable outrage, ensuring coverage in the local press. The artists' visiting card states

CORNFORD AND CROSS
PROBLEMS SOLVED
reasonable rates

with a PO Box number they had registered for one year. But *Camelot* (as the work was called) had highlighted the problem of neglected urban spaces, to the consternation of local councillors. The artists write:

> *Camelot* is a literal interpretation of the 'City Limits' theme; we chose to invite reflection and debate on the physical and social boundaries which often determine the patterns of city life – in this case by denying people access to some small, neglected fragments of public urban land.
>
> (artists' statement)

In a linked exhibition photographic images of the lawns of Oxford Colleges, which are protected by observance of custom rather than security equipment from encroachment, were shown. *Camelot* was a public art commission, selected from a submission of designs; Cornford and Cross used adapted photographs of the site to give a visual impression of their proposal. Their view is that by reinforcing the boundaries of a site with security fencing, as an intentionally excessive display of authority, they sought to relate current debates on security and access to the tragedy of the commons,[45] as they put it, when sites of open mixing are encroached on by those of consumption and routine. The title references the UK national lottery, and the mythicised ideal (chivalric) life of Arthurian Britain, but the work's intention was to interrupt the public realm thereby pointing to its sterility. Later Cross found himself on a local bus when the driver pulled over, pointed to the installation and challenged its status

7.4 • Cornford and Cross, *Camelot*, Stoke-on-Trent. Photograph courtesy of Cornford and Cross

as art. Outrage turned to discussion after Cross admitted to being one of the artists responsible. Cross sees most urban regeneration and associated public art schemes as amelioration in the nineteenth-century mould, masking underlying issues of inequality and injustice. Nevertheless, the outcry caused by *Camelot* levered £70,000 for urban improvements in the area, including provision of a ceramic bench in keeping with the town's industrial (potteries) past. The head teacher of a local primary school asked for and was given the fence to put round her school playground.

At the University of East Anglia in 1997, next to the Sainsbury Centre and as part of the East International exhibition, a grey, steel-and-cladding turkey breeder unit was erected, modified to have no door or windows, with rave sounds played continuously within at high volume. The campus is sedate and picturesque, including a lake and woodland, while the extension to the Sainsbury Centre by Norman Foster is under turf. Yet the shed-like architecture of the Sainsbury Centre building (and of most of the supermarkets of the same name) enabled the shed Cornford and Cross built to have an ambivalent relation to its site. More to the point was the association of East Anglia with industrialised turkey production:

> *New Holland* grew out of a consideration of the relationships between architecture, economic activity, and cultural responses to the landscape in a consumer society. The installation . . . referred to a 'Bernard Matthews' turkey breeder unit . . . The heavy mechanical beat of a blend of rap, house, and garage music from CD compilations could be heard pumping out from darkness.
>
> (Cornford and Cross, 2001: 335)

New Holland was positioned next to a reclining female figure by Henry Moore. Within the Sainsbury Centre modern art is presented next to artifacts from ex-colonial countries. Writing in *Art Monthly*, Sotiris Kyriacou noted that of all the items in the exhibition *New Holland* had received most flak: 'The landscape tradition versus Bernard Matthews turkey breeder units, the patriarchy of Foster versus the matriarchy of Moore, urban versus country contexts, rural idyll versus youthful dissent, were some of the binary oppositions wheeled out by the artists' (*Art Monthly*, September, 1997, p. 209) and concluded it got what it deserved. Yet the work set out to provoke; the shed articulates the implicit and multiple contradictions of the situation. Binary oppositions are more often encountered inside the Sainsbury Centre, where Giacometti stands next to African carving within a narrative of universal human expression. In the Dadaist period a collage might have sufficed to explode this; now more concrete forms are required.[46] The installation by Cornford and Cross, and Norman Foster's Sainsbury Centre are both industrially produced, though the latter is more eloquent and expensive. Yet *New Holland* irrevocably deconstructed its relation to the site. Cross spoke of this project as 'systemic, like weed killer' (conversation, October 16th, 2002), which sounds like Chin's idea of a viral art infiltrating the fat targets of the dominant society. Those targets have their defences and the work was removed after six weeks. It left its mark – the grass inside had grown but become bleached, a ghost presence which was cancelled by re-turfing, which substituted a bright green rectangle.

The railway station at Bournville, near Birmingham, is decorated in Cadbury's purple – a chromatic totality which gives a glimpse of a total benevolence, as a company founded by Quakers takes the form of a Fordist institution. But the regulation of behaviour was there already in the provision of separate recreation areas for male and female employees. The administration of the company site was contextualised by the growth, through the late eighteenth and nineteenth centuries, of a range of institutional building types from the school to the hospital, the workhouse, and the prison. On their first reconnaissance of the site, Cornford and Cross noted an ambiguity in that purple is not only Cadbury's colour, but was used by the suffragettes.

The pool in the women's recreation ground was derelict and had been used as a tipping ground for rubble and waste, but was restored as their work for *In the Midst of Things*, a mixed show of temporary installations on the site in 1999. The pool was cleaned, its surround rebuilt in newly quarried stone, and 37,000 gallons of water pumped into it by diverting the factory's water supply. The artists then collaborated with the firm's Chief Food Scientist to dye the water purple with food colouring:

> Although it was non-toxic, the dye blotted out the light, preventing photosynthesis in a suffocating extension of the corporate identity. The water in

7.5 • Cornford and Cross, *Utopia*, Bournville. Photograph courtesy of Cornford and Cross

> the pond grew dark, translucent so that it was impossible to judge as to depth, and reflective so that its surface mirrored the surrounding garden and viewer.
>
> (artists' statement)

It was called *Utopia*.

Not all projects devised by Cornford and Cross are realised, and they regard rejected ideas as equally worthy of exhibition.[47] In *Coming up for Air* (2001) a grey chimney is proposed to be sited in Chasewater Reservoir in Staffordshire. The first phase of the work, referencing the nineteenth-century practice of funding public statues by subscription, is a prospectus.[48] After centuries of deforestation, the valley of Cannock Chase was dammed in 1797 to form the reservoir – 'This was a period of profound change in the relations between the people who owned natural resources and the people who transformed them into commodities' (artists' statement). The region saw the beginning of industrial-scale coal mining, the first step to climate change. *Coming up for Air* points to a lack of political will to do anything about this while referencing a buried past as the chimney rises over the water like the return of the repressed.

Finally, I turn to *Capitaldonada*, a trans-disciplinary project initiated by the group Extra]muros[in Marvila, a social housing district of Lisbon, in 2001. The title is a reference to Oporto's designation in the same year as European Capital of Culture, and the tendency of such cultural marketing to privilege aspects of cultural consumption associated with affluence. Mario Caeiro, who co-ordinated the event with architects Daniela Brasil and Luis Seixas, geographer Teresa Alves, and residents of Marvila, writes that the project was intended to 'animate a specific social and territorial fabric' and that

> The key element in this strategy in which not only different technical and artistic subjects converge but also many non-professional actors, is the capacity to generate participation, co-responsibility and a real empathy on behalf of common fate. Perhaps following this project, institutions and individuals will see intervention . . . with a new care as well as with new demands.
>
> (Caeiro, 2002: 145)

Marvila is a non-place but not in the sense used by Marc Augé of a space that is a means to an end, such as the airport lounge,[49] in which occupancy is temporary and likely to be structured by text (information which is a key to a displaced imaginary of place as in the departures board, and instruction as in signs for movement in a controlled direction). Marvila is a residential area, but again not in the normative sense of the term to denote middle-income habitation. Marvila is a vast area of concrete blocks separated by waste ground, inhabited by immigrants from rural areas of Portugal and its ex-colonies in Africa. The more recent blocks are of an adequate standard of design and construction, and here and there are survivals of an agricultural past in pockets of densely cultivated green next to an eighteenth- or nineteenth-century farmhouse.

Marvila stands between the historic districts of Lisbon and the Expo-98 site, in preparation for which some blocks visible from the airport road were painted in bright colours – orange, mauve, lime green, blue, yellow. There is a mall within sight, flanked by two concrete towers left unfinished when the money ran out like the skeletons of giant multi-storey car parks, and a metro line beneath. To use the metro requires entry through the mall, which opens after poor people go to work. Employment declined with the transition of the port area on the Tagus to a post-industrial condition in the 1990s (it is now beginning to be redeveloped for bars and night clubs), so not all of them need the bus either. On most maps available to tourists Marvila is a white space, as if to say there is nothing there. To intervene here was to draw attention to the existence of a parallel city to the known Lisbon, capital of empire and city of arcades.[50] Marvila may be blank on maps, but it has patterns of sociation within several distinct neighbourhoods, such as Chelas, Armador and Loios, elects mainly communist representatives to the local authority, and has an active network of tenants' associations based on individual blocks.

The project team collaborated with representatives of the blocks, and enjoyed support from those able to offer it, such as the café owner who provided them with free lunches. In *porque é que existe o ser em vez do nada?* (from Heidegger, why are there things not nothing?), José Maça de Carvalho

7.6 • José Maças de Carvalho, *porque é que existe o ser em vez do nada?* (from *Capital do nada*, Marvila)

photographed twelve individual residents whose images, and in a second wave their cell phone numbers (issued for the project) as well, were reproduced on cards and posters at bus stops and in magazines. The individuals came forward over a year-long research period; de Carvalho sees them as 'heroes of daily life whose activity had a communitarian objective/pursuit' (Extra]muros[, 2002: 157). They acted as communicators for the project, setting up meetings with callers to share their regular activities, for example kickbox training (Francisco), graffiti (Mário), dance parties (Vanesa), music (Beto), and quoits (Sr Casimiro). In a final phase, a 10-metre high poster of one of the twelve, Deborah, was sited on the side of a five-storey block. David Santos (citing Hal Foster) likens de Carvalho's work here to an idea of artist-as-ethnographer,[51] collecting information on site; this helpfully links the project to current debates within ethnography as to the construction of subjects, and relations between peripatetic researchers and the people they encounter in the crevices of society (increasingly now urban society in the affluent world) into which they insert themselves for long enough to understand what happens – but not to be other than observers. There is an element of that, I agree, but the aim of the project is more interesting, problematising the concept of belonging to a place. Irit Rogoff gives a reading of visual culture which is for me closer to this than Foster's idea: 'Visual culture . . . opens up an entire world of intertextuality in which images, sounds and spatial delineations are read onto and through one another' (Rogoff, 2000: 28). Addressing the question 'where do I belong?' she writes:

> It is one of those misguided questions which nevertheless serve a useful purpose, for while it may naively assume that there might conceivably be some coherent site of absolute belonging, it also floats the constant presence of a politics of location in the making. This very act of constant, plaintive articulation serves to alert to the processes by which identity comes into being and is permanently in flux.
>
> (Rogoff, 2002: 14)

The dwellers of Marvila do not *belong* there any more than Marvila belongs in Lisbon. Previous senses of belonging are maintained through extended family structures and group sociation; at the same time they are there whether they belong (or think belonging is constructed in discourse) or not, and their social and cultural experiences begin to construct a sense of being there.

Of several other projects in *Capital do nada* I mention two: *Belcanto* by Catarina Camparo; and *[e] vazao* (*E-vasion*) by Cláudia Taborda and Victor Beiramar Diniz. *[E] vazao* consisted of a series of wooden posts, painted red, black, brown, and white on their four sides, placed at regular intervals in a valley. At one end of the grid were two palms. In October 2001, a holm-oak

7.7 • Claudia Taborada and Victor Beiramar Diniz, [*e*] *vazao* (from *Capital do nada*, Marvila)

was planted to replace one of the stakes. One idea was to replant a large number of trees about to be displaced by a new damn in Alqueva, but it did not happen. *Belcanto* consisted of a series of performances by a pianist and a baritone, using a concert grand piano. At undisclosed locations, a large, white Mercedes van arrived and a piano was unloaded and tuned while the performers, in evening dress, arrived by limousine. The programmes included Mozart, Tchaikovsky, Verdi and Gershwin, in the spaces between blocks, play areas, road verges, a park. The movements of piano-movers, tuners, and performers were choreographed by cell phone. At the end, the piano was loaded back into the white van. The point was not to widen access to elite culture, but, more like a Dadaist performance, to produce banality and at the same time provoke reactions which are themselves a production of culture. Caeiro sees *Capital do nada* as fusing provocation and hope. In a politicised environment, it attracted the support of local representatives including the President of the District of Marvila, for whom it was an attempt to valorize Marvila. Caeiro writes that 'Some people perceived the event as a festival, others regarded it as an opportunity for a collective workshop' (Caeiro, 2002: 134). He accepts that not everything worked in the ways foreseen, and that one event does not change the colossal disparities between Marvila and the Lisbons of the old centre and the new Expo. He cites Kafka in an aphorism that the more horses are harnessed to a stone, the faster you will run: 'You may not be able to move it, but it is possible the belts break and you will obtain an empty and cheerful walk . . . the team that accomplished *Lisbon Capital of Nothing* may not have moved the stone, but feel a kind of suspension of gravity and experience a certain lightness' (Caeiro, 2002: 133).[52]

Where does this bring us? First, to see the limitations of ideas such as artist-as-ethnographer which hold to a separation of art and society which the artist bridges. Where co-producers are involved, the separation weakens. Second, to an insight that the means of production include factors which are structural and systemic such as language and hierarchy, which are open to interruption. Third, to reconsider solidarity in terms of the conditions of post-industrial, multi-ethnic cities; and to think again – in a world of performativity which collapses conventional differentiations of public and private – about what constitutes a sphere of common purpose.

I take up this question in Chapter 9, and will end here by citing Beate Roessler, who finds the lament for the decline of public space in Arendt and Sennett inadequate to present conditions. She argues that diagnoses of an alleged disappearance of domain-boundaries are historically based, and 'in their diagnoses of the decline and fall, are likewise – and particularly – interested in the loss inflicted upon the public sphere by those forms of disintegration' (Roessler, 2002: 38). In a time of mobile phones, reality television, and so forth, the boundaries are probably no longer there, and the content of hitherto bounded spaces is dispersed – but not dead. Roessler does not suggest an absolute shift in the meaning of public and intimate life, but a 'redefinition of the borderline between private and public, a borderline that has never been and is still not definite, but always disputed and constantly in revision' (ibid.). Further, when the personal is political – which nowadays is expressed more often as the political is personal – not only is the baggage of an enclosed feminine realm separated from an open masculine realm (as in the Athenian democracy which influenced Arendt and Sennett) questioned, but attention is directed, too, to structural violence against others. Roessler concludes that 'not every displacement of the conventional border between private and public may be . . . described as dysfunctional' (Roessler, 2002: 45). If this is so, then the site of intervention may not appropriately be the public square or the monumental space; it may be through new awarenesses in domestic and intimate life, or in domains which cannot be categorized according to geographical or architectural terms, such as cyberspace and the nebulous but actual space of informal networks of communication – by word of mouth as much as by advertisement or web site. One extension of the former may be the development of sustainable ways of dwelling (Chapter 8); and of the latter a new resistance in a space as yet unwasted by the increasingly total power of the world's remaining super-state (Chapter 9). In both, analysis of conditions may be conveyed through incremental means – a viral culture, a working away in the cracks, an art of infiltration in the means of producing either tyranny or liberation.

NOTES

1 GP 160 refers to glyco protein 160 molecular weight; Chin cites *The New England Journal of Medicine*, vol. 324, no. 24, June 13th, 1991.

2 For other projects by Chin, see Lacy, 1995: 210–11; Strelow, 1999: 112–13, 204–5.

3 The remark was made in a seminar at the University of Plymouth in October, 1999. See also reference to *killing us softly* in Chapter 6, and discussion of PLATFORM in Chapter 8.

4 In 1985, a group of anonymous women artists put up posters in lower Manhattan asking 'What do these artists have in common?' above a list of white, male artists who show in galleries 'that show no more than 10% women artists or none at all'; and others stating 'These galleries show no more than 10% women artists or none at all' (Guerrilla Girls, 1995: 34–5). They write: 'Unlike the suffragist sisters, the Guerrilla Girls have never taken a real weapon to a work of art. Instead, they use a rapier wit to fire volley after volley of carefully researched statistics at artworld audiences, exposing individuals and institutions that underrepresent or exclude women and artists of color from exhibitions, collections and funding' (Guerrilla Girls, 1995: 7).

5 Guerrilla Girls, 1995: 62. Recent projects have focused on gender and race discrimination in context of the first Gulf War, gay and lesbian rights, living with HIV and homelessness.

6 Deutsche (1996: 70–9) draws on urban geography, citing Smith and Harvey as well as Lefebvre, to emphasise the production of homelessness in gentrification, for which eviction is a more appropriate term. Rosler (1991: 20–31) charts the loss of rental housing and peripheralisation of populations in New York.

7 Kaprow differentiates two avant-gardes: 'one of artlike art and the other of lifelike art' (Kaprow, 1996: 203). In the latter he puts Futurism, Dada, Happenings, Fluxus and Conceptualism, seeing stages of recognition of

secularisation; a shift out of the studio and museum; use of performative modes; connection to natural processes such as seasons and cycles of decay and regeneration; a blurring of boundaries; emergence of a specific public; continuity with daily living; and adoption of a therapeutic purpose 'to reintegrate the piecemeal reality we take for granted. Not just intellectually, but directly, as experience' (Kaprow, 1996: 206).

8 Rosler used the site of a gallery in New York's SoHo during the period of its gentrification to draw attention to eviction. At the beginning of her essay 'Fragments of a Metropolitan Viewpoint', Rosler cites Lefebvre and summarises: 'the city, which at first might appear to be an unplanned welter of heterogeneous structures with streets and avenues threaded throughout, itself encodes an image of the economic realities of the society that produced it' (Rosler, 1991: 15–16).

9 See Bordieu and Haacke, 1995. Haacke uses documentation as follows: German firms supplying arms to Saddam Hussein in *Raise the Flag* (1991); the interests of the Rembrandt Group in South African mines in *Les must de Rembrandt* (1986); New York real estate holdings in *Shapolsky et al: Manhattan Real Estate Holdings, a Real-Time Social System, as of May 1, 1971* (1971) – see Deutsche, 1996: 159–92; and the corporate affiliations of the Guggenheim's trustees in *Solomon R. Guggenheim Museum Board of Trustees* (1974).

10 Bürger cites Benjamin's 1934 text in a note (Bürger, 1984: 120, n. 13), but is more concerned with his essay on the work of art in the age of technical reproduction (Benjamin, 1970: 219–53). Benjamin, Bürger argues, foresees in photographic media a collapse of the separation between object and recipient: 'It is not the break between the sacral art of the Middle Ages and the secular art of the Renaissance that Benjamin judges decisive . . . it is that break that results from the loss of aura. Benjamin traces this break to the change in techniques of reproduction . . . reception characterized by the presence of aura requires categories such as uniqueness and authenticity. But these become irrelevant to an art . . . whose very design entails reproduction. It is Benjamin's decisive idea that a change in reproduction techniques brings with it a change in the forms of perception and that this will result in a change in the "character of art as a whole." . . . the simultaneously distracted and rationally testing reception of the masses' (Bürger, 1984: 28).

11 'Whereas during the period of realism . . . the development of art was felt to lie in a growing closeness of representation to reality, the one-sidedness of this construction could now be recognized. Realism no longer appears as *the* principle of artistic creation but becomes understandable as the sum of certain period procedures. The totality of the developmental process of art becomes clear only in the stage of self-criticism' (Bürger, 1984: 23). See Foster, 2000; and Bürger's Postscript to the second German edition (Bürger, 1984: 95–9).

12 'with the historical avant-garde movements, the social subsystem that is art enters the stage of self-criticism. Dadaism, the most radical movement within the European avant-garde, no longer criticizes schools that preceded it, but criticizes art as an institution, and the course its development took in bourgeois society' (Bürger, 1984: 22).

13 Bürger cites Tzara's 'Pour faire un poème dadaiste' in *Lampisteries précédées des sept manifestes dada* (1963) and Breton's 'Manifeste du surréalisme' of 1924 in *Manifestes du surréalisme* (1963) – Bürger, 1984: 113, n. 19.

14 'A literature whose primary aim is to impose a particular kind of consumer behaviour on the reader is in fact practical, though not in the sense the avant-gardistes intended' (Bürger, 1984: 54). Bloch takes a different view, reading a latent utopian content into popular fiction (Bloch, 1991: 153–68).

15 'The avant-garde artist is conceived as a kind of Promethean adventurer, an individualist and risk taker in a sheepish society, an Overman bringing to the more timid world of the herdman, to use Friedrich Nietzsche's distinction, a new kind of fire'; 'The artist, then, not only can realize himself more than anyone else by reason of his creativity, but is a beacon to these banal others, even a kind of Moses leading them out of their ordinary world of perception . . . In a final mythifying touch, the artist is idealized for the transmutation of value . . . that his perceptual and personal authenticity effect and symbolize'; 'In sum, the myth of the avant-garde artist

involves the belief that he is initiated into the mysteries of primordial experience' (Kuspit, 1993: 1, 2, 5). Kuspit sees art's cult status as Modernist yet reproduced in a new worldliness and seeking after publicity: 'the inherently parasitic character of the neo-avant-garde commodity especially confirms it as a cultification of the avant-garde creation' (Kuspit, 1993: 21). See also Krauss, 1986: 221–42 for a reading of Pollock in which negation (as making work about Nothing) is taken as central; and Krauss, 1986: 23–40 for a reading of Cubist collage as signs without referents.

16 Marcuse refers to non-objective visual art, stream-of-consciousness and formalist literature, twelve-tone composition, blues, and jazz: 'not merely new modes of perception reorienting and intensifying the old ones; they rather dissolve the very structure of perception . . . the familiar object has become impossible, false' (Marcuse, 1969: 45). Bürger cites Marcuse's (1937) 'The Affirmative Character of Culture': 'In bourgeois society, art has a contradictory role: it projects the image of a better order and to that extent protests against the bad order that prevails. But by realizing the image of a better order in fiction, which is semblance (*Schein*) only, it relieves the existing society of the pressure of those forces that make for change' (Bürger, 1984: 50), adding 'In Aestheticism, the social functionlessness of art becomes manifest' (Bürger, 1984: 51).

17 The paper was given at a symposium *Where Art and Society Meet*, SUNY College at Cortland, in March 1988 and published in *Block* no. 14 (1988).

18 'The act of walking is to the urban system what the speech act is to language or to the statement uttered . . . it is a process of *appropriation* of the topographical system on the part of the pedestrian (just as the speaker appropriates and takes on the language); it is a spatial acting-out of the place . . . and it implies *relations* among differentiated positions' (de Certeau, 1984: 97–8). Prior to this de Certeau sets out three aspects of the production of the city: the production of space as rational organisation; the substitution of a synchronic system for the residual traces of everyday tradition; and the universalization of a subject 'which is the city itself' which take over the roles of now subordinate entities (de Certeau, 1984: 94). On the speech-act, see de Certeau, 1997: 11–24, 25–40.

19 A revised version is in Krauss, 1986: 276–90.

20 Krauss appears oblivious also to the art of people of colour: 'Item: 1983: Rosalind Krauss explains to her fellow symposiasts at the NEA Art Criticism Symposium that she doubts that there is any unrecognised African–American art of quality because if it doesn't bring itself to her attention, it probably doesn't exist' (Piper, 2001: 58).

21 Strelow, 1999: 51–2, 210–11.

22 '*Der Autor als Produzent*' – Benjamin, 1983: 85–103; also in *Walter Benjamin: Reflections* (1978) trans. Jephcott, E., New York, Harcourt Brace Jovanovich. For a critical commentary see Leslie, 2000: 92–100. Leslie draws out the pre-awareness of art, a factor in Benjamin's work which may be linked to his conversations with Bloch during the 1920s – 'Art can be prefigurative of social and technical relations to come. Prefiguration is important, for it indicates the extent to which Benjamin is convinced of a dynamic inlaid in technology and the forces of production. To pre-empt that development in art is to glimpse the potential (communist) future in the (capitalist) present' (Leslie, 2000: 92). See also Wolin, 1994: 154–62; and Gilloch, 2002: 144–8. Gilloch reads Benjamin's insistence on a relation within the means of production as adoption of a location within the proletariat, as writer-as-worker.

23 The use of community radio is a tool for solidarity, for instance – see Schelling, 1999.

24 Leslie, citing a note to the essay, reads an emphasis on the rupture of barriers between forms of production: 'Intellectual production is politically useful at the point when it forces an overcoming of separate spheres of competence, between genres, between specialists and lay persons, and between creators and recipients' (Leslie, 2000: 93). On Benjamin's criticism of documentary: 'Crucial to Benjamin's rejection of new objectivist photographic practice is his refusal of its passive model of reception, a model that ensures the potential of technological culture for representing conditions of existence is converted into political paralysis' (Leslie, 2000: 95).

25 See Leslie, 2000: 130–3 on the versions of this essay in German, French, and English, and revision required by Horkheimer for publication. For a link to contemporary cultural production see McGuigan, 1996: 79–81.

26 The paper was delivered in Tirrenia, July 22nd, 1986, to the Italian Communist Party's Women's Festival. See Deutscher, 2002: 23–41.

27 Massey, 1994: 232.

28 For examples: Massey, 2001; Finnegan, 1998; Munt, 2001; Leadbeater and Way, 1996. Agrest deconstructs representation of the city and subject experiencing it: 'The city presents itself as a fragmentary text escaping the order of things and of language, a text to be "exploded", taken in pieces, in fragments, to be further decomposed in so many possible texts, open in a metonymy of desire' (Agrest, 2000: 367). Fat (see below) write: 'The annual arrival of F1 (Formula 1 motor racing) to . . . Monaco provides an alternative model of urban planning . . . a temporary (and recurring) anomaly – an impossible combination of scenarios that undermine the supposedly "natural" condition of the city . . . It is this rupture in the understanding of the city that provides an opportunity to change the relationship between the civic institution and its citizens, offering a possibility that the city is an ephemeral experience' (Fat, 2001: 347).

29 Frascina sees art's major institutions as objects of post-1968 resistance: 'MoMA's image as . . . an oasis of modernist culture was fractured by realisations that it was not only a major manipulator of that culture but also a site of power where . . . the trustee's love of art dissembled the sources and relations that guaranteed their economic capital. Other museums, including the Whitney and the Metropolitan, were also sites for resistance and demonstration both during the "Art Strike" in 1970 and from that year on by women artists' groups such as Women Artists and Revolution (WAR), the Ad Hoc Women Artists' Committee and Women Students and Artists for Black Art Liberation (WSABAL)' (Frascina, 1999: 209–10). See also references to Guerrilla Girls above. On the Museum of Modern Art see Grunenberg, 1994. See also Lippard, 1981, 1995.

30 Leslie, 2000: 134–5.

31 Originally a community arts group undertaking large-scale firework displays and lantern parades (which they still can), WFI has for the past few years been developing ways of working within the textures of everyday lives, as in a provision of alternative rites of marriage and death. They state 'We are seeking a culture which may well be less materially based but where more people will actively participate and gain power to celebrate moments that are wonderful and significant in their lives' (publicity brochure); see http//:www.welfare-state.org.

32 On art criticism and the roots of public art in community politics and urban development, see Miles, 1997: 84–103; on art and intervention in post-industrial cities, see Miles, 2000: 179–202. Lacy arrives at her definition of an art based in public service via a refutation of art as appropriating public sites and commentary on art in the public interest (see Raven, 1993): 'The cannon in the park was encroached upon by the world of high art in the sixties, when the outdoors, particularly in urban areas, came to be seen as a potential new exhibition space' (Lacy, 1995: 21). But: 'An alternative history of today's public art could be read through the development of various vanguard groups, such as feminist, ethnic, Marxist, and media artists and other activists. They have a common interest in leftist politics, social activism, redefined audiences, relevance for communities . . . and collaborative methodology' (Lacy, 1995: 25). For a more detailed discussion of the category, see Lacy, 1995: 171–85. On Lacy's practice, see Roth, 1993; Labowitz-Strauss and Lacy, 2001. See also Felshin, 1995; Frascina, 1999.

33 'The obliteration of art is repudiated by a fixing of capitalist relations of production in the art world and . . . the film industry' (Leslie, 2000: 136).

34 'if power is indeed merely repressive, he [Foucault] asks, then how come power relations are not much more unstable than they are? Translation: the cause of power is its capacity to do something other than repress, just as the cause of the survival of the prison is its capacity to do something other than fail to prevent crime' (Merquior, 1983: 109). Or as Foucault writes, power is 'a silent, secret civil war that re-inscribes conflict in various "social

institutions", in economic inequalities, in language, in the bodies themselves of each and every one of us' (Foucault, M. (1980) *Power/Knowledge: Selected Interviews and Other Writings 1972–77*, Brighton, Harvester Press, cited in Merquior, 1983: 110).

35 'In community arts, workers get their kicks out of joining in, perhaps, but it is the artist-instigator who generally gets the kickbacks . . . The division of labour in an administered market society has as its concomitant the detailed organisation of space . . . "right conduct" becomes identified with knowing one's place' (David Reason, in 'Public Art & Collaboration: an interdisciplinary approach', in the symposium 'Context & Collaboration', Birmingham, April, 1990, papers, p. 57).

36 *Penetration and Transparency: Morphed*, made with video makers Kathy Brew and Roberto Guerra was shown in 'Fresh Kills: the Art of Waste', an exhibition of work by 18 artists at Snug Harbor Cultural Centre, Staten Island, from October 2001 to February 2002. See Carr, 2002; Belascu, 2001; Goddard, 2001. Ukeles has a seven-year Percent for Art commission for the site.

37 Meanwhile sculptor Sergio Furnari was told to remove a life-size monument based on a 1932 documentary photograph of 11 iron-workers from 175 Broadway, which he saw as inspiring workers clearing Ground Zero (*Daily News*, February 5th, 2002, p. 16).

38 'The involvement of users of the building in its design was one aspect . . . but of equal importance in the creation of the finished work was the unusual organization of the design team . . . The team who produced the design for the building included a representative appointed by the Bank to oversee the whole process of building procurement from the client's point of view; the architect; the structural engineer; a building physicist; interior design consultants; acoustic advisers and landscape designers. While this is not an unusual team . . . there was none of the conventional hierarchy that might be found in such a team' (Vale and Vale, 1991: 161–2).

39 Vale and Vale, 1991: 135. On Fathy's mud-brick buildings in Egypt, see Fathy, 1973; Miles, 2000: 105–28. On *casitas*, see Cline, 1997: 21–2, 95. Barefoot architecture at Tilonia, India, is discussed in Chapter 8.

40 I had intended to try to make contact with Mockbee but he died from complications of leukaemia in December 2001. My source is thus Dean and Hursley, 2002. The rural studio is a teaching programme at Auburn University, which entails hands-on construction as well as design, and has undertaken projects at Mason's Bend, Newbern, Sawyerville, Greensboro and Thomaton, and Akron.

41 Sennett, in a lecture in the *CivicCentre* programme, London School of Economics, April 12th, 2003.

42 Till (1998) cites Wates and Knevitt (1987); see also Towers, 1995; Miles 2000: 154–61.

43 Sandercock and Forsyth cite feminist theory including Spender, 1985; Collins, 1990; hooks, 1984; and empirical research including Belenkey *et al.*, 1986. The text is taken here from LeGates and Stout, 2000, but was first published in the *American Planning Association Journal* vol. 58, no. 1 (Winter 1992).

44 I am grateful to Prof. Antonì Remesar of CER Polis at the University of Barcelona for this information, and to AHRB for support in attending meetings in Barcelona. The project is evaluated in a forthcoming publication, *Interventions* (Intellect Books, 2004).

45 The reference is to the North American concept of a site of open, unplanned public mixing as cited by Phillips (1988) and a key underpinning idea for Jane Jacobs (1961). Whether the commons were such sites, or structured by gender, race and class as well as family ties I leave open. All quotes and information in this section are derived from material given by the artists and a conversation with Cross in September 2002 and various talks he has given on his work in Barcelona, Plymouth and London. See also Cornford and Cross, 2001: 332–3 on *Camelot*.

46 'it is important that all our finished installations have a material presence and be experienced in a particular context. Each piece is manifested as physical objects positioned in real space, but each is the result of a process of interaction with a wide range of systems and

organizations, from local turkey barn builders to the National Remote Sensing Centre. This way of working not only gives us a continually changing insight into some of the forces shaping the built, natural, and social environment, but it also exposes our emerging ideas to indifference, criticism, and the test of relevance to "everyday life". We are constantly surprised and reassured at the amount of time given to us by people who have no direct connection to the art world' (Cornford and Cross, 2001: 337).

47 'Cornford and Cross Unrealised: Projects 1997–2002' took place at Nylon, London in 2002.

48 'We propose to build a large industrial chimney, cylindrical in form, perhaps made from smooth, pale concrete or polished steel, and very plain as to detail. The scale of the chimney would be informed by current decision making processes around public health, taking account of the type and quantity of emissions, the physical geography of the location, and the distribution and density of settlements in the fallout area' (Cornford and Cross, 2002, n.p.).

49 'the word "non-place" designates two complementary but distinct realities: spaces formed in relation to certain ends (transport, transit, commerce, leisure), and the relations that individuals have with these spaces. . . . non-places mediate a whole mass of relations, with the self and with others, which are only indirectly connected with their purposes. As anthropological places create the organically social, so non-places create solitary contractuality' (Augé, 1995: 94).

50 The central part of the city was rebuilt after the earthquake and fire of 1755, laid out by military engineers Manuel de Maia, Eugénio dos Santos, and Carlos Mardel according to precedents including London's Covent Garden (Maxwell, 2002: 29–40).

51 Santos cites Foster (1996): 'Foster makes here a type of transition between the "artist as a producer", as in the thesis of Walter Benjamin . . . where he asserted the utopian and modernist sense of a politically committed art that would make art and bourgeois culture disappear in the example of Russian constructivism – and the "artist as ethnographer" of the last few decades' (Santos, 2002: 161–2, citing Foster, 1996).

52 There had been a lightness in Portugal before, on April 25th, 1974, when people placed carnations in the gun barrels of soldiers whose left-leaning officers had mounted a putsch to end the fascist regime of Salazar. Early on April 25th, the national radio broadcast a banned song *Grandola, vila morena* (*Grandola, dark city*) by José Alfonso: 'Land of brotherhood / It is the people who command' (Fremian, 2002: 216). Soviets were set up, strikes organised to break capital, though soon a more familiar pattern was reasserted.

8

2001 (I)

SUSTAINABILITIES

•

In the previous chapter I argued that recent participatory and provocative cultural practices intervened in the production of meaning and social form. That argument is carried forward here in terms of the production of space in post-industrial urban landscapes, contextualised by concerns, in a period of growing resistance to globalisation, for sustainable social environments and ecologies. I look at architecture using low-impact materials while renegotiating the relation of work to domestic spaces in London; at a campus built by barefoot architects in India; at projects which contribute to a reclamation of post-industrial spaces in the UK and USA; and at the work of a London-based artists' group whose work on oil consumption links individual responsibilities and local initiatives to global power. The chapter begins with an account of the World Social Forum in Porto Alegre, Brazil in January 2001.

> Porto Allegre represents the hope that a new world is possible, where human beings and nature are the centre of our concern . . .
>
> We are women and men, farmers, workers, unemployed, professionals, students, blacks and indigenous peoples, coming from the South and from the North, committed to struggle for peoples' rights, freedom, security, employment and education. We are fighting against the hegemony of finance, the destruction of our cultures, the monopolization of knowledge, mass media, and communication, the degradation of nature, and the destruction of the quality of life by transnational corporations and anti-democratic policies. Participative democratic experiences – like that of Porto Alegre – show us that a concrete alternative is possible. We reaffirm the supremacy of human, ecological and social rights over the demands of finance and investors.
>
> ('Call for Mobilization', in Houtart and Polet, 2001: 122)

I THE OTHER DAVOS

The chapter is titled sustainabilities because the term denotes several linked but dissimilar concepts. Sustainable development tends to dominate the thinking of trans-governmental bodies such as the World Bank, and commissions such as the World Commission on Environment and Development (the Brundtland Commission),[1] as well as the environmental agendas of governments. Sustainable development assumes continuing growth in the global economy while managing its impact on the planet's natural resources. If, as an enlightened management of exploitation, sustainable development seems an oxymoron, this is one of the difficulties of current environmental policy. Some of the concept's limitations are indicated by Robert Chambers:

> debates on environment and development have been dominated by values which reflect the 'first' biases of normal professionalism. These start with things rather than people, the rich rather than the poor, men rather than women and numbers rather than qualities. They bear the imprint of interests that are urban, industrial and central in location rather than rural, agricultural and peripheral. Poor rural children, women and men have been treated as residual not primary, as terminal problems not starting points.
>
> (Chambers, 1988: 1)

In opposition to transnational calls for trade liberalisation combined with minor reforms in areas such as carbon emissions, two emphases have arisen: a recognition that local knowledge can contribute to dealing with problems such as deforestation and soil erosion;[2] and a belief that only radical changes to the global economic structures which drive development will end their destructiveness. The latter is enhanced by perceptions that the policies of national and international agencies – which might mediate between development and the needs of poor countries and for environmental protection – are eclipsed by a diminution of regulatory regimes, while some transnational corporations have larger economies than some European states.[3]

An alternative to the concept of development is that of a radical revision of needs in place of the excess consumption generated by market requirements.[4] A case for a reduction in consumption follows, too, from analysis of footprints of urban consumption.[5] And while low-impact technologies can achieve significant damage limitation in environmental terms, sustainability can be seen as depending on an interaction of cultural, social, economic and environmental factors.[6] There is, then, a tension between the privileging of economic data and concerns for cultural and social equity. Peter Worsley writes of a cultural turn in development discourse:

> Increasingly, at least outside the World Bank and the IMF, it seems clear that true development should be measured not just in terms of economic

> criteria . . . but in terms of who gets what, and of whether what they get via the market is really what they need to enrich their quality of life.
>
> (Worsley, 1999: 38)

Appropriate measures include access to health care and education, and the empowerment of citizens. Worsley cites a post-development school[7] for whom development is inherently imperialist. Among them, Arturo Escobar posits a post-structuralist political ecology of new narratives produced in 'the mediations that local cultures are able to effect on the discourses and practices of nature, capital, and modernity' (Escobar, 1996: 65). Richard Peet and Michael Watts argue that societies cohere by 'systems of meaning and representation which organize their natural worlds and establish ways people are socialized' (Peet and Watts, 1996: 267).[8] They propose an environmental imaginary in which the vocabularies of environmental futures are created. The name World Social Forum (counter to the World Economic Forum) thus states an alternative agenda. A Call for Mobilization from Porto Alegre announces: 'We challenge the élite and their undemocratic processes, symbolized by the World Economic Forum in Davos. We came here to share our experiences, build our solidarity, and demonstrate our total rejection of the neoliberal policies of globalization' (Houtart and Polet, 2001: 122).[9]

The Forum introduces a concept of power based on solidarity amid self-determination. It includes a reclamation of rights by the ex-colonised but is more than a reassertion of the South against the North when economic colonialism produces territories of abjection in the affluent and non-affluent worlds.[10] In face of the colonialism of consumption, alliances have emerged using new technologies of communication,[11] and new discourses in which environmental degradation is aligned to human rights abuse.[12]

Just as there are several sustainabilities, there are varieties of environmentalism and ecology.[13] I want to suggest that the concept of ecology as a model of life systems of interdepending elements can be applied to cities as metaphor (rather than in an alignment of biological and social processes) to denote the non-separability of cultural, social and economic conditions. This accentuates the role of narratives in scripting urban change, and the need for counter-narratives; but it does so not from a model of biological stasis but from one of adaptation, compatible with a Marxist position of intervention in the conditions in which narratives are produced, and in a particular direction. Perhaps the imagination of the story of what might be is a quality specific to humans. We cannot say, but Murray Bookchin's concept of social ecology sees human ecologies as differing from those of non-humans.[14] Andrew Light summarises: 'the history of social and natural evolution has become the history of two competing logics: the logic of spontaneous mutualistic ecological differentiation and the logic of domination, which works against everything represented by the other' (Light, 1998: 7). Bookchin polarises organic communities and urban societies; but if some of the organic communities were not Edenic, the *imagination* of such idealised states of sociation is interesting as expression of present disquiet. Bookchin calls for a rebuilding of human relations in a libertarian political economy.

There are many departures from conventional models of settlement, in co-housing, intentional communities and eco-villages.[15] A terrain of resistant politics has emerged in parallel with these departures from mainstream and often from urban living. After road protest, we see demonstrations against globalisation and its neo-liberal organisations, as in Seattle in 1999.[16] Douglas Kellner sees cyberactivists trying to 'carry out globalization from below, developing networks of solidarity and propagating oppositional ideas and movements' (Kellner, 2003: 189). The new networks offer dissemination outside the global news media.

For the Association for the Taxation of Financial Transactions in the Interest of the Citizen (ATTAC), it is now a question of 'taking back, together, the future of our world' (ATTAC, 2001: 71).

II WORKING AND DWELLING

To do that, local initiatives are a beginning; but the term local can be understood in two ways: local to place and a specific set of problems and conditions; or local to a cultural or social form, or specific practice. I turn to low-impact architecture as a case of the latter. There are many cases of projects using renewable energy sources and recycled materials, restructuring patterns of mobility in multi-function zoning, and reintroducing high density living in forms other than the tower block.[17] But I look now at 9/10 Stock Orchard Street in north London, by Sarah Wigglesworth and Jeremy Till, completed in December, 2000 on the site of a forge and outbuildings where automobile springs were tempered (which replaced the original, nineteenth-century house). The building combines work and domestic space, and demonstrates a range of low-impact building technologies. The site is adjacent to a rail track, and previously belonged to British Rail. It was auctioned in 1994 and planning permission for the house and workspace granted in 1997. In one part of the L-shaped building are the studio and office of Wigglesworth and Till's practice, and in the other a bedroom, living room, kitchen and bathroom. Above the living room at the studio end is a library tower to house tiers of books rising to a room at the top in which a day-bed and desk offer a space for contemplation, or to look out over London. The living room roof supports a meadow, and the land around the house is cultivated to provide herbs and flowers in organic plots. On open days in 2002, more than 1,500 people visited the studio-house.[18] Its technologies, such as straw-bale walls and recycled concrete in gabions, have potential for use in a spectrum of public- and private-sector buildings, in social housing for instance, where cost is a key factor.

The method of design enacts a fusion of working and living. In 1998, Wigglesworth and Till asked rhetorically: 'Faced with a blank sheet of paper and a couple of buildings to design, where do you start?' (Wigglesworth and Till, 1998b: 31). Their design began from the transition from a perception of order, disorder and reorder on a dining table (through a dinner party), charted in a series of drawings tracing the placing and displacing of objects.[19] The same table was used at other times for office meetings in their previous house. This intersection of domestic life and work is carried through in the new building, where the dining room is a hinge between office and domestic spaces: 'This final condition became an emblem for the plan of the house, a seemingly disordered collection of objects set on a plane – but in fact a collection which allows the passage of time and domestic life to pass through it in a relaxed manner' (Wigglesworth and Till, 2001: 16). The window is set low, at a height level with that of the trains passing by outside.

The interior spaces of 9/10 Stock Orchard Street are simple but not without luxury, and decidedly urban: the dividing wall which can be pulled across to close off the studio from the dining room is of Douglas Fir, with a rich natural variegation; the living room has the extent and light of loft living, and there are paintings and pieces of designer furniture which denote cultural capital. At the same time, the adobe larder cooled naturally by air, and use of recycled materials and those which ensure an even interior temperature through the day, denote an economy of means. As the architects say: 'Our tactic is not that of the hair-shirt puritan; we aim to seduce you with the gloss, and slip the world of the everyday in through the back door' (Wigglesworth and Till, 1998a: 7). There is an arresting inversion of convention: the living room has industrial-scale windows to give maximum light and views (in and out); the office is enclosed, its slit-like windows framed by railway sleepers to peer out through a sandbag wall as if under siege from the passing trains, or as a place to silently watch them. The domestic zone is open to, and the work zone enclosed from, the visible world.[20] The building has an ambivalent relation, too, to Modernism: the wide glass windows of the living room suggest the transparency of the houses designed for the Bauhaus masters in the 1920s; the emphasis on horizontal windows in the office suggests Le Corbusier; the library tower is reminiscent of that of the Villa Stein à Garche into which Le Corbusier retreats in his film *L'architecture d'aujourd'hui* (1929).[21] Yet the building is clearly of now in its accommodation of a tension between order and disorder which is a refusal of Modernist purity; and the references in materials and surfaces are, despite some pre-industrial nuances, to post-industrial urban conditions and a post-modern contingency: '[It] will remain permanently incomplete, as what is both already there and not yet there is continually reinvented, adjusted and played with' (Wigglesworth and Till, 2001: 2).

Beginning at the entrance: the gate is made of willow hurdles in a galvanised steel frame.[22] Next are the gabions filled with roughly broken up, recycled concrete which support the base of the office floor above. This material is in

8.1 • Sarah Wigglesworth and Jeremy Till, *9/10 Stock Orchard Street*, London, gabions containing recycled concrete

cheap and plentiful supply[23] and has a weight tolerance in excess of that required here. Above the gabions the office wall uses sand, cement, and lime packed in sand-bags which will gradually erode to leave a wall of rippling forms; and on its other side – referencing the fusion of domestic and work spaces in a metaphor of comfort which is also gendered – the office outer wall is clad in a cloth of silicone-faced fibreglass 'puckered and buttoned like a domestic quilt' (Wigglesworth and Till, 2001: 6) with an insulating layer and inner lining.[24] Under the building, two tanks collect rain-water for use in clothes-washing and to irrigate the roof meadow. There is a compost toilet, a solar panel and a wood-burning stove (though heavy insulation reduces the need for heating). The bedroom is encased by a straw-bale wall, which continues on the north side of the building.[25] While in vernacular buildings straw tends to be rendered with lime plaster, here it is encased in a rain-screen of galvanised steel, ventilated to allow moisture to escape. A section is visible behind transparent polycarbonate. Inside, lime plaster is applied directly to the straw, with small metallic strips bridging the wooden frame to prevent cracks. Recollecting on the catalogue of materials and processes used, the architects

8.2 • Sarah Wigglesworth and Jeremy Till, *9/10 Stock Orchard Street*, London, straw-bale window

write that they 'form a repertoire of technologies from the everyday, raiding the techniques of other disciplines for inspiration' and that their approach 'has been to borrow and adapt technologies from outside the normal . . . architectural canon' (Wigglesworth and Till, 2001: 12).

Thinking again of Benjamin's idea of the artist (architect) as producer I see this building as intervening in the production of categories such as domestic space and work space. Learning from an architectural everyday,[26] beginning with people not things, its form is not an engineering solution to a visual concept but follows from its means of production. This refuses modernity's privileging of visuality (but does not mean it is visually unrewarding), suggesting that in a new architecture the act of building and engagement with materials might be as important as design, and not relegated to a secondary tier or delegated to technicians and construction workers.[27]

Wigglesworth and Till write of mainstream architecture set like More's Utopia on a remote island:

> We first came to the everyday from the furthest shores of architecture. Conceived of as an island, this architecture concerns itself with internalised notions of form and style. Aesthetics and technology enter into an unholy alliance which allows the self-contained and self-referential language of architecture . . . Occasionally boats arrive at this island, bringing with them fresh supplies of theory, geometry and technique which inject the flagging body of architecture with new life. It is not surprising that the architecture which is thereby created is obsessed with notions of the iconic, the one-off, the monumental. It privileges the final produce over the process, the perfected moment of completion over the imperfections of occupation.
>
> (Wigglesworth and Till, 1998a: 7)

Yet there were no difficulties in gaining planning permission for 9/10 Stock Orchard Street, nor in insurance or mortgage cover. So why is this kind of architecture not taken up more widely? Wigglesworth thinks that in the bureaucracies of health and social housing which might be obvious clients, few are prepared or encouraged to take risks. Within its profession, architecture remains 'structured around a cult of novelty', and alternative architecture is identified as either conceptual or fit for rural situations (conversation, September 4th, 2002). Yet it is in cities that low-impact forms of high-density housing and hybrid spatial categories can produce sustainable forms of settlement.

I turn now to a case that might seem out of place in a book on urban avant-gardes: a rural campus built by barefoot architects at Tilonia, near Jaipur, India.[28] I include it because – apart from its intrinsic interest – it demonstrates more than a technology for sustainable settlement. The buildings on the college site and in surrounding villages denote new social and cultural possibilities, and new forms of power relations, which have implications for urban living in the affluent world.

The campus comprises a clinic with dispensary, a library and dining hall, guest houses, an amphitheatre, residential blocks, craft centre, workshops and administrative spaces. It was created in local stone by Bhanwar Jhat with 12 barefoot architects and local labour. In the surrounding area, 250 or so homes have been built for homeless people by 60 barefoot architects; a rainwater harvesting system was installed to collect rain from rooftops and ensure local control of its use, by Laxman Singh assisted by Ram Karam, Kana Ram and Ratan Devi. Geodesic domes made by Rafeek Mohammed and seven barefoot architects are used for a clinic, telephone exchange, teaching rooms and guest rooms.[29]

The social architecture – the structures of relation between people – of the Barefoot Campus is as engaging as the visual:

> The college aims to demonstrate that village knowledge, skills and practical wisdom can be used to improve people' lives – an attitude of self-help drawn from Mahatma Ghandi's example. Since rural people already have tried and

8.3 • The Barefoot Campus, Tilonia, India. Photograph courtesy of Bunker Roy

> tested traditions, the college believe that all that is needed is a little training and upgrading combined with respect for local skills, which are slowly dying out because people are migrating to the cities to look for jobs.
>
> (Aga Khan Award for Architecture, 2001: 78)[30]

The ecological footprint of Tilonia's Barefoot College would be very small – its local materials and locally harvested water, and the use of solar power throughout, combined with a simplicity of dwelling, does not require imported energy – and its education programmes contribute to maintaining a proactive community.

The relation to tradition of the Barefoot Campus is like that of Hassan Fathy's mud-brick architecture, which he saw as a no-cost solution to Egypt's housing problems.[31] Fathy writes of tradition that it is based in material culture: 'a tradition need not date from long ago but may have begun quite recently. As soon as a workman meets a new problem and decides how to overcome it, the first step has been taken in the establishment of a tradition' (Fathy, 1973: 24).

Similarly, it is a set of responses to conditions, which tend outside industrial society to be localised. At New Gourna in upper Egypt, Fathy facilitated the partial building of a model village in the 1940s using mud as a plentiful and free local material (before construction of the Aswan High Dam). Wood was scarce and expensive. Mud-brick vaulting (which masons in Aswan could still produce) avoided the use of large timbers, while doors were made decoratively from small sections. Wood is in short supply, too, at Tilonia, so the geodesic domes use scrap metal available from discarded agricultural tools and machinery.

Fathy sees acquired vocabularies as part of how a colonial power assimilates the elite of a subject state to its norms.[32] The import of high-cost, high-energy, high-maintenance technologies is equally debasing and ensures a relation of dependence on the part of the recipient, but I would argue that this does not mean local technologies are beyond adaptation. The water harvesting system at Tilonia exemplifies this: unlike standard water schemes in the non-affluent world it does not pump up water from below ground (which is expensive and could produce brackish water here) but collects surface water in a traditional but updated way, gathering it from flat roofs and channelling it into tanks:

> In an arid region such as Tilonia, water and its adequate storage and supply are critical to the very existence of the community. The new system is inexpensive and provides a year-round supply, even when the monsoon rainfall is low . . . in several rural primary schools the attendance of girls has improved because they do not have to spend hour walking several kilometres to collect drinking water. The system has also led to wasteland reclamation.
>
> (Aga Khan Award for Architecture, 2001: 80)

Land reclamation, primary education, and basic sustenance are, from this account, integrated in a response to one problem which has outcomes for the others. This organic process contrasts with the separatist methods conventionally employed by state agencies in planning and development aid, at home or abroad. Nabeel Hamdi and Reinhardt Goethert describe the laborious quest for data which precedes any intervention;[33] and the orthodox cycle of survey, analysis and plans using statistical modelling to forecast future trends – 'It will be full of good intention . . . five years into the project, the whole will be neatly put onto the shelves of the planning office, until the time comes to revive the plans' (Hamdi and Goethert, 1998: 26, 27). A magazine report of delays in the rebuilding of Bhuj in Gujurat after an earthquake confirms this gloomy view:

> A masterplan to rebuild the devastated centre of earthquake hit Indian city Bhuj has been delayed by six months . . .
>
> The delay comes after [the planning consultancy] was told its five month study to produce a masterplan for Bhuj – accepted by the Gujurat urban development corporation – was not detailed enough.
>
> . . . [the firm] said the original brief did not demand the detail now requested . . . It has been asked to produce an extra plan of the walled city including the location of individual land plots, roads and open spaces.

> Some residents are reported to have started rebuilding their homes . . . without official permission. But some of these rebuilt homes could be in areas set aside for new roads or open spaces. [An official] . . . confirmed that houses rebuilt without permission would be demolished.
>
> (*New Civil Engineer*, January 17th, 2002, pp. 12–13)

Hamdi and Goethert see the planning assumptions that followed imperialism as reflecting the slow, continuous urbanization of northern Europe and North America; and a self-referential system in which planning practices bid for projects designed by other planners, 'driven by the demand to promote productivity through increased mechanization and transfer of technology . . . rather than by a concern with encouraging local resourcefulness' (Hamdi and Goethert, 1998: 29). But amid this failure they see a growth of new paradigms, relationships among actors and understandings of the responsibilities and liabilities of experts. I wonder if there could be scope for reverse technology transfer, so that settlements in affluent countries develop technologies equivalent in their terms to those of the Barefoot Campus.

In barefoot architecture the categories of design and production, and of production and reception, are collapsed, as they are in Fathy's work at New Gourna; the architect ceases to design for others to build, and uses her or his status to drive through projects, contributes specific expertise in a process which gives equivalent value to the expertise of dwellers on dwelling, and mediates between conflicting interests which may not be resolved in a single solution. Lefebvre sees no purity in technology:

> A dialectical contradiction . . . presupposes unity as well as confrontation. There is thus no such thing as technology or technicity in pure or absolute state, bearing no trace whatsoever of appropriation. The fact remains, though, that technology and technicity tend to acquire a distinct autonomy, and to reinforce domination far more than they do appropriation, the quantitative far more than they do the qualitative.
>
> (Lefebvre, 1991: 392)

Perhaps from the non-affluent world's dwelling in contingencies the affluent world can learn to redress the balance between order as control and as negotiation, and begin to renegotiate our categories – architect and dweller as co-producers, for instance.

Walter Segal writes that when people build their own homes it makes for a better use of resources;[34] for the time-rich and fit of post-industrial cities it is an option (if facilitated by local authorities and mortgage lenders). In South Africa, the building of social housing is now partly handed over to township dwellers, who have ample skills and can make buildings more adaptable to future needs than those of government schemes.[35] But if it is naive to think self-build will replace mass housing, it is instructive to ask why. It is not for technical reasons because projects using Segal's design exist in Lewisham and Brighton, and there is a small literature of handbooks for self-builders higher up the income scale.[36] Neither is it in the economic interests of dwellers, since self-build removes from the cost of housing the profits of the volume house builders. So it is for cultural reasons: because there is a resistance to learning from the non-affluent, or the urban imaginary has no vocabulary for it.

If so, then small-scale, local initiatives may be crucial in creating a new socio-environmental imaginary as prerequisite of the confidence for widespread action. For example: in 1997, at Quaking Houses in County Durham – a mining village in a rural landscape – local people made a wetland to treat pollution in a local stream. It is a local project addressing a problem that has global import. The stream carried aluminium hydroxide producing white foam, acidity causing red-orange iron oxyhydroxide deposits, and aluminium giving a milky colour to the water; and was unable to support life. Oppositional action began in Quaking Houses with moves against open-cast mining, in 1989, after the Thatcher regime's attack on the deep mining industry, with the founding of an Environmental Trust. In 1995, digging of a pilot wetland was begun by local volunteers working with Paul Younger, a water engineer with experience of self-help water cleansing schemes in Bolivia, and of mine water treatment in Cornwall. The wetland was completed in 1999, has two pools and uses a layer of horse and cow manure mixed with straw and composted garbage, with a cell of limestone cobbles mixed with substrate under an aerobic flow of water. Bacteria in the compost combined with dissolution in the limestone cause pollutant to settle in the mud and reduce acidity. Aluminium is deposited in the wetland as alkalinity rises. Water quality is monitored and the wetland has achieved its aims in terms of removing most of the pollutants while adding a new focus to the surrounding landscape, its banks planted by local volunteers.

The project was not without conflicts, particularly between Younger's team and artists introduced to the project through Artists Agency in Sunderland (now Helix Arts, relocated to Newcastle). There seemed a perception that an artist would produce illustrative material as part of a popular understanding of science agenda, contrasting with a desire to collaborate in the design process – inserting a walkway over the site to allow public access – and be a full partner. The addition of an arts budget to a low-impact project was also problematic. Artist Helen Smith nevertheless devised a series of solar-powered listening posts with a pre-recorded mix of local voices and water data, available through headphones. She held IT workshops with local young people, and worked with local people of all ages in broadcasts on Sunderland University's Radio Utopia.[37]

Looking to the literature of development studies,[38] the use of appropriate technology, and of both local and professional knowledges, resembles work in the non-affluent world. Quaking Houses is anyway a pocket of non-affluence. In the non-affluent world there are no artists-in-residence in such projects, yet no shortage of creativity within local cultures. That might be the point: cultures

8.4 • Quaking houses, County Durham. Chas Brooks by the second pool of the wetland

in traditional societies are repositories of the stories through which a society coheres, while in industrialised societies the anthropological sense of culture is replaced by the elitist and exclusionary categories of high art and mass culture. I think Smith did contribute to a communicative network which is a post-modern equivalent for a modern popular understanding of science. But it may be that in otherwise well-intentioned art projects one sort of culture displaces the other. So what can an artist add? Can they lend visibility through their professional skills, or can that visibility be co-produced? Could the arts become a form of cultural development work, renegotiating the narratives which normalise power, or creating an imaginary for a post-industrial society?

Pittsburgh was a steel town and is now in the midst of the US rust-belt. One of the benefits of its post-industrial condition is the opening to public access of its water front, where paths and cycle trails – one goes to Washington, D.C. – replace the closed spaces of mills and railway yards. The city ambivalently embraces nature through restorative planting along the waterfront, much of it carried out by citizens' groups such as Friends of the Riverfront, and at the same time still considers mining residual coal deposits in city hills in order to produce further industrial development. This is contextualised by a growth in cultural industries centred on the Warhol Museum, but also by a population drain and tendency of graduates from the city's universities to go elsewhere. A negative image is curiously imparted by sports teams such as the Pittsburgh Steelers whose names conjure an industrial past, and by physical traces of industrial hegemony, as when for more than fifty years the steel industry dumped slag in a 240-acre site of what is now lifeless artificial rock, a continuous escarpment the height of a ten-storey building. The valley below is thickly wooded and wild turkeys sometimes fly out of the trees. In 1910, Frederick Law Olmsted Jun. pronounced it to be an ideal site for a public park with playing fields and shaded walks between cool resting places.[39] Steel-industry slag is not highly toxic but is unable to retain moisture and so has no means to support growth. Only where building rubble has been dumped on the slag can plants and trees set roots to provide that shade now.

The slag site was marked for housing development in the 1990s,[40] and the stream bed in the valley written off as dead. The first plan was to culvert it. Adjacent to this valley, called Nine Mile Run, is Frick Park, a sign of the city's philanthropic past, another being the Carnegie institutions. At Carnegie Mellon University, the Fine Arts School houses a unique research facility, the Studio for Creative Inquiry, in which artist-fellows undertake medium-term projects of artistic and contextual innovation. Tim Collins and Reiko Goto worked with Richard Pell, Bob Bingham, John Stephen and others including researchers in the natural sciences, landscape and urban planning to facilitate alternative plans for the valley – in which public access would be combined with intervention to promote biodiversity. Collins and Goto were clear that the aim was not to restore the stream bed to its pre-industrial state, which would be impossible without removal of the millions of tons of slag, but to research the ecologies of the stream and stream-side green areas, to clear invasive plant species and selectively plant those, from the evidence of sites such as Frick Park, most likely to flourish and merge with indigenous vegetation, to investigate pollution, and publicise the stream and its ecosystems.[41]

The sterility and uniform grey of the slag led people to see Nine Mile Run as a convenient place to dump anything, from rubble and old tires to fridges and cars. One of the first tasks undertaken by the Studio was to remove junk; later a pilot scheme greened part of the slope to aesthetically recode the site. Another tactic was to take people on walks along the valley to see and hear for themselves the diversity of wildlife it supports. Much of the pollution was sewage from leaking sewer lines and illegal connections to storm-water drains. This dry weather condition is aggravated by currently legal combined sewer overflows which spew wastes in periods of heavy rain. Investigation identified sources and responsibilities, while an education project in local schools

8.5 • Tim Collins and Reiko Goto, Nine Mile Run Greenway, Pittsburgh, collecting junk from the slag site

encouraged a revaluing of the site. In 2000, agreement was reached with all parties that a third of the area will remain green, at which point the project team made arrangements to hand over implementation to a local partnership of interests.

The project's core was a series of workshops (charettes) known as the Community Dialogues in which local people met experts, officials, and developers to argue through specific issues concerning the valley. All had access to available data, and some participants added local knowledge to the imported knowledges of experts. This led to visualisations produced by the Studio, but more important was their mediation between different authorities and languages: stakeholders, developers, planners, engineers, biologists, environmentalists . . . and artists; and languages of public administration and academic disciplines as well as everyday life. So where is the art? It does not matter. More interesting is the agency of the Studio to draw in other experts, to treat dwellers as experts, to use their professional and academic status to talk to public authorities and private-sector interests, and to raise funds (the Heinz Endowments being the primary sponsor). A second project is now in development by Goto and Collins: *Three Rivers, Second Nature*, which extends ecosystem analysis and social dialogue to a series of River Dialogues on public access to and uses of the Monongahela, Allegheny and Ohio rivers. A boat has been purchased for water monitoring, and interfaces set up with groups ranging from environmental to development interests in post-industrial communities outside the city.[42]

Collins and Goto identify the philosophies informing their work as social ecology, eco-feminism, and deep ecology:

> These three philosophies, with their spatial commitment to city, town, country, or wilderness, and their political commitments to humanity, post-dominion humanity and the intrinsic rights of nature itself, provide a broad intellectual foundation for the eco-artist . . . the strategic points of engagement . . . are interface, perception and human values. It is through human values that we know nature, it is through human values that artists act upon culture to change the perception of nature.
>
> (Collins and Goto, 2003: 142)

They advocate legal rights for nature,[43] but the last sentence hints at an idea of cultural development. If cultural production and reception change perceptions of nature, then perceptions of society and economy are equally open to mediation. An emphasis in post-modern discourses on representation asserts the potency of signs in maintaining both domination and resistance, as Sharon Zukin argues, or can 'isolate cultural factors . . . from their material context' (Zukin, 1996: 224). But if cultural production is a form of critical and sensuous activity, it can ground signs in the material and bodily worlds they interrogate.

Codes, then, which have been produced can be reproduced or re-produced. Even the model of art-work as non-productive time questions the myth of productivity. Tim Hayward argues that

> the sustainability of liberal democracies with capitalist economies is premised on an assumption of the indefinite sustainability of economic growth – an assumption which appears seriously questionable from an ecological perspective. . . . the emphasis on growth, as opposed to what it is for, combined with the domination of market values, means there can also be certain important interests people have whose effective expression is systematically blocked.
>
> (Hayward, 1998: 162)

He adds that even interests advocated by environmentalist groups are vulnerable to liberal democratic re-framing. But the framing devices, too, are vulnerable to exposure in cultural critique. The market is adept at manipulation,[44] but citizens who gain confidence and expertise through participatory democracy – which will usually be at a local level – may be harder to manipulate.

III OIL AND HONEY

PLATFORM promotes creative processes of democratic engagement in order to advance social and ecological justice. Founded in 1983, its core members are Jane Trowell, James Marriott, and Dan Gretton;[45] it works also with a range of other individuals, among whom Emma Sangster and Greg Muttitt are long-term collaborators. PLATFORM initiated the production of renewable energy using a water turbine in the river Wandle in south London; and since 1996 has developed projects engaging with Londoners' consumption of energy and the operations of the global oil industry, collectively called *90% Crude*.

I focus on the latter as a case of creative work at the interface of cultural production and environmental campaigns. But after the discussion in Chapter 7 on art and society, I must avoid dualities. It would be easy to make them, between art and environmentalism or at the next level between mainstream and alternative arts, or environmental campaigns and activism; but rather than situating PLATFORM at a cusp of what then becomes a polarisation, I could say there are creative tensions among the agendas and tactics of those engaged in differing ways in art and environmentalism, and that incisive cultural work can be done in the spaces between such categories; and that other categories are interrogated in the process (such as production and reception in Benjamin's terms – discussed in Chapter 7 – or personal and political). I return at the end of the chapter to aspects of these questions in terms of the kinds of knowledge produced in differing modes of engagement; here I observe only that some things in the material world suggest our categories are too neat. Honey, for example, defies the categories of liquid and solid, having qualities of both. Oil, too, can take the form of a liquid or a solid mass, and is closely related to gas. In terms of concepts, some contemporary cultural processes operate in a dynamic and active tension between the anthropological idea of culture as a way of life and the bourgeois notion of art (itself in tension with mass media); and in another way there is a tension between instrumentality and spontaneity, like organised revolution (Leninism) and anarchic rupture (or a millenarian state of grace). The example of the global oil industry, as PLATFORM reveal in some of their work, is of excess instrumentality, while local initiatives to counter its effects may adopt more consensual and organic means – though this does not preclude grassroots instrumentality. But tactics embody ends, and as Audre Lorde argued in 1979: '*the master's tools will never dismantle the master's house*. They may allow us temporarily to beat him at his own game, but they will never enable us to bring about genuine change' (Lorde, 2000: 54). How, then do we forge new tools to take back the idea of a common wealth which the masters of, say, oil have stolen from us (and from themselves because master *and* slave are dehumanised)?

A leaflet distributed by PLATFORM in 1993 stated the following:

> PLATFORM is a meeting place for desire and acts of change . . . described as many things – an art group, a forum for political dialogue, an environmental campaign – but, in essence, it is an idea, a vision of using creativity to transform the society we live in; a belief in every individual's innate power to contribute to this process.
>
> (PLATFORM, leaflet, 1993)

In a more recent formulation they describe their work under six headings: catalyst for change; individuals not representatives; practical and poetic; interdisciplinary creativity; here and elsewhere; and infectious visions (PLATFORM, leaflet, 2003). In the ten years between the two formulations they have continued to place emphasis on, and set aside time for, critical self-evaluation of, and feedback from participants in, their work. This embodies the collaborative aim it serves, and enables refinement and extension of a theorised practice. The group makes decisions collectively, while members sometimes develop individual works, and accepts a hand-to-mouth funding position as giving the freedom to set agendas other than those of arts funding bodies.

In 1990, planning began for *Still Waters*, a project on London's buried rivers;[46] in 1992 several events took place, including ritual dowsing walks and establishment of an Effra Development Agency in an empty shop in Herne Hill in south London (a spoof before *Ignite* in 1996–7), and the pilot use of a water-turbine to power a school's music room in 1993.[47] The demonstration of an alternative technology can be compared with Mel Chin's *Revival Field* (1989–92) in St Paul-Minneapolis which used hyper-accumulating plants such as maize, bladder campion and pennythrift to leach toxins from a waste site.[48] Chin's aim was to test the technology, which worked but would be a slow if organic means to cleanse brownfield sites, and is happy for the idea to be adopted freely.[49]

90% Crude includes practical and material aspects – thinking is a practice – but is closer to a campaign (though not exactly so) than to the technological model of *RENUE*. It has seven elements apart from *RENUE* (which began within *90% Crude*), beginning in 1996 and planned to culminate in 2006. They are as follows: *Funding for a Change*, a national network of 200 people and groups engaged in five debates on the ethics of global corporate activity, and the relations of the voluntary sector to corporate sponsorship; *Ignite*, a spoof newspaper distributed free at London rail termini; *Crude Operators*, a two-day international gathering of activists, campaigners, and groups concerned for the negative social and ecological impact of global dependency on oil (with subsequent publication); *Vessel*, a proposal for conversion of an ex-industrial Thames-going craft for the use of artists and community activists; *Agitpod*, a solar-powered, pedal-propelled video/slide projection vehicle on a Brox quadricycle base, designed with students at Southwark College – which, like the water turbine, demonstrates a technology; *Carbon Generations*, a performance by Marriott using images, stories, and objects to convey histories of his family and the oil industry in context of global warming; and *killing us softly*, a lecture-performance by Gretton, and research project on the psychologies of contemporary bureaucratic behaviour and the historic proximity of corporations to genocide (see Chapter 6). An extension of *Carbon Generations* has in effect become a new project from around 2002, called *Unravelling the Carbon Web*. It is focused on Shell and BP, and includes elements such as *Gog and Magog* – a journey which takes participants into the imagined minds of the two giants of the City of London, Shell and BP – and *Freedom in the City*,

8.6 • PLATFORM, The Agitpod, a solar-powered, pedal-propelled, image-projection vehicle, made with design students at Southwark College, 1997. Photograph courtesy of PLATFORM

8.7 • PLATFORM, *Freedom in the City*, 2003. Photograph courtesy of PLATFORM

a series of guided walks in the financial district of London, devised to raise buried histories, specifically that of the East India Company. PLATFORM has also participated in a detailed study of the Baku–Ceyhan pipeline (see Muttitt and Marriott, 2002), and has produced a booklet to distribute to universities, titled *Degrees of Capture*, on the relation of universities in the UK to graduate recruitment in the oil industry.

One long-term aim is to nudge the consciousness of graduates likely to be recruited by major companies, so that to work for an oil company would be not something to admit socially. This is an ambitious aim, and significant aspects of *90% Crude* use small-scale encounters, and personal histories and voices, to communicate intimately rather than rhetorically. But *Ignite* represents an appeal to the non-specific range of people who use London's main rail stations.[50] In relation to this, Trowell uses the metaphor of a viral particle 'slipping a proposition into the blood-stream under the guise of a safe publication' (Trowell, 2000: 107). *Ignite* appropriated its masthead from *Tonight*, a recently defunct evening paper. Readers would have quickly seen, however, that *Ignite* published stories ignored by other papers (apart from *Socialist Worker* or *The Morning Star*). On the front-page it announces 'Shell police accused of torture'; a city-page feature 'Lessons in crude PR' compares the involvement of BP and Shell in corrupt regimes – 'Don't forget that despots are good for business, although the cocktail parties get a bit tedious when you have to turn a deaf ear to tales of another hanging' (*Ignite*, 1996: 1, 2, 8). A competition offered a chance to win a developing country: 'Imagine owning a country of your very own. Imagine the joy of running the government, the economy and the military . . . being in complete control . . . only multinational companies are eligible to enter this draw' (*Ignite*, 1996: 1).

The lead story concerns BP's dealings with the military rulers of Colombia, and the death on October 24th, 1996 of farmers Carlos Arrigui and Federico Ascensio after protesting about the loss of their land. The editorial states:

> *Ignite* is here to tell you the stories behind the stories . . . about our love for oil and our need for the transnational corporations which deal in it. It's about how London's life is addicted to this commodity. It's about how every single aspect of our daily lives depends on this addiction.
>
> (*Ignite*: 1996: 9)

A two-page spread covers the suppression of a government report on pollution in the North Sea; an image of the tanker *Sea Empress* is overlaid with a caption 'Built in Spain; owned by a Norwegian; registered in Cyprus; managed from Glasgow; chartered by the French; crewed by Russians; flying a Liberian flag; carrying an American cargo and pouring oil onto the Welsh coast . . . who takes the blame?' (*Ignite*, 1996: 6). 'Flares still the fashion in Niger Delta' heads a story of gas flares in Ogoni land;[51] and Delia Spliff offers a recipe for stuffed lungs from ground level ozone, nitrogen dioxide, sulphur dioxide, carbon monoxide, volatile organic compounds, particulates and your own lungs (if you do not need them, presumably).

The second 'Smogbusters' issue followed an apparent change of regime in the UK general election of 1997, and was distributed between November 26th and December 12th, 1997, during the Kyoto summit. Its status as art is denoted by a credit to the Arts Council and London Arts Board (a change from the first issue). PLATFORM describe its distribution as a series of 10,000 2-second art actions involving a gift to passers-by: 'The gift will seem at first glance to be a new 20-page tabloid London evening paper . . . [but] will tell a specific story – the story of London and Londoners' compulsive addiction to oil, and its relation to climate change (PLATFORM, leaflet, 1997). There are spoof adverts: Thomas Sucks announce cancellation of all skiing holidays due to lack of snow caused by global warming; and Automobile Anonymous (AA) offers day and night support for people unable to give up their cars. A section on addiction includes an advert for a tobacco industry lobbyist:

> In this challenging post, you will ensure that we continue our legitimate, legal and economically vital business without interference from scientists, liberals or the public . . . You will be handsomely rewarded, at least in this life, though your children will loathe you and your grandchildren will tell their friends you died many years ago . . . We are an equal opportunities employer. Women and ethnic minorities may apply if they wish.
>
> (*Ignite*, 1997)

Features concern Premier Oil (UK)'s links to the Burmese junta, Greenpeace's occupation of Rockall and Blair's support for tobacco sponsorship in Formula 1 motor racing.

PLATFORM has moved since to more intimate investigations of political responsibility.[52] This began with *Carbon Generations*, a performance in which Marriott interweaves narratives of global warming denoted by Meteorological Office charts – the globe is coloured redder as the generations pass – and the story of his family over seven generations including his unborn children, from 1870 to 2080. Oil is used in motor cars owned by family members, in aircraft used in foreign holidays, and in central heating. Marriott recalls a tank for heating oil behind the house in Sussex in which he grew up, and the seductive smell of the wood on which it rested and dripped – passing a piece of it around the audience, with a school-book from the 1970s showing an early diagram of global warming. The links are precise: a member of Marriott's family was a colonial official in pre-1914 Nigeria; an aunt had one of the first cars in the Surrey town in which she lived; and Marriott and Trowell fly to Europe. The issue of climate change is relocated to their everyday lives. The intention is not to imply guilt but, just as the personal is political, to show the political to be comprised by personal responsibilities.[53]

But it is comprised also by a selectivity in public debate which the informational aspect of *90% Crude* seeks to upset.[54] *Some Common Concerns*, co-authored by Marriott with Greg Muttitt as part of *Unravelling the Carbon Web*, gives an account of the plan to build a pipeline from Azerbaijan through Georgia to Turkey. Pipeline stories do not make happy reading; neither does this: 'In 1999, Azerbaijan's labour law was . . . amended with a new article which states that workers may be fined or jailed for protesting' (Muttitt and Marriott, 2002: 150). The authors note that when something goes wrong like an ecological disaster or exposure of human rights abuses, BP finds individual scapegoats to avoid discussion of a systemic abuse of human and environmental rights for which responsibility is dispersed; meanwhile corporate power outsources even key functions in a decentring of its own power which 'enables the corporation to lay the blame . . . at the door of the sub-contractors' (Muttitt and Marriott, 2002: 181–2).

In *killing us softly*, this nebulous control, which has deathly outcomes, is investigated with an invited audience, while *Ignite* tended more to set a counter-agenda for a less (or un-) specified readership. In the elements of *90% Crude* as a whole, PLATFORM use contrasting ways to approach the complex webs of responsibility in which a globalised industry, governments, groups and individuals all have places: *RENUE* demonstrates low-impact technologies; like Chin's use of hyper-accumulating plants in *Revival Field*, or the wetland at Quaking Houses, this future is available to others to adopt. In the case of the wetland, as in the work at Nine Mile Run, there is a social engagement as well as a technology. Such projects state the power of local action to make a change, and this decentres power (while its own self-dispersal retains a fixed zone of membership). Once accomplished in one field, the idea may spread virus-like to others, even to democracy. Other projects invite a change of heart, may contribute to shifts in personal, group or public consciousness. In *Ignite*, dominant narratives are exposed as false by counter-narratives; in *killing us softly*, parallels are drawn between different situations in which a compartmentalisation that allows decisions that are literally murderous is encountered.[55] But in Lanzmann's *Shoah* it is the ordinariness of those interviewed which connects: the Holocaust was administered by those with power of decision but operated by bureaucrats of the petit-bourgeois class, workers, and its victims – people like the invisibles whose ash is mixed with that of the CEOs at Fresh Kills landfill (Chapter 7) – and I think this raises a question as to whether decision-making from a position of power is the most valid concern, or whether it is the culture which allows this to be acceptable which requires interrogation – a question the performance and discussion after also raise, no doubt. But I want to dwell on the ideas of demonstration and invitation, and narratives and documents, because it might be that the spaces in which people produce kinds of knowledge are being configured in different ways, and will produce different kinds of knowledge.

Andrew Jamison writes of four forms of cognitive practice, based on work by Bronislaw Szerszynski at the Centre for the Study of Environmental Change in Lancaster, UK:

> He distinguishes between purposive and principled action, the first aiming to change political decisions or achieve direct political results, the second concerned to modify values or behaviour. Szerszynski further distinguishes between counter-cultural and mainstream forms of syntax.
>
> (Jamison, 2001: 150)

This gives a monastic piety (as Szerszynski calls it), which characterises the principled and counter-cultural activity of groups living eco-lifestyles; a sectarian piety which is purposive and counter-cultural, as in activism; a churchly piety which is purposive but established, as with campaigning organisations such a Greenpeace; and the folk piety of principled action which is safe but less organised, as in green consumerism. The four are not exclusive, particularly in the trajectories of individual lives, and Jamison notes co-operation and collective identity as well as competition, but the type of cognitive praxis that characterises the varieties of environmentalism seems important to consider. PLATFORM is counter-cultural, but is it purposive or principled? Trowell says its core members are 'from the dominant class' but intent on exercising a responsibility to shift the values embedded in culture 'from the patriarchal, the imperial, the disdainful, the erasive and the extractive to the co-operative, the consensual, the vigorously debated, and the maintained' (Trowell, 2000: 108). This is principled. But the narration of *killing us softly* underlines, as well as the oil industry's violence, that transnational corporations are managed by people. Is the aim to change their values, or those of the graduates who are the next generation of executives, or to change the decisions of the corporations which are the quasi-governments of a globalised economy? It may be both; but does the centrality of Gretton's narration imply a purposive attitude, in contrast to the principled shift (in values among the invited audiences) which might be seen as the aim? Does he, too, deconstruct the power he critiques, or reproduce it? Does the silence followed by exchanges of feelings and ideas produce monastic piety? Or is it more sectarian, a space of withdrawal into a ritual of dissection of corporate mentalities, in which the narrator plays a central part in the middle of a semi-inverse panopticon?[56] I am not sure.

Trowell sees a power of listening as 'vital to the long-term success' of PLATFORM's work (Trowell, 2000: 103). I imagine that power of listening to be like the content of Paolo Freire's pedagogy of liberation. Freire worked initially through adult literacy classes in Brazil before the 1964 coup, enabling participants to tell their own stories, to name their own agendas, rather than have set texts:

> How can the oppressed, as divided, inauthentic beings, participate in developing the pedagogy of their own liberation? Only as they discover themselves to be 'hosts' of the oppressor can they continue to the midwifery of their liberating pedagogy. As long as they live in the duality

> where *to be is to be like*, and *to be like* is *to be like the oppressor*, this contribution is impossible.
>
> (Freire, 1972: 25)

Freire sees two stages of realisation: a change in the perception of oppression, then expulsion of its myths. The first is principled in Szerszynski's sense but leads to actions which change, not the decisions of the state, but the state itself. The differentiation of principled and purposive action then ceases to matter (as the state withers away in a classical Marxist account). Of course, revolutions tend to fail, and this draws attention to the separation of principled individuals from a state, or corporate quasi-state, the purposiveness of which is violently destructive of human and environmental rights.

NOTES

1 WCED define sustainable development as development that meets the needs of the present without compromising the ability of future generations to meet their own needs. Meadowcroft sees two key concepts: 'the concept of "needs", in particular the essential needs of the world's poor, to which priority should be given; and the idea of limitations imposed by the state of technology and social organisation in the environment's ability to meet present and future needs' (Meadowcroft, 1999: 13). See also the Independent Commission on International Development Issues, 1982: 'The industrialization of developing countries, as a means of their overall development efforts, will provide increasing opportunities for world trade and need not conflict with the long-term interests of developed countries' (p. 287).

2 Rangan, 1996: 215–17; Jarosz, 1996; Pradervand, 1989: 37–9, 138–9; Zimmerer, 1996; Reij, 1988.

3 Elliott cites statistics on the GDPs of states and transnational corporations: Exxon ($110 billion) and Shell ($109.8 billion) are larger economies than Norway ($109.6 billion) or Poland ($92.8 billion), and more than three times the size of Nigeria ($30.4 billion) (*New Internationalist*, 1997, no. 296, in Elliott, 1999: 29). See also Bauman, 1998, 55–76.

4 Papanek, 1984, 1995; and, for instance, the magazine *Eco Design*. See Illich, 1990 on productivity, discussed in Miles, 2000: 206–7.

5 'Ecological footprint analysis is an accounting tool that enables us to estimate the resource consumption and waste assimilation requirements of a defined human population or economy in terms of a corresponding productive land area' (Wackernagel and Rees, 1996: 9). Analysis includes extents of the import to a defined zone (such as a city of a given population) of energy and resources such as food, water, energy and raw materials; and export of unassimilated waste. See also Chambers, Simmons and Wackernagel, 2000.

6 'as with all aspects of sustainability, basic needs must be seen as an integrated whole. Tackling basic needs on a piecemeal basis . . . is a self-defeating policy. The provision of an improved primary health care system, for example, is futile if those being treated . . . are unable to meet the nutritional needs on which improved health depends . . . in practice the poor cope with their own needs as best they can on their own, with variable but diminishing assistance from the state, and minimal input from the private sector (Drakakis-Smith, 1990: 141–3).

7 Crush, 1995; Escobar, 1995; Ferguson, 1990, Rahnema and Bawtree, 1997; Sachs, 1992; and Tucker, 1997 (Worsley, 1999: 39).

8 Peet and Watts cite Blaikie (1985) as attempting to refine political economy in the ambit of political ecology: 'to make the causal connections between the logics and dynamics of capitalist growth and specific environmental

outcomes rigorous and explicit' (Peet and Watts, 1996: 9). They outline approaches seeking to integrate a discourse of political action in political ecology; undertaking analysis of social institutions; and dealing with pluralities of perception and definition of environmental resources. Calling for a critique of classical Marxism and citing Castoriadis (1991) on the self-production of society, they arrive at the term 'liberation ecology' to denote a field between Marxism, post-structuralism and environmental determinism: 'the intention is not simply to *add* politics to political ecology, but to raise the emancipatory potential of environmental idea and to engage directly with the larger landscape of debates over modernity, its institutions, and its knowledge' (Peet and Watts, 1996: 37).

9 A Global Forum for Alternatives met in Cairo in March 1997 to call for a reversal of global economic policy to serve the interests of people rather than capital; The Other Davos met in Zurich on January 28th–29th, 1999, with a press conference in Davos on January 30th – the Movement of Landless Farmworkers (Brazil); PICIS, a Korean trades union group; the National Federation of Farmworkers' Organizations (Burkino Faso); the Womens' Movement (Quebec); and the Movement of the Unemployed (France) were represented. It aimed to amplify protest against structural injustices; raise awareness of alternative scenarios; and set up networks of solidarity (Houtart, 2001: 79).

10 I use the term 'affluent' for the industrialised countries in which globalised capital tends to be located; and 'non-affluent' for the pre-industrial, ex-colonial, and impoverished countries often seen as cheap labour markets and sources of materials. In the case of, say, a gated residential compound in Delhi one world penetrates the other (Seabrook, 1996: 210–16).

11 See Johnson, 1999 on local resistances, citing Scott, 1985 on the idea of everyday resistance in non-privileged settings; Longo, 1998 on coalition-building on environmental issues; and Meikle, 2002 on new communications technologies and resistance.

12 Development Studies addresses the problems of non-affluent countries. The following references indicate the terrain in terms of urbanisation: Gilbert and Gugler, 1992; Beall, 1997; Fernandez and Varley, 1998; Hamm and Muttagi, 1998; Berg-Schlosser and Kersting, 2003. Kabeer, 1994 considers gender issues in development work; Hamdi, 1996 deals with education and training for development work with an emphasis on planning.

13 On diversity within environmentalism see Pepper, 1996. On social theory and environmental issues and processes, see Goldblatt, 1996; Barry, 1999. On ecologisms, see Hayward, 1998: 1–18; on deep ecology, see Naess, 1989. On radical ecology and the desire for nature, see Heller, 1999.

14 For Bookchin animals live in communities but not societies: 'they do not form those uniquely human contrivances we call institutions' (Bookchin, 1982: 357, in Light, 1998: 7).

15 See Barton, 2000 on eco-villages; and Schwartz and Schwartz, 1998 on a spectrum of alternative settlements.

16 Lee, 1995; McKay, 1996; Wall, 1999; Jordan and Lent, 1999; Jordan, 2002. Further protest occurred at the meeting of the World Bank and International Monetary Fund (IMF) in Washington in April 2000, in Prague and Melbourne the same year, in Quebec in April 2001, and in the violently suppressed demonstration in Genova in July 2001. There were festive days in Firenze in 2002 when global companies closed their outlets but local shop-keepers gave away food, café-owners allowed demonstrators to use wash-rooms, and guards at museums gave them free entrance.

17 An example of low-impact settlement design is Bed-Zed, designed by Bill Dunster and developed by the Peabody Trust and BioRegional (an environmental charity), in Sutton, Surrey, UK. Of the 82 units, a third are set aside as low-rent homes. Arranged in a series of terraced buildings, the development uses a combined heat and power plant fuelled by tree waste, which also fuels a fleet of electric cars for local journeys. Materials for buildings are locally sourced, where possible recycled. On co-housing, see Schwartz and Schwartz, 1998: 26–42; Sangregorio, 1998. For cases and methods of alternative building, see Pearson, 1989; Vale and Vale, 1991; Kennedy, Smith and Wanek, 2002.

18 Wigglesworth contrasts the engaged and supportive responses of these visitors, many of whom were not involved in architecture or related professions, to that of critics and practitioners within architectural fields. She observes that 9/10 Stock Orchard Street does not correspond with the category of a traditional vernacular; for ecological purists it 'does not use straw bales properly and upsets them as well'; neither is it a case of self-build – 'a vernacular of DIY – B&Q architecture' (conversation, September 4th, 2002).

19 Wigglesworth and Till, 1998b: 32.

20 'the house is expansive and embraces the exterior, the office is inward focused, a place for concentration' (Wigglesworth and Till, 2001: 18). Cf. 'the four walls of one's private property offer the only reliable hiding place from the common public world' (Arendt, 1958: 71).

21 Colomina, 1996: 283–336. See Chapter 3.

22 'hurdles are made to agricultural tolerance . . . steel frames manufactured in millimetres. Twenty drawings and countless telephone calls late, the two sensibilities were reconciled' (Wigglesworth and Till, 2001: 2). This level of involvement was characteristic, a team of six craftspeople producing most of the non-industrial elements.

23 'It costs less to bring a lorry load of recycled concrete to this site than to take away a lorry load of spoil. In the developed world the construction of buildings accounts for 50% of the consumption of raw materials, and the production of building materials accounts for 22% of manufacturing energy consumption. There is thus an imperative to find ways of building with materials that reduce environmental impact both in terms of toxicity and embodied energy' (Wigglesworth and Till, 2001: 4).

24 The cladding can be replaced with other coverings: 'a provisional architecture resisting the demands for eternity, fixity and progress' (Wigglesworth and Till, 2001: 6).

25 Wigglesworth and Till (2001: 9) cite the use of straw bales for building in Europe for three centuries, but state that 9/10 Stock Orchard Street is the first such UK dwelling to receive building control approval. The Straw was baled in the Cotswolds, 550 bales costing £825 delivered. See Passchich and Zimmerman, 2001: 53–70 on straw bale use in a demonstration eco-house near Albuquerque: 'In New Mexico, straw bale construction has become so trendy that it's almost a cliché' (p. 1). See also Daglish and Thepaut, 1993; Morgan, 2000. See also Jones, 2002; Magwood and Mark, 2000; Myrhrman and MacDonald, 1999.

26 Hill, 1998; Wigglesworth and Till, 1998a; Miles, 2000: 153–78.

27 Wigglesworth regrets the absence of interest in materials and on site architectural education (conversation, September 4th, 2002).

28 The Barefoot College Campus was given an Aga Khan Award for Architecture, 2001 for 'its integration of social, ecological, cultural and educational elements to aid rural development while promoting the architectural traditions of the region' (Aga Khan Award for Architecture, 2001: 78). It was founded by Bunker Roy as a place where urban professionals would immerse themselves in the actualities of rural life, in a joint venture with local people. It began as the Social Work Research Centre, later coming to be known as the Barefoot College after the Chinese experiment in rural health care of the 1960s, and today emphasises self-help for the rural poor, disregarding barriers of caste, education, and gender. Collective decision-making is used, and austerity taught in thought and action. The College has administration and training facilities, offering programmes on water, health, education of children, women's empowerment, rural industry and solar energy. A development plan for the surrounding area using appropriate technologies has produced several building schemes, of which the campus is the largest.

29 The domes vary in size: 3 metres for dispensaries, telephone exchange and other small facilities; 6 metres for class rooms and housing; 10 metres for a 100-person meeting hall (Aga Khan Award for Architecture, 2001: 79, illus. 82).

30 On Ghandi's influence on environmental debates and discourse, see Guha and Martinez-Alier, 1997: 153–68.

31 See Fathy, 1973, 1986; Steele, 1988, 1997; Miles, 2000: 105–28. Fathy, 1973 is a reprint of the original edition published by the Egyptian Ministry of Culture, 1969, 2nd edition 1989. See also Chapter 3.

32 'the work of an architect who designs, say, an apartment house in the poor quarters of Cairo for some stingy speculator, in which he incorporates various features of modern design copied from fashionable European work, will filter down, over a period of years, through the cheap suburbs and into the village, where it will slowly poison the genuine tradition' (Fathy, 1973: 21).

33 'The final document (typically 250–300 pages long) with its ten annexes will include project justification and development objectives, project inputs and outputs, risk considerations, and exit strategies, deadlines and reporting schedules. It will be primarily concerned with ensuring proper accountability amongst the consultants' (Hamdi and Goethert, 1998: 26).

34 Segal, 1980: 174. See also Gilbert and Gugler, 1992: 114–54.

35 Umenyilora, 2000.

36 For example, Armor and Snell, 1999 (16th edition); pp. 265–77 deal with community self-build. The authors note that in the UK housing market crash of 1988 and rise in interests rates to 15 per cent, more than 70 self-build groups failed to meet loan repayments. The first group self-build was in Brighton in 1948, by ex-servicemen on land provided by the local authority (p. 266).

37 The wetland and the radio station were part of *Visions of Utopia*, co-ordinated by Artists Agency and involving over 40 local projects in the northern region. I visited the wetland a number of times and interviewed Younger and Smith, and Lucy Milton and Esther Salamon from Artists Agency. I am unsure what to think about the conflicts; the arts funding system drives arts organisations to constantly seek new projects, so that any social or environmental project can be seen as 'needing an artist' and the artists and arts managers are necessarily peripatetic. A second project at Skinningrove in Cleveland led to a situation in which the artist, Jean Grant, was supported by some local people but the art project became secondary to the needs of flood prevention after a catastrophic flood in the first months of the project in 2000. See Miles, 2002.

38 See Camacho, 1998 for a range of cases; Callicott, 1994 on ecological awarenesses in different cultures; and Carmen, 1996 on empowerment in relation to environmental questions.

39 Simony, Brott and Prior, 1998: 14.

40 Plans for a mall in 1982 met with opposition. In 1995 the city acquired the site, appointing a developer and setting up an Urban Redevelopment Authority, a private–public partnership.

41 For explanation of restoration ecology and its relation to histories of landscape conservation and preservation in North America, see Collins and Goto, 2003: 134–8. They state the democratic potential of restoration ecology as working in three stages: setting a goal which reflects local aesthetics; action which involves local people bodily; and a monitoring period in which citizens gauge effectiveness (p. 136).

42 At the time of writing the project is in progress, and will include a conference bringing artists, critics and the city's cultural industries together to reconsider the relation of art and ecology, as well as charettes. The goal is 'to conduct an analysis of the **green infrastructure** which provides social, aesthetic, ecological and economic benefit to the Three Rivers Region. Green infrastructure, when identified and integrated into an ongoing program of urban redevelopment, can provide significant multi-benefit returns on investment. The program will complement . . . efforts to implement innovative technical and institutional solutions to "grey" infrastructure problems (stormwater and sewer systems) and the wet weather discharges which soil our rivers. Combined, these grey and green programs will reawaken the public interest in the natural benefits which sustain, define and complement life in the cities of the Three Rivers Region' (unpublished project document).

43 Hayward emphasises consensual recognition: 'political theorists have to recognize that relations with non-humans can in principle, and should, be normatively regulated. In practice this is only likely to happen

by being harnessed to and picked up by (human) interest groups. But what can be done is to allow such groups particular right: to alter the burden of proof and rule of standing in relation to animal rights cases . . . to favour and promote institutions aimed at inculcating "care" for nonhumans' (Hayward, 1998: 160).

44 'The US public relations firm, Mongoven, Biscoe and Duchin . . . divides opponents into four categories: "opportunists", "idealists", "realists", and "radicals" . . . The strategy is to isolate the radicals, cultivate and educate the idealists into becoming realists, and co-opt the realists into agreeing what industry had already decided' (Cornerhouse, 1998: 5).

45 PLATFORM is an artist-led group, and has maintained a small size despite encouragement to enlarge. It has also retained a link to London, where it has a project base near Tower Bridge. Marriott and Gretton come from upper-middle and middle-class backgrounds (respectively) in the home counties, Trowell from a mining family. Trowell is an art educationalist and historian; Marriott studied history and then sculpture in London and was assistant to Suzi Gablik in the late 1980s; Gretton studied literature and was a founder of Cambridge Student Campaign for Nuclear Disarmament. A fourth member, John Jordan, left in 1995 to work with Reclaim the Streets. Marriott and Gretton met through 'a shared desire to fuse the campaigning aspect of political activism with the imagination of art' (annotation to draft of an earlier text), and in 1987 went to Germany to meet followers of Joseph Beuys. Marriott refers to Beuys as revealing 'an invisible boundary between art which decorates a political process and art which *is* a political process (conversation, March 18th, 1999); and sees Beuys' concept of invisible sculpture as seminal: 'thinking *as* sculpture, in itself' (conversation, July 5th, 2000), as well as his concept of direct democracy.

46 The Fleet (buried in the eighteenth century); the Walbrook (buried in the fifteenth century); the Effra (buried in the nineteenth century); the Wandle (above surface): 'At best, London's rivers have become toxic, rubbish-filled streams, robbed of most plant and animal life. At worst, they've disappeared under roads, reduced to pipes carrying human waste. Like blocked arteries in a human body, the destruction of London's rivers has caused great damage to the Thames valley ecosystem. . . . Platform hopes to inspire discussion about London's hidden rivers, and begin a process whereby the resurrection and re-enchantment of London's lost rivers becomes an inevitability' (PLATFORM, leaflet, 1992).

47 This grew into a larger project, *RENUE* (Renewable Energy in the Urban Environment), with Millennium Commission funding, linked to the London Borough of Merton. Trowell has reservations on the Effra Development Agency, which proclaimed the imminent conversion of the river to public recreation, seeing *Ignite* as having a clearer relation to its content and difference from its model (conversation, July 5th, 2000). *London Cyclist* reported 'The public was led to believe windsurfing down Brixton Road, fishing by the Oval or paddling through West Norwood could soon become a possibility. In the midst of lively public meetings and excited coverage in the press, the Effra Development Agency silently disappeared' (August/September 1997, p. 25).

48 Lacy, 1995: 210–11. See also *Agricultural Research*, November 1995, p. 9.

49 Conversation, June 20th, 1999, Aachen.

50 The first issue of *Ignite* (10th December, 1996) was edited by Cindy Baxter, ex-Press Officer of Greenpeace, and the second (November/December, 1997) co-edited by her with Emma McFarland, PLATFORM's administrator. Fifteen thousand copies of each were printed and distributed; those not handed out at stations were given to environmental campaigning organisations as educational material.

51 '300,000 Ogoni peacefully protested against Shell's operations . . . 2000 have been butchered and countless others raped and tortured by the Nigerian military. In November 1995, Ogoni leader Ken Saro-Wiwa was executed, framed by the Nigerian authorities. While Shell denied any complicity with the Nigerian regime, it has since admitted paying the Nigerian military "field allowances" on occasion' (*Ignite*, 1996: 18). See also Midnight Notes Collective, 1992: 87–90; 91–106; Okome, 2000.

52 Between the two editions of *Ignite*, PLATFORM contributed ideas to the production of *Evading Standards* (April 11th, 1997, published by Reclaim the Streets – see Chapter 9). Taking the masthead design from the *Evening Standard*, the stock was to be distributed before an event organised by Reclaim the Streets in support of striking Liverpool dock workers, in the run-up to the general election; but confiscated by the Metropolitan Police and returned without charge three weeks later. The headline states 'GENERAL ELECTION CANCELLED' after a meeting between party leaders at which it was agreed that voter cynicism made postponement of the election unavoidable. A photograph shows City dealers in panic as share prices tumble; there is a lottery for a dream home (10 Downing Street, London), and an offer of national leadership from the Spice Girls (details p. 9). *Evading Standards* has eight pages. It seems oddly prophetic but mistimed – voters were only too keen in 1997 to oust the old regime. Blair is shown handing out burgers at MacDonalds, which might today be credible, as might the headline in 2005 or 2006.

53 On visits to North America and Azerbaijan in 2001 and 2003, Marriott travelled by boat and train, respectively. On global warming, see Christianson, 1999.

54 'One problem is that power analyzed solely in terms of individual decision-making fails to capture those aspects which lie outside observable decision-making processes. A broader view of power would focus not only on the enactment of decisions, but also on exclusion of certain issues from the decision-making agenda . . . Power in this view no longer rests only in the ability of some actors to initiate, decide and veto decisions, but also in their ability to confine decision-making to "safe" issues' (Kabeer, 1994: 225).

55 I question how far the comparison can go, if the Third Reich was an aberration in capitalism, a perverse chiliasm: 'Capitalism had no other choice than that which it has excellently made with fascism up till now; yet it would certainly prefer old liberalism to romantic "anti-capitalism" (without which business could admittedly no longer be done in Germany). The blood myth, and intoxication as a whole, is not the most desirable servant of capitalist *reason*' (Bloch, 1991: 55).

56 I say semi-inverse panopticon because the participants retain what is perhaps the key aspect of mutual isolation in adjacent 'cells', but instead of being observed by a central watcher, they watch a central narrator. After several conversations on this with Gretton and Marriott, I still have reservations as to whether the means are appropriate to the aim, or whether Gretton's central role is too much that of interpreter. Two models come to mind: first, Dr Nicolaas Tulp in Rembrandt's *The Anatomy Lesson* (1632, den Haag, Mauritshuis), presiding over the dissection of Aris Kindt (Barker, 1984: 73). Tulp has power of interpretation as well as office, which contrasts with a power of listening (see Trowell, 2000: 103); second, Courbet painting himself at the centre of his representation of the Fourierist *phalanstere* in *The Studio* (see Chapter 1).

9

2001 (II)

COSMOPOLIS

•

I frame this chapter by reflecting on the attacks on the Pentagon and the World Trade Centre, New York on September 11th, 2001; and more specifically on their impact on the terms of engagement with power. I begin by asking whether there are continuities in the emergence of a security state, as well as a rupture in the codes of contestation; and end by asking whether there is a possibility for a new understanding of cosmopolis – despite a prevailing mood of crisis management designed to inhibit such aspirations. In a context of contingency and recoding, and keeping in mind the axis introduced in the previous chapter between work which aims to transform the values of those who take part in it, and that which seeks to impact public policy, I look at cases of activism; then at guerrilla and squatter gardening, and uses of cyberspace and robotics for subversive ends. Although the technologies range from the most basic to the most advanced, the cases share an intention to infiltrate the spatial production of the dominant society, so that growing vegetables or launching spoof advertising on the World Wide Web are both forms of subversion. In the chapter's second section I return to the site of cultural dissemination, the art gallery or museum, to consider the work of Slovenian artist Marjetica Potrč. This work, however, brings into the spaces of elite culture the images and material reconstructions of modes of dwelling encountered in marginalised situations, such as the squatter camp. By juxtaposing survival architecture in the non-affluent world to the paraphernalia of a survival mind-set in the affluent society Potrč draws attention to the brittle quality of boundaries.

I begin, though, at 9–11. The following is from Vincent Cornell's 'A Muslim to Muslims: Reflections after September 11':

> Shortly after September 11 Christiane Amanpour of CNN interviewed apologists for Osama bin Laden at an exclusive secondary school in Pakistan. For the children of the elite, geopolitics followed the plot of a Marvel Comics book. Bin Laden was tough, Bin Laden was cool. Bin Laden gave

> the West what it deserved. He struck blows against the Empire in the name of all who resist Zionism and Western imperialism. But what did the west do to harm these young cricket-players? Without the social and economic changes wrought by colonialism and postcolonial imperialism, their parents could never have earned the tuition that allows them to indulge in fantasies of empires lost and then redeemed. The west they claim to despise even wrote the curriculum of the school in which they study. In terms of core values, they have more in common with Londoners than with their own people. It is ironic that the majority of Pakistanis do not share Bin Laden's views. In the face of grinding poverty, they still refuse to trade their common sense for a false ideology.
>
> (Cornell, 2002: 329–30)

I GRASSROOTS GLOBALISATION

Early on the morning of September 12th, I heard a report on the BBC World Service of a crowd in Islamabad carrying a banner saying (in American) 'Think America, why are you so hated around the world?'. Going home the previous evening I had joined a crowd watching identical replays of the Twin Towers falling down on multiple screens in the window of a television shop. Live news was framed as history and coded as apocalypse.[1] The war was scripted, if not in detail, and perhaps had been for some time.

Now the Pentagon has been repaired and the debris cleared from Ground Zero – a site worth too much money to leave vacant. A new high-rise structure more elegant than the Twin Towers will reassert New York's status as a world city, and will no doubt be engineered to withstand terrorist attacks. The Twin Towers *were* designed to survive the impact of an aircraft, as it happens, but a 737 not a 757 with full tanks. The financial services industry, too, had arrangements for contingency, and re-connected within hours to its information super-highways using back-up sites. Was 9–11 a watershed, then, or not? Noam Chomsky observes that it was the first time since 1812 that the national territory of the United States has been attacked;[2] and Susan Buck-Morss writes that September 11th 'ruptured irrevocably' the context in which public intellectuals speak (Buck-Morss, 2002: 2). Yet she points out that history does not have clean breaks, that after the end of the Cold War and through the electronic media revolution the signifiers of Enlightenment modernity remained in place to state a self-evident, rational freedom. Both the dominant powers and struggles for national liberation, armed or not, tended to adopt these values. Campaigns for social and environmental justice tend similarly to appeal to a perception of irrationality in global capital's hegemony, implicitly assuming a universal rationality. Terrorists who hijacked aircraft and took hostages before 9–11 generally sought to trade a meeting of demands for their release in a transactional terrorism which assumed a code of sorts.

9.1 • Graffiti, Barcelona, 2003

The new struggle for power is carried out by a super-power which constructs demands it knows to be unanswerable, and by terrorists who do not make demands at all. On September 11th there was no thought that either hijackers or victims would survive, hence no code for negotiation:

> It was not only, or even primarily, to Americans that the act was addressed. Indeed to Americans the aim was less to communicate than to explode understanding, a weapon of sabotage with devastating effect because . . . to receive the communication had the consequence of destroying the code.
>
> (Buck-Morss, 2002: 3)

Perhaps it was not terror, though, which tends to be associated with marginalised groups, but intended to look like war, the prerogative of states; perhaps a strike of such magnitude and scale of casualties 'could not be mistaken as being other than an act of war . . . couldn't be responded to within the area of policing alone' (Gupta, 2002: 52). And perhaps the absence of a national identity is now no inhibition to such a declaration of hostilities.

Suman Gupta sees the choice of targets in New York and Washington as a claim on the part of the attackers to equivalent status to that of the superpower, making the specifics irrelevant.[3] He notes the attackers' blurred identity: 'since the enemy wasn't going to declare themselves and thereby become obvious, the enemy had to be retrospectively constructed' (Gupta, 2002: 53), but sees this blurring as produced in part by the global news media.[4] The new enemy can thus be imprinted in various ways, and meets the security state's need to fill the vacuum left by the collapse of the Soviet Union as prime target. It also conveniently limits the scope for dissent at home and legitimates increased military expenditure and the undertaking of overseas adventures. Other targets than Afghanistan and Iraq will be (have been) named as threats to national security. The list is written, and the ends are predictable.[5]

For Buck-Morss, the war on terror supports a state within the state: a 'national security state' called into existence in a state of emergency, 'a wild zone of power, barbaric and violent, operating without democratic oversight' (Buck-Morss, 2002: 6).[6] If September 11th marks a watershed in terms of a new kind of confrontation, there is also a continuity between the wild zone of power and the military-industrial complex of the Cold War. What is new in this fusion of the agendas of capital and armed force is the excess, the degree of overt violence the new security state is prepared to use, and the shift (as from liberalism to neo-liberalism) from a pretence of allegiance to Enlightenment values to the construction of a permanent sense of emergency which gives *carte blanche*, in which all options are permanently open.

The state of permanent crisis management is not entirely new, however, and extends practices in business management to that of public life. Jérôme Bindé argues that globalisation privileges time as just-in-time over a sense of time unfolding in reflection:

> The keener sense of emergency stems from both the primacy of real time and the absence of any reference to a collective aim. What has to be done, therefore, is to invert the logic of emergency . . . it is not the emergency of problems which prevents the formulation of long-term plans but the absence of any plan that subjects us to the tyranny of emergency.
>
> (Bindé, 2001: 91)

Permanent emergency has no beginning, middle, or end – hence no clean slate – but requires permanently increasing vigilance. It is *Waiting for Godot* speeded up as public life, and has to be so because total uncertainty is the only condition in which the means used by capital to ensure its hegemony will be tolerated. At the same time, certain contradictions are subsumed in the emergency state: for instance that technologies regarded in modernity as value-free become overtly ideological; and that the project of free trade – a global free-market economy to produce maximum profit for capital – operates through a mix of deregulation of overseas markets and protection as required for the industries of the super-power.

The wild zone of power was prepared, too, in popular culture. It reconstitutes another wild west in the wake of fantasies of New York's destruction,[7] and obliquely reflects the anti-urban (to my mind anti-social) implications of counter-cultural retreats to wilderness.[8] And as the myths return, so the money system operates in the usual uneven-handed way in the rules for compensation after 9–11:[9] business as usual; shopping.[10]

Carried forward as well is a practice of recoding. But if in the French Revolution time started again in a new calender while space was produced anew in rational units of measurement,[11] now recoding is a stock in trade of global brands.[12] In the terms of the super-power it becomes a blunt instrument, requiring an absolute polarisation – for or against with no transitional or trans-active zones, no interstices in which to hide, no grey areas in which to negotiate. Hence anti-globalisation protestors are reclassified as terrorists,[13] and in Genova in 2001 subjected to the tactics of the fascist squads of the 1920s.[14] In a less violent but equally controlling way, in New York and London in 2002, their movements were choreographed by the police to isolate them from the city's publics.[15] Yet in exposing the extent of violence the state will use to protect the interests of market economics, demonstrators test the limits of public acceptability in a system which rests nominally on elections.[16] Protest must therefore find imaginative ways to communicate.

In most of the global news media, unsurprisingly, anti-roads and anti-globalisation protest has received mainly negative coverage. The experience of such exposure as has been achieved is in any case of misrepresentation During the main period of anti-roads protest in the UK in the 1990s,[17] for example, when tactics such as living in trees marked for felling and digging labyrinthine tunnels under road construction sites attracted wide coverage, Daniel Hooper, known as Swampy (a tunneller at an anti-roads camp in east Devon), briefly became a folk hero: 'The media celebrated roads protest and crystallized the actions of thousands of grassroot campaigners into the form of a single personality. Swampy became an icon, better known, at least briefly, than many TV presenters and cabinet ministers' (Wall, 1999: 91). He was offered a recording contract. Media attention brought anti-roads protest to a mass public, but the protestors were presented as unique individuals despite their own refusal of a cult of personality: 'direct action was transformed from the "real" to a media spectacle, something for heroes rather than adults and children in local communities' (Wall, 1999: 92). Was (or is) publicity the aim of anti-roads (or

anti-globalisation) protest anyway? Taking Szerszynski's counter-cultural categories of a monastic piety of groups whose own values are transformed through par-ticipation, and a sectarian piety of efforts to change the policies of those in power, the anti-roads movement seems to cross the categories. In one way, the quasi-tribal identities of protest groups such as the Dongas indicates an evolution of alternative values within the alternative social formation; in another, the explicit aim is to stop the building of roads, or to limit the reach of capital, which are policy matters.[18]

I want now to reconsider this axis (which I see as a creative tension to be observed and not a polarisation to be resolved) in terms of direct action and the work of Reclaim the Streets (RTS), who became known during anti-roads protest in London in the mid-1990s. In 1995, for instance, RTS spread sand on the road outside Goodge Street station in London, set out deck chairs, and held a beach party to stop the traffic. In 1996, 7,000 people joined them at the M41 near London; under the voluminous skirts of carnivalesque dancers on stilts, activists dug up the tarmac with road drills and planted trees. On June 18th, 1999, a day of activist carnival in London and other cities worldwide, 10,000 people joined street parties in London's financial district, or produced their own resistance.[19] At first, direct action seems aligned with interruption of the dominant society's routines, often in symbolic gestures that appeal to the humour and imagination of diverse groups of people and might be expected to widen a base of popular support for alternative policies or anti-capitalist practices of consumption. But it may equally, or more, be a means through which the values of participants evolve into a new social consciousness – in effect the new consciousness foreseen by Marcuse (see Chapter 4) if by other means. First impressions may in this case be misleading: if the pattern of activity in direct action is sporadic, while participation in anti-roads camps is continuous, direct action may still create formations which are no less coherent in their common interest (but are spatially dispersed, defined by the day of action not the site of dwelling) than those of more evidently close-knit groups. The solidarity of direct action may extend through networks into which participants come and go at different levels of commitment, but this dis- (or de-) organisation is not a state of disorganisation.[20]

The case for a monastic piety, so to speak, in direct action is enhanced by a leaflet distributed by RTS in 1996, in which direct action is claimed as a process of participatory empowerment:

> London RTS uses direct action. This is not, as many commentators would suggest, a clever technique to gain media exposure . . . Direct action is about perceiving reality, and taking concrete action to change it yourself. It is about working collectively to sort out our own problems, doing what we thoughtfully think is the right course of action . . . It is about pushing back the boundaries of possibility, about inspiration, about empowerment. It is about thinking and taking, not asking and begging.
>
> (cited in Jordan, 2002: 62)[21]

This raises at least three complex issues. First, it may be that direct action is not primarily aimed at changing policy (though it may open public debate, particularly when there is an over-reaction by the authorities to its playful tactics); it may be that the ephemeral site of direct action is instead a social formation co-present with (while rejecting the values of) the dominant society; and that the key aim is to build an alternative value structure among participants, so that the means is the end and what is seeded is likely to be long-lasting. Second, the model of disorganisation is not exclusive to direct action but found in a systematic way in the operations of large corporations which out-source functions to become power networks rather than power centres; hegemony and resistance thus mirror each other's tactics within a common condition. Third, actors in any situation act as it allows. Solidarity and imagination extend a capacity to act, but it remains bounded; yet is bounded, too, for those in power if they wish to retain a semblance of acceptability.[22]

As the war on terror unfolds and mass publics are increasingly disenfranchised from conventional politics, a new political form may emerge informed by (if not fully adopting) direct action and single-issue campaigning in a reassertion of autonomy. But it will not be a rehash of the individual subject's autonomy in modernity, but in terms of group autonomies – to which I return in the final paragraphs of the chapter.

I move now to a range of cases of how such autonomies are applied in practices on the edges, or within the crevices, of mainstream society, but which construct differing relations with power and its bureaucracy. These practices include guerrilla and squatter gardening, and potentially subversive uses of new technologies of communication.

Guerrilla gardening developed in the UK during the late 1990s from reclaiming the streets. In 2000, informal gardens appeared in London, Bristol, Brighton, Bradford, Manchester and Sheffield. Ben, a guerrilla gardener in Bristol, says: 'People go out and plant stuff where there are no trees . . . Willows are simple to do on a large scale: you push willow sticks into the ground and you get a 40% success rate. Some people just drift through, plant and then move on' (Jones, 2000). For RTS, it 'is not just green; it is also symbolic of taking back the land under people's feet' (in Jones, 2000). On May Day, 2000 RTS initiated guerrilla gardening in London and other UK cities under the name Resistance is Fertile, for instance digging up and replanting a grassed area near the Houses of Parliament. A similar action had taken place in Washington, D.C. during the International Monetary Fund meeting in April 2000 – a leaflet stated: 'We must do it creatively enough that they cannot credibly label us as terrorists, and militantly enough that we cannot be ignored' (Jones, 2000).

There are several ways of taking back the land. While the event on May Day 2000 enabled members of various networks to reconnect and make a symbolic and carnivalesque gesture, there was no attempt at long-term occupation. The Land is Ours, in contrast, reclaimed a redundant brewery site in Wandsworth on May 1st, 1996 as an incipient eco-village.[23] The site became known as Pure Genius from a sign left by the brewer, but raises some awkward questions. The occupation could be seen as a squat following many previous occupations of buildings by homeless people in cities such as London and New York, except that the shelters here were built from material salvaged, donated, or found in skips, and many of the initial occupiers were anti-roads protestors or new age travellers. Yet it was not a protest camp either, its open access attracting street-dwellers as well as campaigners, and in the later stages a significant proportion of people with mental health and other problems, the abject of Thatcher's Britain who need more than an empty site to support recovery.[24]

To me (but I was not there) there seems a confusion between one approach which is about making colourful or spectacular gestures likely to achieve wide media coverage, and another, which is closer to the aims of alternative or intentional communities, such as eco-villages, but without a long-term viability; and a third which is more opportunist or survivalist, a making-do on a site which happens to be there, for a while. It remains interesting, though, that about six weeks before bailiffs cleared the site on October 15th, George Monbiot wrote in a national newspaper, 'What Pure Genius achieved does not end on this piece of blasted land. All over Britain, communities are waking up to the fact that neither government nor conventional developers are going to help them' (*Guardian*, September 4th, 1996, cited in Schwartz and Schwartz, 1998: 64). Monbiot adds whimsically that 'While the Government watches the dead leaves blown across the water, beneath the surface Britain has begun to flow the other way' (ibid.). But has it? As one of the occupiers pointed out, 'You can't build an ecological house . . . for less than £20,000 downpayment . . . For the working class, ecology is a luxury' (Dave, in Schwartz and Schwartz, 1998: 64). It could also be objected that only a small minority of people will resort to direct action, while the stages in which its imaginative and symbolic gestures are translated into more everyday aspects of a culture of empowerment outside such events – in the wider society – remain unspecified. It might be that spectacular events add to a vocabulary of moments building towards a new consciousness in which things that seemed impossible become viable aspirations. Equally, however, the clearance of a site of major commercial development potential was probably inevitable (and foreseen by the initiators). Perhaps crystallising attitudes of sympathy or rejection, this is likely to be a disincentive to future such action, or to action on a mass scale as implied in a counter-current – unless it is so romantic as to be an underground. In this respect The Land Is Ours has little if anything in common with a longer history of squatter (or avant-) gardening in New York's Lower East Side.

The squatter gardens which began in Lower Manhattan in the 1970s were semi-permanent, and produced not by in-comers but by dwellers in adjacent tenement blocks. There was mobilisation, but on a local basis. Sarah Ferguson compares these squatter gardens to community gardens cultivated under US federal welfare schemes in the 1930s, but argues that those of the post-war period reflect official neglect rather than support.[25] The move to green abandoned plots began when artist Liz Christy, with a group of activists, or Green Guerrillas, reclaimed a garden on Elizabeth Street in 1973, which was immediately paved by the authorities. Another garden was made on Bowery and Houston 'where a few months earlier a couple of homeless men had been found frozen to death in a cardboard box' (Ferguson, 1999b: 83). This attracted media attention and public support, leading to a lease from the authorities in 1974. A federal urban garden project was set up in Brooklyn in 1976. In 1977 Operation Green Thumb leased plots to squatter gardeners for $1 a year, but on a temporary basis that retained control for the City bureaucracy, and only to those able to deal with the application procedure. That aside, the activity of gardening vacant lots became part of the cultural identity of the Lower East Side, and part of its self-produced social cohesion. Superficially, this experience supports an argument that public policy, at least locally, is responsive to the tactics of dwellers. But in the late 1980s gentrification and the violent eviction of homeless people from Tompkins Square Park[26] changed the situation radically. Community gardens were seen by developers as merely vacant lots ripe for redevelopment. Ferguson cites a case of summary eviction from licensed gardens on Avenue B in 1997 to enable speculative redevelopment.[27] Despite a previous accommodation of squatter gardening, then, once the economic agenda moves, it seems the political agenda follows, reflecting an uneven distribution of power that was legitimated by the licences.

Perhaps radical democracy remains a dream. In '¡Viva Loisada Libre!', Bill Weinberg conjures a Lower East Side Autonomous Zone: a moratorium is declared on new commercial enterprises whose prices or aesthetics offend the working-class grain of Loisada; and yuppie taxation funds tenement repairs:

> We have a vision that one day in the near future, the residents of the Lower East Side will start to meet and talk with each other in our tenements, on our blocks, in our gardens; Puerto Ricans and Dominicans and Central Americans, Poles and Ukrainians and Slovaks, Bengalis and Chinese and Korean, Blacks and Jews and Italians, punkers and hipsters and homeboys; artists and activists and squatters; and decide to find our common interest in reclaiming our neighbourhood from the occupying forces of speculators, developer, landlords, organized crime, police and automobiles.
>
> (Weinberg, 1999: 38)

The yuppies, meanwhile in reality, have become a new bohemian class in post-industrial cities. Unlike the bohemians of the nineteenth century they are mainstream not counter-culture, and are attracted to the frisson of living in proximity to multiple deprivation.[28] The outcome, with many other factors at work, is a new layering of the city no longer delineated as discrete zones and quarters, but in which enclaved excesses of power, wealth and consumption are juxtaposed with residual abjection.[29]

Autonomy is still an enduring term, however, as Christiania in Copenhagen demonstrates. Initially squatted in the 1970s when it was a disused army barracks, it now constitutes an enduring free zone semi-integrated in the city, if today it is also a tourist destination appealing to an exoticism of alternative living in the view of the mainstream society. Other departures from that mainstream, which have quietly grown, include co-housing, first developed in Denmark and the Netherlands,[30] and efforts to construct alternative economies. A case of the latter is Solidair in Utrecht, a network of organisations, housing cooperatives and individuals who build grassroots and human-scale collaboration for mutual benefit outside the main economy.[31] There is also the network of Local Economy Trading Schemes (LETS) first established in Canada and now widespread in the UK and Australia among other countries.[32] And in Amsterdam a subculture of alternative living based in part on houseboat dwelling – as an alternative to the city's expensive land housing – has evolved since the late 1960s.

The above, taken together despite the differences between each form of departure from the main society, may constitute a new society within, or alongside, the old. Individual projects and networks have differing degrees of integration with or separation from the main society and economy, but the effect is to suggest that the arena for change is no longer the public spaces of a city, but, in one way, the domestic and work spaces of citizens who co-operate for mutual benefit; and in another the ephemeral networks of activism. Both seek to build new systems of value within new social formations, contesting but generally not in open confrontation with the dominant society.

There is, too, an arena not in geographical space but in the spaces of new technologies of communication developed by global capitalism, which are open to infiltration. One term to describe this new resistant cultural production is tactical media:

> Tactical media are media of crisis, criticism, and opposition. This is both the source of their power . . . and also their limitation. Their typical heroes are: the activist, Nomadic media warriors, the prankster, the hacker, the street rapper, the camcorder kamikaze. They are the happy negative, always in search of an enemy. But once the enemy has been named and vanquished, it is the tactical practitioner whose turn it is to fall into crisis.
>
> (Garcia and Lovink, 2001: 90)

David Garcia and Geert Lovink note attacks by the right on political correctness, but also that concepts from the left, such as identity politics, become dated. This gives rise to a totality of contingency, of formations which emerge and fade within other equally contingent formations, and to an aesthetic of 'poaching, tricking, reading, speaking, strolling, shopping, desiring', which is articulated in 'polymorphic situations, joyful discoveries, poetic as well as warlike' (Garcia and Lovink, 2001: 91). As well as hacking, the production of spoof advertising,[33] and anti-brand campaigns,[34] tactics include spoof web sites and corporate structures; etoy.CORPORATION, for instance, describes itself as 'a corporate structure officially incorporated 1994 in Zurich. Etoy is a typical early mover (on-line since 1994) and developed rapidly into a controversial market leader in the field of experimental internet entertainment and art.[35] The group's core agents appear at symposia dressed as corporate executives, with company ties and dark blazers, and claim to be serious business-people, not artists. The corporation's parodic work is mainly concerned with mapping its own decentred structure, and the apparatus of a company that makes only its own reputation. Peter Hill, based in Melbourne, has for some years maintained a spoof contemporary art museum, the Museum of Contemporary Ideas, New York.[36] Hill began by drafting press releases on fictional art groups and sending them to the art press, sometimes finding that an item would be taken up and given a paragraph. From this he developed performative fictions, *The Artfair Murders*, and the Museum. The outcome is a critical-ironic reflection on structures specific to a post-modern art world but which reflect nonetheless those of the contingent and value-fugitive (rather than value-free) world of post-industrial urban economies and their global culture and media industries.

Hill and etoy mix web-based and performative means. In contrast, Critical Art Ensemble, a cyber-artists' and theorists' group based in various US universities, sees the arena of electronic communications as replacing that of conventional contestation. This is more than a shift to new media; it is also recognition that old categories of struggle cease to hold currency:

> In the postmodern period of nomadic power, labor and occupation movements have not been relegated to the historical scrap heap, but neither have they continued to exercise the potency that they once did. Elite power, having rid itself of its national and urban bases to wander in absence on the electronic pathway, can no longer be disrupted by strategies predicated upon the contestation of sedentary forces. The architectural monuments of power are hollow and empty, and function now only as bunkers for the complicit and those who acquiesce . . . As with all monumental architecture, they silence resistance and resentments by the signs of resolution, continuity, commodification, and nostalgia. These places can be occupied, but to do so will not disrupt the nomadic flow . . . The electronic valuables inside the bunker, of course, cannot be taken by physical measures.
>
> (Critical Art Ensemble, 1994: 23)

Hence the means of cultural resistance must extend to the electronic: 'Just as authority located in the street was once met by demonstrations and barricades, the authority that locates itself in the electronic field must be met with electronic resistance' (Critical Art Ensemble, 1994: 24). This, however, is also a field in which the structures of subject and autonomy are refigured, beyond both a subordination to the state, which is the target of the political right, and a subordination to consumption, which is attacked from the left; and in a state of information overload.[37] This raises several dilemmas, among them a claim to privacy in face of data overload which goes against a grain of decentring the subject; and a suspicion of technologies of data gathering alongside a potential re-utilisation of systems.

Finally in this section, I note two groups which demonstrate a potential to critically re-use robotic technologies: the Bureau of Inverse Technology coordinated by Natalie Jeremijenko; and the Institute of Applied Autonomy (anonymous).[38] In both cases the level of technology used is accessible and affordable. The Bureau of Inverse Technologies has piloted, for example, a robotic pet in the shape of a dog for use by community groups, which can among other versions be fitted with equipment for detection of toxic waste in land-fill sites (where it would be dangerous for humans to go, and where data is not freely available from polluters). The Institute for Applied Autonomy (IAA) has produced a robotic leaflet dispenser, its caricature of a mechanical voice cutely inviting passers-by to take literature from its metallic pouch; and a robotic graffiti-writer based on the technologies of a remote-control model car and a dot-matrix printer (adapted for industrial spray-cans as used for road markings), with encoder and micro-controller. The graffiti-writer prints a dot-matrix text on the street and moves at up to 15 kph.[39]

II THE PURSUIT OF HAPPINESS

I turn now to the work of Marjetica Potrč. This includes gallery installations and large, full-colour ink-jet prints, and a web-based work made in collaboration with Aisling O'Beirn in Belfast, *Virtual Urban*.[40] Potrč was trained as an architect,[41] and much of her work re-presents the marginal territories of informal settlements, or the lives of those human and non-human creatures who live in the cracks of the dominant society. Francesco Bonami sees it as 'an anthropological urbanism that tracks not only the failure of the modern planned city, but also those aspects of that failure that have allowed unintegrated human beings to create spaces not just of survival but of development' (Bonami, 2001: 3). The squatter shacks of what used to be pejoratively called shanty towns are read, then, as locations of human social development rather than of an abjection which can be documented or romanticised but is beyond the pale. Bonami continues: 'Her shelters do not preach or blame, but rather stress the idea that multiple levels of existence and economies can exist around an urban context' (Bonami, 2001: 4). Eleanor Heartney likens Potrč's work to that of Krzysztof Wodiczko (see Chapter 5): 'Both artists are interested in legitimizing the means employed by people left out of the larger economy to create viable living spaces' (Heartney, 2001).[42] But perhaps this applies the logic of art's institutional structure – in which work is validated as art by its acceptance in that structure, rather than by a self-evident aesthetic quality – to the culture, in an anthropological sense, of others. If anything is legitimised it is the *reconstruction* of shelters as art, not the act of shelter dwelling. In any case, to think of legitimation is to apply the terms of the affluent world's cultural apparatus to the non-affluent world's everyday life, while dwellers in informal settlements have made no such request, and are more likely to seek legalisation than cultural legitimation.

The reconstruction of shelters using authentic materials in a gallery, however, asks other questions: first, throwing the question back to those in whose terms it is written, what the affluent society can legitimate (or not)

9.2 • Marjetica Potrč, *Travellers*. Reproduction courtesy of the Max Protetch Gallery, New York

9.3 • Marjetica Potrč, *House for Travellers*. Reproduction courtesy of the Max Protetch Gallery, New York

in its state of excess by looking to a culture and economy of necessity; second, how sustainable (or not) is a widening division of affluent from non-affluent worlds, and whether another quite different society is produced in non-affluence to demonstrate an autonomy within its limits, as well as ingenuity; and third, what are the specifics of difference in the cultures of borderland and marginal places. These are fairly obvious, and I do not imply that the work occupies any moral high ground. Its point seems more to give material form to difference, to the alterity of the cultures it represents. But there is a more than latent concern for a right to the city, and questions as to how it might be exercised with less impediment. In this respect the work is politicised,[43] has an awkward relation to its museum site, and refuses to look on informal architectures with the gaze of the technocratic rational planner or the tourist (though tours to favellas now occur.[44]) The shelters are rebuilt as they are in their original settings, while their re-contextualisation by not being in those settings but in the rarefied art space of the affluent society emphasises the questions they put to that society.

Potrč reconstructs shelters using the materials and principles (such as what constitutes basic provision) of each case.[45] The work is mediated unavoidably by its location in art space but not in its material form. *Kagiso: Skeleton House*, for instance, adopts the floor, roof and plumbing system which are the elements of subsidised housing schemes in South Africa in which dwellers build the rest of the structure themselves. The skeletal frame was exhibited at the

9.4 • Marjetica Potrč, *Kagiso: Skeleton House* (2000–1). Photograph courtesy of the Max Protetch Gallery, New York

Guggenheim, New York in 2000 alongside a small structure of cast-off materials more typical of squatter shacks in unregulated townships. At one level, the contrast between the two, in scale and regularity of form, exhibits the adaptation of a planned housing economy in recognition of the self-build skills of township dwellers. At another, it relays a more individual story, the work's point of departure for Potrč, of a family who moved their squatter shack piece by piece to the site of the subsidised base they were allocated to guard its toilet against theft. The shack bears a satellite dish, putting the technology of global media next to that of sanitation in a reminder that priorities are established in different ways in different places.[46] In *Nerlidere: the 24-Hour Ordinance*, to take another case, Potrč references a Turkish law which allows the construction of what can be built in a day. The shelter looks half-finished, with a tarpaulin for part of its roof and a mix of cement block and brick in its construction, and is a day's work in the gallery as on site.

Potrč is not, however, concerned only with informal settlements and recycling economies of the non-affluent world, and observes 'similar strategies . . . of individual initiative and an emphasis on private concerns, as well as sustainability in first and third world countries', adding that 'Though many of my locations are in distressed urban landscapes, these are today in many ways accepted as mainstream locations' (e-conversation, May 12th, 2002). The contexts for her work include the mobilisation of marginalised groups in the affluent as well as non-affluent worlds (itself a factor in the fluidity of

9.5 • Marjetica Potrč, *Nerlidere: the 24-Hour Ordinance* (1999). Photograph courtesy of the Max Protetch Gallery, New York

boundaries between wealth and deprivation, and in the drawing of other kinds of lines than those on maps); and that of development work in the non-affluent world since the 1970s. In relation to the former, Leonie Sandercock writes of voices newly heard (but which may have spoken longer) from border territories within the dominant society as much as on its peripheries or beyond:

> These . . . belong to people who dwell in cultures of displacement and transplantation, to cultures with a long history of oppression, to people who have been marginalized for hundreds of years, but who are now insurgent, and who are turning their very marginality into a creative space for theorizing. They challenge familiar/dominant notions of theory and practice, of epistemology and ontology, of what we know and who we are.
>
> (Sandercock, 1998a: 111)[47]

In relation to the latter, the growing literature of development studies (references in Chapter 8) puts recognition in terms of legalisation of informal settlements (as an effective way to meet the housing needs of migrants into rapidly expanding urban zones) in terms of reassertions of autonomy and reclamations of a right to build autonomous social as well as built architectures.[48]

The materials used by informal builders vary according to local conditions, and range from wooden crates and metal oil drums to bricks, cinder-blocks and corrugated iron sheeting. The detritus of advanced consumerism is mixed pragmatically with materials of an incipient vernacular. But Potrč is not concerned with a truth-to-materials in the sense of Modernist sculpture of the 1970s, but with the uses and combination of materials in acts of construction which produce both housing and a means to identity:

> Her work demonstrates that a shelter is not simply a temporary and nomadic dwelling, but has the power to generate different and autonomous identities. Such identities would counter the segregation of people in prefabricated ghettoes, and would call attention to the fact that this segregation involves, more than just social and economic isolation, a kind of 'temporary solution' in which 'civilized' societies attempt to remove the decaying elements of the social structure.
>
> (Bonami, 2001: 4–5)

In a period of ethnic cleansing, urban cleansing operates to sort residual populations into those accepted for assimilation and those rejected by it to become ghettoised or peripheralised, in a strategy not unlike that adopted at times in the Soviet Union to relegate abstract art to the bad-lands of mental instability (on the basis that a society which has found its objectively given form does not produce such abberations from the empathetic relation to form of classicism).

To put shacks in an art museum takes on added significance as a denial of programmed invisibility, though in some cases shelters represent a progressive policy on the part of authorities who upgrade rather than demolish informal settlements. An example is the East Wahdat Urban Upgrading scheme in Amman, Jordan – where a quarter of the population of the urban region live in informal settlements. Potrč has taken this as the basis for *East Wahdat*, which uses a core unit of block-wall, roof, door, roof-mounted water container, drain, and electricity connection (as used in the upgrading scheme) and reconstructed it in several gallery locations, handing over the process of specific design to the gallery's or museum's team of installers:

> I like the idea that time makes the work. I don't need to control the production. I don't have to be there to build the work and to agonize over materials or colors. If a team decides to put up a core unit, they take care of all these decisions themselves. The core unit has a life of its own. It changes and grows or disappears. As for the city space, if I build a core unit or an attachment to a house, these structures change, too. A core unit is just a functional entity. It's meant to be upgraded in the first place. People add walls to it, incorporating it in a new building. There is nothing wrong with letting a situation take over. A dialogue is always more productive than a monologue.
>
> (Potrč, 2001: 40)[49]

In a gallery space, painted pink or bright yellow and with satellite dishes added, the structures refuse categorisation as sculpture or architecture, and are not documentary either. Separated from their practical site but no less material in their exhibited forms, the core units sit awkwardly in the ambience of art as excess consumption. They are also decidedly non-Modernist, that is, neither value-free in the convention of the white-cube gallery – quite the opposite, loaded with significances relating to worlds outside the art museum – nor utopian:

> I don't feel that utopia makes any sense today. For me, the present time is about self-reliability, individual initiatives and small-scale projects. A few small-scale water turbines work better in the long term than one large dam. People still do build big dams – it's a slow process to change mentalities . . . City planning is another kind of big project. I visited Chandigarh years ago. It was really hard to find my way around, which was a joke, since the city was planned [by Le Corbusier – see Chapter 3] according to a serious rationale. The building materials were wrong too. The houses were made of concrete, which is terrible in the Indian climate because it's very cold in the winter and very hot in the summer. On top of that you had to navigate through the grid structure.
>
> (Potrč, interviewed by Hans Ulbrich Obrist, http://www.potrc.org/obrist)

9.6 • Marjetica Potrč, *This Then That* (2001), pepper-spray canister. Photograph courtesy of the Max Protetch Gallery, New York

But it is not so simple as to say these reconstructed shelters are post-modern either, with the implications of eclecticism and avoidance of engagement which might be part of post-modernity's mainstream cultural baggage. The same goes for the images of travellers and urban animal sightings, or exhibition of objects which denote a fixation on security in the affluent world, such as the canister of a pepper spray for use against bears.

Potrč exhibited the canister of such a spray in 2002, asking a friend in the US to buy it for her in New Jersey. The bears which have replaced muggers as the target for such means of personal defence – an inverse image of the survival tactics of the non-affluent world – forage in inner city dustbins or sit by swimming pools, or are electrocuted while climbing a pylon; and the alligator which might once have been a pet is now an outcast in suburbia. The presences are transgressive, making overt the conditions in which previously accepted boundaries between civilisation and wilderness break down.[50] Under the heading 'The Pursuit of Happiness', Potrč writes:

> Urban bears and nuisance alligators lead us into suburbia and sometimes even into the inner city, as is the case with a coyote that got trapped in an elevator in downtown Seattle. You might say that wild animals don't belong in cities. However, there they are, alive and well, adapting fast to a new lifestyle.

> For me, animal sightings prove that city space is constantly negotiated. Furthermore, they draw attention to the creativity of border spaces in general. The US/Mexican border, for instance, is a good example. Border cities like Tijuana and San Diego flourish because of the border. As for actual border barriers, no matter how high you build a wall, immigrants will always find a way to cross it in order to seek a better life. What would you do if one morning you woke up and an immigrant was pumping water from your swimming pool? . . .
>
> I read someplace that by 2100 national states will have yielded to city states. If you can imagine that, it becomes clear that the idea of border space will change too. It will become a close proximity experience, all due to survival instincts, and of course, the selfish pursuit of happiness.
>
> (Potrč, 2001: 23)

Running through the work is a problematization of the boundaries between the affluent and non-affluent societies, between art and cultures in an anthropological sense, and between spectators and representations of material cultures which are not their own. The abject are seen to be resourceful, not the residual humanity of charity,[51] and to have a culture which, as I suggested above, requires no validation by the affluent society's institutions but takes place regardless of them, almost despite them. The cultures of informal settlements, that is, are perceived as alternatives (if involuntary) to that of a mainstream unprepared for the idea of alternatives, more attuned to assimilations.

David Sibley writes that moral panics break out when boundaries are questioned. Boundaries which are porous can have the effect, as well as leakage which undermines the dominant society's self-image, of accentuating difference. When 'those who are usually on the outside occupy the centre and the dominant majority are cast in the role of spectators' (Sibley, 1995: 43), the centre appears no longer to hold and this brings to the surface deeper elements in the formation of a subject among the objects which comprise its world. Drawing on object-relations theories in post-Freudian psychoanalysis, as in the work of Melanie Klein, Sibley links social and individual patterns of relation. He adds, too, drawing from cultural theory, that inversions of power relations can expose the norms of oppression.[52] Powerful symbolic representations which shape processes of social change are sometimes located in marginal domains where they continue to contest the rule of the dominant class – as in carnival.[53]

There is not much sense of carnival in an informal settlement. It is everyday not feast-day life which goes on there; yet may entail moments, in Lefebvre's sense, of liberating consciousness. If so, the moment of liberation does not depend on assimilation of these cultures of everyday life into the dominant society, nor even of approval (to ask for which is already subordination) – which is not an argument against the provision by the affluent and official world of basic provision such as water, sanitation and energy. The question, though is less of a power supply than of a power relation. If informal settlements are a threat, in their very existence as well as the visibility they begin to receive in the dominant world's cultural production, to the established order, what is most threatening to it is, as Sibley argues, their demonstration of porosity in boundaries fixed to safeguard the urge to perfection of the dominant society. And perhaps equally threatening is the fact of autonomy – that the poor can organise themselves and do not need legitimation by the rich for their social formations.[54]

III CONCLUSION

After 9–11, politics in the anglophone part of the affluent world has been shelved for the most part in favour of the war on terror. In the US a president of dubious electoral legitimacy makes statements about the need for democracy (for which read opportunities for construction contracts and oil extraction) in far-away places seen as suitable to be annexed to the globalised economy. In the UK, despite the largest mass demonstration of anti-war sentiment ever seen, public policy is relentlessly driven by trans-Atlantic ties. The uneven distribution of power and uneven access to wealth that characterises the globalised economy is now provided with a mask of necessity – a case for the abolition of reflective commentary and of the idea that there are always alternatives to the adopted position. One page of a UK daily newspaper, picked more or less at random, gives a cameo of the post-political sphere: on the day a colossal statue of Saddam Hussein is toppled by a US army of occupation in Tikrit, it is announced that a former dictator charged with genocide is running for president of Guatemala; Palestinians in the Israeli-occupied West Bank resort to begging when trucks of food sent to their villages are unloaded at Israeli roadblocks; and a South Korean woman with a credit card debt she could not repay jumped with her three children from the balcony of a high-rise apartment (*Guardian*, July 19th, 2003).

But it might be that the question is not the traditional 'what is to be done?' but more 'what already happens?' Looking back to the cases of activism, avant-gardening, and cyber-interruption of the mainstream culture, and to representation of informal settlements by Potrč, a possible deduction is that activities currently taking place in the interstices and on the edges of the dominant society are not direct contestations of its power – that is, not primarily aimed at shifting public policies (though that may be a by-product) – but are alternative, more or less autonomous, and radically different structures of empowerment.

This draws me back to the basic flaw in the concept of an avant-garde: that acts of interpretation assume a power (or gaze) over what is interpreted and a subordination of the audience (and their kinds of knowledge) who receive that interpretation to the voice (and knowledge) of the interpreter. Perhaps it is time, as Freire argued (see Chapter 8), to refuse passive reception, to reclaim the terms of dialogue. Or, as Buck-Morss writes, citing Walter Benjamin, 'If the war is brought to the homeland, let *us* be the ones who wage it – not with terrorist violence whereby the ends justify the means, but with divine violence as Walter Benjamin, a Jew and a Marxist, conceived it: collective political action that is lethal not to human beings, but to the mythic power that reigns over them' (Buck-Morss, 2002: 8). The interruption of the rituals which enshrine high culture in its position of privilege which Benjamin sees in the mechanical reproduction of images through the lens is one application of this thought. It supposes, too, a collapse of the separation between producer and receiver, echoed in the fusion of making and dwelling in informal or self-build architecture, which can be extended to the formation of identities and the structuring of categories in verbal language.

I do not try to put the applications in any order (which would be to reproduce the model of hierarchy), and simply say that if culture in the affluent society has a potential to contribute to a newly cosmopolitan society, or diversity of social formations, it may be in exactly this fracturing of boundaries, in a refusal of the old categories and a viral-like infiltration within the dominant society which seeds quite other ways of looking at the world, and at the constitution of a subject alive and conscious in it. This occurs in context, I think, not of a chasm between the world pre- and post-September 11th. Buck-Morss emphasises the continuities of the war on terror and previous efforts to annexe territory: 'the military action that George W. Bush calls "the first war of the twenty-first century" looks remarkably similar to US military actions in the past' (Buck-Morss, 2002: 8) – this despite the lack of any clear geopolitical alignment of the new enemy with the territories of annexation, and the terrorists' perception of a world which no longer has an inside or an outside of dominance. Buck-Morss concludes that

> The true nightmare is that, under the terror produced by a total and unlimited war on terror, a US-led alliance of powers . . . will develop in a way that protects the global mobility of capital and its interests, but not that of the multitude and the interests of its public sphere.
>
> (Buck-Morss, 2002: 9)

She sees this as a reactionary cosmopolitanism in contrast to the similarly inadequate reactionary radicalism of Al-Qaeda. She adds that, if radicalism and cosmopolitanism were to converge, this would produce a global public sphere.

It might. But what would it be like? It is no longer the domain of public space; given the growth of alternative political and environmental campaigning in face of a parallel diminution of a conventional politics of representation, the site of dialogue is more likely to be outside conventional locations – again in the crevices – but is equally likely to have an incremental effect. That is why, to me, it seems important to recognise that groups taking direct action (of which I am not part) are in practice a new society evolving alongside the existing

9.7 • Barcelona – flyposting

society, and that there are many levels of contestation, but also of reclamation of an autonomy no more seen as a power to own, but experienced as a power to speak. Perhaps, even, there is a power in silence, as a cessation of the noise of a world constructed, at great expense to the majority of its human and non-human inhabitants and their built and natural habitats, by the project of capital (now in a global phase) to process the planet into profit. If the residue of that project is dust, it is timely to reconsider the ways each citizen participates in it. Taking the hint of the reference to silence and ending here, I do not conclude but offer the thought that in critical cultural practices of the kinds I have described there may be a beginning of something which embodies *Eros*.

NOTES

1 For discussion of the apocalyptic associations of this coverage in context of a literature of urbanism, see Miles, 2003.

2 'Many commentators have brought up a Pearl Harbour analogy, but that is misleading. On December 7, 1941, military bases in two US colonies were attacked – not the national territory, which was never threatened . . . Exactly what this portends, no one can guess. But that it is something strikingly new is quite clear' (Chomsky, 2002: 11–12).

3 'Since the aim of terrorism is behaviour modification of the enemy and/or public and not of the immediate victims, a certain arbitrariness in the election of the instrumental targets is characteristic of the terrorist form of violence' (Schmidt and de Graaf, 1982: 15, cited in Gupta, 2002: 53).

4 Chomsky sees a negatively defined identity: 'they draw support from a reservoir of bitterness and anger over U.S. policies in the region' (Chomsky, 2002: 13). Gupta sees the news media as constructing a blur and then selectively defining an identity within it: 'the mass media have more or less *created* a fragile frame by a method of constant fittings and refittings, by repetition and adjustment, by blurring distinction and then drawing out their particularities from the blur, by a process of throwing up a surfeit of connected but incoherent information and then adding to these and gradually letting a few ideas take dominance, and most importantly by constantly trying to gauge and accommodate to the pulse of political opinion and consumer demand within the West' (Gupta, 2002: 21).

5 Gupta sees the US response to September 11th as predicted by the attackers: 'The military action against bin Laden and the Taliban . . . was widely anticipated. Clearly, the United States was simply going through motions and routines long ingrained by the experience of *initiating* state terrorism. But this time the United States is not the *initiator*, this time the United States has simply gone mechanically and with clockwork-like predictability into its well-worn routine at the prodding of an *outside* hand' (Gupta, 2002: 93).

6 Buck-Morss dates her political maturing to September 11th, 1973 when the US government carried out murderous acts in support of a coup against the democratic government of Chile; and draws an analogy between the US school of the Americas (a training camp for terror against leftist republics) and al-Qaeda training camps in Afghanistan (Buck-Morss, 2002: 6). Chomsky characterises the US as a terrorist state: 'The most obvious example, though far from the most extreme case, is Nicaragua . . . It is worth remembering . . . that the US is the only country that was condemned for international terrorism by the World Court and that rejected a Security Council resolution calling on states to observe international law' (Chomsky, 2002: 44). Gupta, citing Chomsky, observes a routinisation of US terror: 'The United States has so much experience of perpetrating state terrorism and sponsoring state terrorism with international effect, while maintaining an apparent role of promoting democracy as ideological principle with international effect, that a certain routinization of procedure has taken place' (Gupta, 2002: 91).

7 Page puts the attack on the Twin Towers in context of the destruction of buildings in redevelopment, and the fantasised destruction of New York in mass culture – in the Japanese animation film *Final Fantasy* (2000), in *Planet of the Apes* (1968) and in Joaquin Miller's *The Destruction of Gotham* (1886): 'Only when Manhattan has "burned and burned to the very bed-rock" is the apocalypse complete' (Page, 2002: 169, citing Miller, J. (1886) *The Destruction of Gotham*, New York, Funk and Wagnalls, p. 232).

8 The tendency to adopt quasi-tribal identities is encountered in road protest (the Dongas at Twyford Down – McKay, 1996: 134–48), the Rainbow Warriors of contemporary millenarianism (Buenfil, 1991), and the departure to a hunter-gatherer lifestyle of two artists in 1992 (Gablik, 1995: 59–83).

9 'What might have been a first step toward national, no-fault insurance, or at least a fair and equitable disaster-relief policy, turned into yet another iteration of market tyranny. In choosing a method of disbursing money, Ken Feinberg . . . chose a market model: basing compensation on the victim's earnings and potential earnings . . . had they lived' (Page, 2002: 181).

10 'we were told to shop. Shop to show we are patriotic Americans. Shop to show our resilience . . . because in consumer capitalism shopping is the only way we can participate. Contrary to our president's call to shop, many Americans chose, instead, to give blood as eucharistic bonding of our life and body with those stricken and maimed' (Willis, 2002: 377).

11 Toulmin quotes Robert Darton in the *New York Review of Books*, February 1989: 'the revolutionaries stamped their ideas on contemporary consciousness by changing everything's name' (Toulmin, 1990: 175–6).

12 A Coca-Cola executive asked about coca-colonisation added a new tang to reification: 'all the thing wants to do is refresh you, and it's willing to understand your culture, to be meaningful to you . . . In Japan, that means one kind of thing, and in Brazil another. And Coke acknowledges those differences. But Coke stands for friendship' (Bayart, 2001: 313). Coca-Cola is less friendly to rival brands, as is picked up in a Benetton image showing two children using a home-made telephone consisting of two cans, one Coke the other Pepsi, joined by a piece of string: 'pretend this is a message from Pepsi *and* Coke' (from *Colors* magazine (no further details), in Franklin, Lury and Stacy, 2000; fig. 5.17; see also p. 173).

13 'A global strategy to counter our movement, even before 9–11, has been to portray us as threatening terrorists and the police as "saviours" of the people' (Starhawk, 2002: 225).

14 'The black bloc suddenly appears in the midst of a square that is supposed to be a safe space for peaceful gatherings; the police gas and beat the women and the pacifists and let the bloc escape. We are having a quiet lunch in the convergence center by the sea, when suddenly tear gas canisters are flying into the eating area and a pitched battle begins directly outside . . . The police rationale for the attack on the school was the supposed presence of members of the black bloc – but they never attacked the actual black bloc encampment, and by the night of the attack most of the black bloc had left the city' (Starhawk, 2002: 112).

15 'A reversal occurred at the New York City World Economic Forum protest. The protestor were no longer a decentralized mass covering a city, attempting to breach a fortress of elites. After all, it was always those big ugly fences, the perimeter defence, that made it look like those on the inside really did have something to hide. The prior decentralization of the protestors throughout a city had made the protests seem democratic . . . The violence of the fences and teargas and nightsticks exposed a real brutality . . . Quebec was turned into a fortress. The WTO ran to Qatar. But here, the strategy inverted. Here it was the protestor who were contained, in a massive military-police-media trap. . . . Now . . . it was the protestors who were the infiltrators into an otherwise peaceful city. Now it was the protestors who must be contained, not the elites who must be defended. A new defensive strategy emerged, the strategy of the security state (now visible and expanding): a strategy of encompassment . . . in which all people participate only as the representational images allotted to them' (Tillet, 2002). I am grateful to Nicola Kirkham for this source.

16 'No system of domination can afford to use force to control every aspect of its functioning. Instead, it engages our fear and our hope. We comply with its decrees because we fear punishment or retaliation if we resist. Or because we hope for some reward, some benefit' (Starhawk, 2002: 224–5). A further argument would be that consumers have a capacity to act through refusal to purchase goods and services from companies with bad records of social or environmental exploitation; or to reduce consumption to levels which are sustainable – see Lodziak, 2002: 150–60.

17 Cases include mass trespass on Twyford Down in 1993 (McKay, 1996: 134–48; Wall, 1999: 65–73; Field, 1999: 79); resistance to extension of the M11 at Claremont Road, east London in 1993–4 (McKay, 1996: 150–3; McCreery, 2001); and a symbolic excavation of the M41 by Reclaim the Streets in 1996 (Field, 1999: 77).

18 Jamison cites 'reclaiming the streets' as a case of sectarian piety. Szerszynski aligns sectarian piety with radical groups in the English Revolution of the 1640s who were urgent witnesses against evil (Jamison, 2001: 150). But Wall (citing Megill, 1985; Laclau and Mouffe, 1985) sees anti-roads protest as cultural in character, with a loose organisational structure and emphasis on lifestyle rather than political issues (Wall, 1999: 38).

19 RTS initiated 43 events between 1995 and June 18th, 1999; for June 18th, see http://www.infoshop.org/j18_reflections.html. On their surveillance by the police, see Wall, 1999: 127–8.

20 'Reclaim the Streets . . . is a non-hierarchical, leaderless, openly organized, public group. No individual "plan" or "masterminds" its actions and events. RTS activities are the result of voluntary, unpaid, co-operative efforts from numerous self-directed people attempting to work equally together' (cited in Jordan, 2002: 69; see http://gn.apc.org/rts/disorg.htm).

21 See http://www.gn.apc.org/rts/streetpolitics.htm. A leaflet distributed in 1996 states: 'Direct action enables people to develop a new sense of self-confidence and an awareness of their individual and collective power. Direct Action is founded on the idea that people can develop the ability for self-rule only through practice . . . Direct action is not just a tactic, it is individuals asserting their ability to control their own lives and to participate in social life without the need for mediation or control by bureaucrats or professional politicians' (cited in Wall, 1999: 192).

22 Rose argues that 'We must jettison the division between a logic that structures and territorialises "from above" according to protocols that are not our own, and a more or less spontaneous anti-logic "from below" that expresses our needs, desires and aspirations . . . our present has arisen as much from the logics of contestation as from any imperative or control' (Rose, 1999: 277). Derrida reads all nation states as founded in violence, with the conceptual legitimation following the act, but sees, too, a new factor in the subordination of heads of state to a newly international justice: 'what appears singular and new today is the *project* of making States, or at least heads of state in title (Pinochet), and even of current heads of state (Milosevic), appear before universal authorities. It has to do only with projects or hypotheses, but this possibility suffices to announce a transformation: it constitutes in itself a major event. The sovereignty of the State, the immunity of a head of state are no longer in principle, in law, untouchable' (Derrida, 2002: 57).

23 Schwartz and Schwartz, 1998: 54–65.

24 Schwartz and Schwartz see the fluctuating and open access of the site as a disabling factor: 'The community has endless hassle from its fluctuating and uncontrollable population: there are no enforceable rules in their illegal settlement, no power to exclude the unwelcome' and 'A hundred serious demonstrators settled here in the first week; in the second week these were replaced by the curious, the homeless, the far-out, the drugged, the unstable and the dishonest' (Schwartz and Schwartz, 1998: 57, 58). They also note criticism of the 'intellectuals' such as George Monbiot who set up the squat and then 'after a week of fanfare in the media, left it in the lurch' (p. 58).

25 'During the Depression, the City's welfare department and the federal Works Project Administration sponsored nearly 5,000 "relief" gardens on vacant lots and city parks'; but: 'By 1977, there were more than 25,000

vacant lots in New York. Littered with trash and rats, these open sores became magnets for drugs, prostitution, and chop shops for tripping down stolen cars' (Ferguson, 1999b: 82–3).

26 Smith, 1997: 3–29; Weinberg, 1999: 46. In 1998, the Green Thumb gardens were transferred from the Parks Department to the Assets and Sales Unit of the Department of Housing, Preservation, and Development (Ferguson, 1999a: 71).

27 98 duplex condominiums called Del Este Village were planned for the site, 71 displacing community gardens hitherto licensed by the City: 'It was as if we didn't exist . . . our garden and the three others were listed as "vacant, blighted lots". This despite the fact that folks at Little Puerto Rico had been tending their plots for over 10 years' (Ferguson, 1999a: 61).

28 Wilson cites Anderson (1998) that while the collapse of cultural categories, such as the conservative bourgeois aesthetic against which the nineteenth-century bohemians revolted, could produce a new plebeianisation, it is more likely to mean that the cult of celebrity extinguishes the possibility of revolt (Wilson, 2003: 242).

29 See Marcuse, P., 2002.

30 Co-housing is designed as 'a new type of cooperative housing . . . [which] integrates autonomous private dwellings with shared utilities and recreational facilities such as kitchens, dining halls, workshops, and children's play facilities. Cohousing residents comprise an *intentional community*. They *choose* to live together and to share property and resources. They develop a rich social life that includes regular shared meals. They aspire to meaningful social relations and a strong "sense of community" ' (http://www.aiid.bee.qut.edu.au/~meltzer/content.htm). See also http://www.cohousing.org for the Co-Housing Network of North America.

31 www.solidair.nl. Member organisations include Ana Maria Fonds, a charitable foundation able to lend money to ecological businesses; SamSam, an investment association; Committment, an association of businesses and housing projects; AMF Onroerend Goed, a foundation managing property for common benefit; Dissident, an association of individuals able to invest in solidarity projects; and Reonans, an association of businesses and housing cooperatives. A leaflet (2003) states: 'The Solidair association . . . has expressly chosen to bring an alternative to the market economy into practice. It wants to create the conditions and circumstances in which you can create your own housing, working and living based on solidarity . . . This also involves entering into confrontation with whoever and whatever threatens this. Developing your own alternative is a condition of achieving the desired effect: welfare for everyone.' Among the contributing organisations is Vof de Veranderung, which offers organisational support and project development, for instance in alternative and democratic forms of property development and the setting up of new, alternative businesses.

32 See http://www.communities.org.uk/lets; see also Schwartz and Schwartz, 1998: 209–14.

33 For instance Adbusters, a Vancouver-based media foundation founded in 1989 following a refusal of media companies to sell air-time to an anti-logging campaign – http://www.adbusters.org; see Meikle, 2002: 129–39 on Adbusters; and pp. 113–26 on the category of tactical media, referencing de Certeau's concept of tactics in urban routines.

34 See Mosbacher, 2002; see also http://www.socialaffairsunit.org.uk.

35 From symposium papers, *Artist as Engineer*, Institute of Digital Art & Technology, University of Plymouth, 2003. See http://www.interrupt-symposia.org; and http://www.etoy.com.

36 See http://www.mociny.com – the museum is organised like the new galleries which appeared in the 1990s in New York's SoHo, especially on Broadway, using several floors of ex-commercial space. Different shows are presented on each floor, with a menu resembling the information given in a museum elevator, and there is a basement bar.

37 'Rather than stopping the flow of information, far more is generated than can be digested. The strategy is to classify or privatise all information that could be used by the individual for self-empowerment, and to bury the

useful information under the realms of useless junk data offered to the public. Instead of the traditional information blackout, we face an information blizzard – a whiteout. This forces the individual to depend on authority to help prioritize the information to be selected' (Critical Art Ensemble, 1994: 132). See also Critical Art Ensemble, 1996.

38 Jeremijenko is a design engineer and technologist who graduated in mechanical engineering from Stanford University, and in the history and philosophy of science from the University of Melbourne. She directs the Engineering Design Studio at Yale, and is associated with the Media Research Lab/Center for Advanced Technology at New York University. Her work includes interactive digital, electro-magnetical and bio-technological projects. The Institute for Applied Autonomy was founded in 1998 as an anonymous collective of graduate engineers, designers, artists and activists, seeking to develop technologies to extend activism through performative acts in public spaces and through its web site. See http://www.interrupt-symposia.org, http://cat.nyu.edu/natalie and http://www.appliedautonomy.com.

39 In a pamphlet, *Contestational Robotics: Theory and Practice*, published in collaboration with Critical Art Ensemble, IAA state: 'Since the notion of the public sphere has been increasingly recognized as a bourgeois fantasy that was dead on arrival at its inception in the 19th century, an urgent need has emerged for continuous development of tactics to reestablish a means of expression and a space of temporary autonomy within the terrain of the social' (undated, p. 3).

40 '*Virtual Urban* is both a journey and a dialogue. It aims to document how cities are decoded and read today. Readings are presented using the material and its sequencing as a metaphor for navigating contemporary space. The surfer can navigate endlessly, much like visiting a city for the first time. It is easy to get lost in this site and almost impossible to retrace your steps . . . Dialogues also have these qualities. Recounted narrative gains a new layer of meaning with each retelling. The ability of the surfer to approach information on this site from different directions each time gives the experience of the site another dimension' (http://potrc.org/vu/background.htm). See also an interview by O'Beirn, *Circa*, Autumn 1997, pp. 26–9.

41 Potrč lives in Ljubljana, where she is Associate Professor at the Academy of Fine Arts; In 2000 she participated in the International Studio Programme of the Künstlerhaus Bethanien, Berlin, and shows regularly in New York.

42 Heartney continue: 'Potrč flirts with the danger . . . of romanticizing the poverty that forces the destitute to erect these temporary cities. Nevertheless, her small exhibition served as a tonic in the context of a museum culture that embraces the ethic of conspicuous consumption and seemingly endless expansion' (Heartney, 2001).

43 'Potrč's Slovenian background seems to have helped form her strongly political stance; it also seems to have suggested a way to address the issues on which her work focuses from a more "ancient" point of view. Her small shacks, that is, sometimes feel like the constructions surrounding a medieval castle. Slovenia itself, of course, is a kind of new shelter, located at the bottom of the walls of the fortress called Europe' (Bonami, 2001: 6).

44 'I found out there are tours organized in the favellas of Sao Paolo, which would have been unthinkable ten years ago . . . In Israel, tourists can take a tour of Gaza's tent cities. I read about this in the *Rough Guide* (Potrč interviewed by Hans Ulrich Obrist, http://www.potrc.org/obrist.

45 'Potrč's attention to the details of materials used by outcast individuals to build their spaces expresses the need to preserve a connection with the reality that produced those materials' (Bonami, 2001: 7).

46 Potrč was awarded the Hugo Boss Prize (2000) for an installation at the Guggenheim Museum including this work; other short-listed artists were Vito Acconci, Maurizio Cattelan, Michael Elmgreen, and Barry Le Vau. A further work references the Barefoot College, Tilonia (Chapter 8), exhibited at Max Protetch Gallery, New York, April–May 2002. On work based on shelter, see http://www.potrc.org/home/projects/strategies/index.html.

47 Sandercock cites hooks, 1990: 152, where she calls for a new 'politics of location . . . marginality as a site of resistance' (in Sandercock, 1998a: 111–13); and West, 1993 on a politics of audacious hope (ibid.).

48 See Turner, 1976 for studies comparing informal and planned housing in central America, in which the former achieve better outcomes for health, family structures and economic viability; and for arguments concerning citizen participation in housing programmes. Turner advocates planned provision of utilities, especially piped water, but continues: 'But the application of the same principles to the dwelling environments within the area defined will inhibit local and personal initiative and therefore deprive society of a major part of the resources available for development' (p. 152). See also Hamdi, 1995: 43–8 for a comparison of Turner's emphasis on self-help and self-build within a context of land tenure, and Habraken's emphasis on support structures for development; and Hamdi, 1995: 75–128 on the practices of enabling dweller participation, appropriate design and sensitivity to conditions in housing projects for non-affluent countries. See Fernandez and Varley, 1998 on the legalisation of informal settlements, and the translation of Lefebvre's concept of a right to the city into legal procedures.

49 A House for Travellers was constructed on the outskirts of Ljubljana in 2000, with migrants not allowed to work more than eight hours per week, who recycle material from a nearby garbage dump to generate living costs (see exhibition catalogue, 2002, Berlin, Künstlerhaus Bethanien, p. 17; and http://www.potrc.org/home/projects/strategies/index.

50 A set of Potrč's ink-jet images of urban animal sightings is included in *The City Cultures Reader*, 2nd edition, Routledge, 2003.

51 See Sibley, 1995: 49–71 on advertising images trading on concepts of the defiled, extending the precedents of imaginary geographies in which minorities were represented as imperfect human forms, or as polluting, to contemporary mass culture.

52 Sibley cites early modern broadsheet illustrations of a World Upside Down – beggars give alms (Cf. reference to Courbet's *The Beggar's Charity at Ornans* in Chapter 1), the blind lead the sighted, and children discipline fathers; and the Rebecca Riots against turnpike roads in Wales and south-west England, in which men dressed as women to challenge the state (Sibley, 1995: 44).

53 Sibley continues that Caribbean carnivals are grudgingly accepted by a British state which is nominally multicultural but has tended to police them heavily: 'The appeal of the exotic for the white majority mixes uneasily with the images of black criminal stereotypes which have informed the responses of the control agencies' (Sibley, 1995: 44–5); and that legislation increasingly contains carnivalesque activity, including that of new age travellers. He notes that in a few cases the marginalised are able to reclaim a site, as with the gypsy pilgrimage to Saintes Maries de la Mer in the Camargue, southern France on May 24th–25th and in October: 'Although the pilgrimage has now been given a tourist gloss . . . it is still a subversive event which expresses the collective but highly circumscribed power of European Gypsies and expresses the long history of racism to which they have been subject' (Sibley, 1995: 45). On Bakhtin's work on the transposition of carnival to literature – its import into high culture – see Brandist, 2002: 133–55.

54 Iris Marion Young argues against the assimilation of minority groups into a mainstream culture: 'Groups experiencing cultural imperialism have found themselves objectified and marked with a devalued essence from the outside, by a dominant culture they are excluded from making. The assertion of a positive sense of group difference by these groups is emancipatory because it reclaims the definition of the group by the group, as a creation and construction, rather than a given essence' (Young, 1990: 172). In a more recent book Young differentiates three levels of sociation: private association, public association and political association. Of self-organising association she writes that it enables people who feel marginalised to seek out each other to develop a political voice; and continues that although some forms of association are or become hierarchical or authoritarian, many associations are democratic, and offer empowering participation: 'the self-organizing activities of civil society contribute to self-determination, and, to a lesser degree, self-development, by supporting identity and voice, facilitating innovative or minority practice, and providing some goods and services' (Young, 2000: 165). See also Sandercock, 1998a: 185–7; and Miles, 2003.

BIBLIOGRAPHY

•

Editions listed are those used. Where the date of a first edition affects the understanding of its context, it is given in square brackets.

Abu-Lughod, J. (1980) *Rabat: Urban Apartheid in Morocco*, Princeton, N.J., Princeton University Press.

Adam, P. (1987) *Eileen Gray: Architect/Designer*, New York, Abrams.

Adorno, T. W. (1978) 'Culture and administration', *Telos*, no.37, Fall, pp. 100–1.

Adorno, T. W. (1981) *Prisms*, Cambridge, Mass., MIT.

Adorno, T. W. (1990) *Negative Dialectics*, trans. Ashton, E. B., London, Routledge.

Adorno, T. W. (1991) *The Culture Industry: Selected Essays on Mass Culture*, ed. Bernstein, J. M., London, Routledge.

Adorno, T. W. (1994) *The Stars Down to Earth and Other Essays on the Irrational in Culture*, ed. Cook, S., London, Routledge.

Adorno, T. W. (1997) *Aesthetic Theory*, London, Athlone.

Adorno, T. W., Benjamin, W., Bloch, E., Brecht, G. and Lukács, G. (1980) *Aesthetics and Politics*, London, Verso.

Adorno, T. W. and Horkheimer, M. (1997) *Dialectic of Enlightenment*, London, Verso.

Affron, M. and Antliff, M., eds (1997) *Fascist Visions: Art and Ideology in France and Italy*, Princeton, N.J., Princeton University Press.

Aga Khan Award for Architecture (2001) *Modernity and Community: Architecture in the Islamic World*, London, Thames and Hudson and the Aga Khan Award for Architecture.

Agrest, D. [1993] (2000) 'Architecture from without: Body, logic, and sex', in Rendell, Penner and Borden (2000), 358–70.

Ainley R., ed. (1998) *New Frontiers of Space, Bodies and Gender*, London, Routledge.

Albrow, M. (1997) 'Travelling beyond local cultures: Socioscapes in a global city', in Eade (1997), pp. 37–56.

al-Khalil, S. (1991) *The Monument: Art, Vulgarity and Responsibility in Iraq*, Berkeley, University of California Press.

Alloula, M. (1986) *The Colonial Harem*, trans. Godzich, M. and Godzich, W., Minneapolis, University of Minnesota Press.

Amato, J. A. (2000) *Dust: A History of the Small and Invisible*, Berkeley, University of California Press.

American Council for the Arts (1987) *Public Art Public Controversy: The Tilted Arc on Trial*, New York, ACA Books.

Angus, I. (2000) *Primal Scenes of Communication: Communication, Consumerism, Social Movements*, Albany, N.Y., State University of New York Press.
Antliff, M. (1997) '*La Cité Française* George Valois, Le Corbusier, and fascist theories of urbanism', in Affron and Antliff (1997), 134–70.
Appadurai, A. (1990) 'Disjuncture and difference in the global cultural economy', in Robbins (1993), pp. 269–95.
Appadurai, A., ed. (2001) *Globalization*, Durham, N.C., Duke University Press.
Arendt, H. (1958) *The Human Condition*, Chicago, Chicago University Press.
Armor, M. and Snell, D. (1999) *Building Your Own Home*, 16th edition, London, Ebury Press.
Aronowitz, S. (1995) 'Literature as social knowledge: Mikhail Bakhtin and the reemergence of human sciences', in Mandelker (1995), pp. 119–35.
Ashcroft, B., Griffiths, G., Tiffin, H., eds (1995) *The Post-Colonial Studies Reader*, London, Routledge.
Ashton, D. (1972) *The Life and Times of the New York School*, Bath, Adams & Dart.
ATTAC (2001) 'Simply a question of taking back, together, the future of the world', in Houtart and Polet (2001), pp. 69–71.
Augé, M. (1995) *Non-Places: Introduction to an Anthropology of Supermodernity*, London, Verso.
Bangma, A., ed. (2002) *Territorial Invasions of the Public and Private*, Rotterdam, Piet Zwart Institute of the Willem de Kooning Academy.
Barber, S. (1995) *Fragments of the European City*, London, Reaktion.
Barker, F. (1984) *The Tremulous Private Body: Essays on Subjection*, London, Methuen.
Barnett, J. (1986) *The Elusive City: Five Centuries of Design, Ambition and Miscalculation*, New York, Harper & Row.
Barry, J. (1999) *Environment and Social Theory*, London, Routledge.
Barton, H., ed. (2000) *Sustainable Communities: The Potential for Eco-Neighbourhoods*, London, Earthscan.
Bauman, Z. (1989) *Modernity and the Holocaust*, Cambridge, Polity Press.
Bauman, Z. (1998) *Globalization – the Human Consequences*, Cambridge, Polity.
Bauman, Z. (1999) *Culture as Praxis*, London, Sage.
Bayart, J.-F. (2001) 'The paradoxical invention of economic modernity', in Appadurai (2001), pp. 307–34.
Beall, J., ed. (1997) *A City for All: Valuing Difference and Working with Diversity*, London, Zed Books.
Beardsley, J. (1996) *Earthworks and Beyond*, New York, Abbesville Press.
Becker, C., ed. (1994a) *The Subversive Imagination*, London, Routledge.
Becker, C. (1994) 'Herbert Marcuse and the subversive potential of art', in Becker (1994a), pp. 113–29.
Beckett, S. (1963) *Proust and Three Dialogues with Georges Duthuit*, London, John Calder.
Beckett, S. (1965) *Waiting for Godot*, 2nd edition, London, Faber and Faber.
Beckett, S. (1967) *No's Knife, Collected Shorter Prose 1945–66*, London, Calder & Boyars.
Beckett, S. (1975) *The Unnamable*, London, Calder & Boyars.
Beecher, J. and Bienvenu, R. (1983) *The Utopian Vision of Charles Fourier*, Columbia, Miss., University of Missouri Press.
Belascu, D. (2001) 'Artist Mierle Ukeles finds great potential, and newfound sadness, in Fresh Kills', *The Jewish Week*, October 29th, 2001 [http://www.thejewishweek.com].
Belenky, M., Clinchy, B., Goldberger, N., Tarule, J. (1986) *Women's Ways of Knowing: The Development of Self, Voice, and Mind*, New York, Basic Books.
Bell, M.M. (1998) 'Culture as dialogue', in Bell and Gardiner (1998), pp. 49–62.
Bell, M. M. and Gardiner, M., eds (1998) *Bakhtin and the Human Sciences*, London, Sage.
Belsey, C. (1988) *Critical Practice*, London, Routledge.

Benhabib, S., ed. (1996) *Democracy and Difference: Contesting the Boundaries of the Political*, Princeton, N.J., Princeton University Press.
Benjamin, A., ed. (1989) *The Problems of Modernity: Adorno and Benjamin*, London, Routledge.
Benjamin, W. (1973) *Illuminations*, ed. Arendt, H., Harmondsworth, Penguin.
Benjamin, W. (1983) *Understanding Brecht*, trans. Bostok, A., London, Verso.
Benjamin, W. (1985) *The Origin of German Tragic Drama*, intro. Steiner, G., London, Verso.
Benjamin, W. (1997a) *Charles Baudelaire*, trans. Hoare, Q., London, Verso.
Benjamin, W. (1997b) *One-Way Street*, trans. Jephcott, E. and Shorter, K., London, Verso.
Benjamin, W. (1999) *The Arcades Project*, trans. Eiland, H. and McLaughlin, K., Cambridge, Mass., Harvard University Press.
Bennett, S. and Butler, J., eds (2000) *Locality, Regeneration & Diversities*, Bristol, Intellect Books.
Bentley, E., ed. (1965) *The Theory of the Modern Stage*, Harmondsworth, Penguin.
Berg-Schlosser, D. and Kersting, N., eds (2003) *Poverty and Democracy: Self-Help and Political Participation in Third World Cities*, London, Zed Books.
Bergson, H. [1910] (1971) *Time and Free Will*, trans. Pogson, F. L., London, George Allen & Unwin.
Berman, M. (1983) *All That Is Solid Melts into the Air: The Experience of Modernity*, London, Verso.
Bernal, M. (1987) *Black Athena: The Afroasiatic Roots of Classical Civilization,* vol. 1, New Brunswick, N.J., Rutgers University Press.
Bernstein, J. M. (1992) *The Fate of Art: Aesthetic Alienation from Kant to Derrida and Adorno*, Cambridge, Polity Press.
Beuys, J. (1973) 'I am searching for a field character', in Tisdall (1974), p. 48.
Bird, J. (1988) 'The spectacle of memory', in catalogue *Michael Sandle*, London, Whitechapel Gallery.
Bird, J. (1996) 'Art history and hegemony' in Bird *et al.* (1996), pp. 68–86.
Bird, J., Curtis, B., Mash, M., Putnam, T., Robertson, G., Stafford, S. and Tickner, L., eds (1996) *The BLOCK Reader in Visual Culture*, London, Routledge.
Birkett, J. (1986) *The Sins of the Fathers: Decadence in France 1870–1914*, London, Quartet.
Blackman, W. S. (1927) *The Fellaheen of Upper Egypt*, London, Harrap.
Blaikie, P. (1985) *The Political Economy of Soil Erosion*, London, Methuen.
Bloch, E. (1980) 'Discussing expressionism', in Adorno *et al.* (1980) pp. 16–27.
Bloch, E. (1986) *The Principle of Hope*, trans. Plaice, N., Plaice, S. and Knight, P., Cambridge, Mass., MIT.
Bloch, E. (1991) *Heritage of Our Times*, Cambridge, Polity Press.
Bloch, E. (2000) *The Spirit of Utopia*, Stanford, Calif., Stanford University Press.
Bonami, F. (2001) 'Promised land: Shelters and other spaces in the work of Marjetica Potrč', in exhibition catalogue *Marjetica Potrč*, Berlin, Künstlerhaus Bethanien.
Bookchin, M. (1982) *The Ecology of Freedom*, Palo Alto, Calif., Cheshire Books.
Borden, I. (2001) *Skateboarding, Space and the City: Architecture and the Body*, Oxford, Berg.
Borden, I. (2003) 'What is radical architecture?', in Miles and Hall (2003), pp. 111–21.
Borden, I., Kerr, J., Rendell, J. and Pivar, A., eds (2001) *The Unknown City: Contesting Architecture and Social Space*, Cambridge, Mass., MIT.
Bordieu, P. and Haacke, H. (1995) *Free Exchange*, Cambridge, Polity Press.
Bown, M. C. and Taylor, B., eds (1993) *Art of the Soviets: Painting, Sculpture and Architecture in a One-Party State, 1917–1992*, Manchester, Manchester University Press.
Brailsford, H. N. (n.d.) *Shelley, Godwin, and their Circle*, Home University Library of Modern Knowledge, London, Williams & Norgate.

Brandist, C. (2002) *The Bakhtin Circle: Philosophy, Culture and Politics*, London, Pluto.

Braunstein, P. and Doyle, M. W. (2002) *Imagine Nation: The American Counterculture of the 1960s and '70s*, London, Routledge.

Briggs, A., ed. (1996) *Fins de Siècle: How Centuries End 1400–2000*, New Haven, Conn., Yale.

Brookner, A. (1971) *The Genius of the Future: Studies in French Art Criticism*, London, Phaidon.

Brown, N. O. (1990) *Hermes the Thief: The Evolution of a Myth*, Great Barrington, Mass., Lindisfarne Press.

Buber, M. (1996) *Paths to Utopia*, Syracuse, N.Y., Syracuse University Press.

Buck-Morss, S. (1991) *The Dialectics of Seeing: Walter Benjamin and the Arcades Project*, Cambridge, Mass., MIT.

Buck-Morss, S. (1997) 'The City as Dreamworld and Catastrophe', in Paetzold (1997), pp. 97–115.

Buck-Morss, S. (2002) 'A global public sphere?', *Radical Philosophy*, no.111, pp. 2–10.

Buenfil, A. R. (1991) *Rainbow Nation Without Borders: Toward an Ecotopian Millennium*, Santa Fe, N. Mex., Bear & Co. Publishing.

Bürger, P. (1984) *Theory of the Avant-Garde*, Minneapolis, Minn., University of Minnesota Press.

Burgess, E. W. [1925] 'The growth of the city: Introduction to a research project', in LeGates and Stout (2000), 2nd edition, pp. 153–61.

Burgin, V. (1986) *The End of Art Theory: Criticism and Postmodernity*, Atlantic Heights, N.J., Humanities International Press.

Byrne, D. (1997) 'Chaotic places or complex places: Cities in a post-industrial era', in Westwood and Williams (1997), pp. 50–70.

Byrne, D. (1998) *Complexity Theory and the Social Sciences: An Introduction*, London, Routledge.

Byrne, D. (2001) *Understanding the Urban*, Basingstoke, Palgrave.

Caeiro, M. (2003) 'Lisbon capital of nothing', in Miles and Kirkham (2003), pp. 133–46.

Callicott, J. B. (1994) *Earth's Insights: A Multicultural Survey of Ecological Ethics from the Mediterranean Basin to the Australian Outback*, Berkeley, University of California Press.

Callinicos, A. (1989) *Against Postmodernism: A Marxist Critique*, London, Pluto.

Camacho, D. E., ed. (1998) *Environmental Injustices, Political Struggle: Race, Class, and the Environment*, Durham, N.C., Duke University Press.

Campanella, T. (1887) 'City of the sun', in Morely (1887) pp. 217–63.

Campanella, T. (1981) *La Città del Sole: Dialogo Poetico. The City of the Sun: a Poetic Dialogue*, trans. Donno, D. J., Berkeley, University of California Press.

Carey, J., ed. (1999) *The Faber Book of Utopias*, London, Faber and Faber.

Carmen, R. (1996) *Autonomous Development: Humanizing the Landscape*, London, Zed Books.

Carr, C. (2002) 'Fresh kills becomes an urban artwork', *The Village Voice*, May 28th, p. 43.

Casciato, M. (1996) *The Amsterdam School*, Rotterdam, 010 Publishers.

Castagnary, J., 1892, *Salons (1857–79)*, Paris, Charpentier.

Celan, P. (1996) *Collected Poems*, ed. and trans. Hamburger, M., Harmondsworth, Penguin [previously published by Anvil Press, 1988].

Çelik, Z. (2000) 'Le Corbusier, orientalism, colonialism', in Rendell, Penner and Borden (2000) pp. 321–31.

Chalfant, H. and Prigoff, J. (1987) *Spraycan Art*, London, Thames and Hudson.

Chambers, I. (2001) *Culture after Humanism: History, Culture, Subjectivity*, London, Routledge.

Chambers, R. (1988) 'Sustainable rural livelihoods: A key strategy for people, environment and development', in Conroy and Litvinoff (1988), pp. 1–17.

Chambers, N., Simmons, C., and Wackernagel, M. (2000) *Sharing Nature's Interest: Ecological Footprints as an Indicator of Sustainability*, London, Earthscan.
Chase, J., Crawford, M., and Kaliski, J. eds (1999) *Everyday Urbanism*, New York, Monacelli Press.
Chasseguet-Smirgel, J. and Grunberger, B. (1986) *Freud or Reich? Psychoanalysis and Illusion*, London, Free Association Books.
Chin, M. (1999) 'I see . . . the insurgent mechanics of infection', in Marras (1999), pp. 68–79.
Chinedu, U. (2000) 'Empowering the self-builder', in Hughes and Sadler (2000), pp. 210–21.
Chomsky, N. (2002) *9–11*, New York, Seven Stories Press.
Christianson, G. E. (1999) *Greenhouse: The 200-year Story of Global Warming*, London, Constable.
Christie, I. and Warburton, D. (2001) *From Here to Sustainability: Politics in the Real World*, London, Earthscan.
Chu, P. t-D., ed., (1992) *Letters of Gustave Courbet*, Chicago, University of Chicago Press.
Cilliers, P. (1998) *Complexity & Postmodernism: Understanding Complex Systems*, London, Routledge.
Clark, T. (1997) *Art and Propaganda*, London, Weidenfeld and Nicholson.
Clark, T. J., (1973) *The Absolute Bourgeois: Artists and Politics in France 1848–51*, London, Thames & Hudson.
Clendinnen, I. (1999) *Reading the Holocaust*, Cambridge, Cambridge University Press.
Cline, A. (1997) *A Hut of My Own: Life Outside the Circle of Architecture*, Cambridge, Mass., MIT.
Cole, I., ed. (1999) *Gustav Metzger Retrospectives*, Oxford, Museum of Modern Art.
Cole, I. and Stanley, N. (2000) *Beyond the Museum: Art, Institutions, People*, Oxford, Museum of Modern Art.
Collins, P. (1990) *Black Feminist Thought: Knowledge, Consciousness and the Politics of Empowerment*, Boston, Unwin Hyman.
Collins, T. and Goto, R. (2003) 'Landscape, ecology, art and change', in Miles and Hall (2003), pp. 134–44.
Colomina, B., ed. (1992a) *Sexuality & Space*, New York, Princeton Architectural Press.
Colomina, B. (1992b) 'The split wall: Domestic voyeurism', in Colomina, ed. (1992a) pp. 73–128.
Colomina, B. (1996) *Privacy and Publicity: Modern Architecture and Mass Media*, Cambridge, Mass., MIT.
Conroy, C. and Litvinoff, M. (1988) *The Greening of Aid: Sustainable Livelihoods in Practice*, London, Earthscan.
Coombes, A. (1991) 'Ethnography and the formation of national and cultural identities', in Hiller (1991), pp. 189–214.
Coombes, A. (1994a) *Reinventing Africa: Museums, Material Culture and Popular Imagination*, New Haven, Conn., Yale University Press.
Coombes, A. (1994b) 'Blinded by science: Ethnography at the British Museum', in Pointon (1994), pp. 102–19.
Cooper, D., ed. (1968) *The Dialectics of Liberation*, Harmondsworth, Penguin.
Cooper, R., Friedman, J., Gans, S., Heaton, J. M., Oakley, C., Oakley, H. and Zeal, P. (1989) *Thresholds between Philosophy and Psychoanalysis: Papers from the Philadelphia Association*, London, Free Association Books.
Corbin, A. [1982] (1996) *The Foul & The Fragrant: Odour and the Social Imagination*, London, Macmillan.
Cork, R. (1995) 'Message in a bottle', *Modern Painters*, Spring, pp. 76–81.
Cornell, V. J. (2002) 'A Muslim to Muslims: Reflections after September 11', in Hauerwas and Lentricchia (2002), pp. 325–36.

Cornerhouse, the (1998) 'Same platform, different train: The politics of participation' [Briefing Paper 4, February], Sturminster Newton, Corner House.
Cornford, M. and Cross, D. (2001) *Coming Up for Air* [project proposal, not paginated], Stoke-on-Trent, Staffordshire University.
Costello, D. (2000) 'The work of art and its "Public": Heidegger and Tate Modern', in Cole and Stanley (2000), pp. 12–26.
Cottington, D. (1998a) *Cubism*, London, Tate Gallery.
Cottington, D. (1998b) *Cubism in the Shadow of War: The Avant-Garde and Politics in Paris 1905–1914*, New Haven, Conn., Yale University Press.
Courbet, G., (1861) 'Letter to young artists', *Courier du dimanche*, Paris, December 25th, 1861.
Crane, D. (1987) *The Transformation of the Avant-Garde: The New York Art World, 1940–1985*, Chicago, University of Chicago Press.
Crary, J. (1990) *Techniques of the Observer: On Vision and Modernity in the Nineteenth Century*, Cambridge, Mass., MIT.
Crawford, M. (1992) 'The world in a shopping mall', in Sorkin (1992), pp. 3–30.
Crawford, M. (1999) 'Blurring the boundaries: Public space and private life', in Chase, Crawford and Kaliski (1999), pp. 22–35.
Cresswell, T. (1996) *In Place, Out of Place: Geography, Ideology, and Transgression*, Minneapolis, University of Minnesota Press.
Critical Art Ensemble (1994) *The Electronic Disturbance*, New York, Autonomedia.
Critical Art Ensemble (1996) *Electronic Civil Disobedience and Other Unpopular Ideas*, New York, Autonomedia.
Crowley, D. and Reid, S. E., eds (2002) *Socialist Spaces: Sites of Everyday Life in the Eastern Bloc*, Oxford, Berg.
Crush, J., ed. (1995) *The Power of Development*, London, Routledge.
Curtis, B. (2000) 'The heart of the city', in Hughes and Sadler (2000), pp. 52–65.
Curtis, K. (1999) *Our Sense of the Real: Aesthetic Experience and Arendtian Politics*, Durham, N.C., Duke University Press.
Daglish, J. and Thepaut, P. (1993) 'Straw Houses', *Permaculture Magazine*, vol. 1 no.4, Autumn, pp. 11–12.
Daniel, J. and Moylan, T. (1997) *Not Yet: Reconsidering Ernst Bloch*, London, Verso.
Darke, J. (1991) *The Monument Guide to England and Wales*, London, MacDonald.
Darier, E., ed. (1999) *Discourses of the Environment*, Oxford, Blackwell.
Davis, J. C. (1993) 'Formal utopia/informal millennium: The struggle between form and substance as a context for seventeenth-century utopianism', in Kumar and Bann (1993), pp. 17–32.
Davis, M. (1990) *City of Quartz: Excavating the Future in Los Angeles*, London, Verso.
Deakin, N. (2001) *In Search of Civil Society*, Basingstoke, Palgrave.
Dean, A. O. and Hursley, T. (2002) *Rural Studio: Samuel Mockbee and an Architecture of Decency*, New York, Princeton Architectural Press.
de Boer, L. (2000) 'Making sense of matter: An interview with Jackie Brookner', *Earthlight*, Fall, pp. 21–5.
de Certeau, M. (1984) *The Practice of Everyday Life*, Berkeley, University of California Press.
de Certeau, M. (1997) *The Capture of Speech and Other Political Writings*, Minneapolis, University of Minnesota Press.
Degen, M. (2002) 'Regenerating public life? A sensory analysis of regenerated public place in El Raval, Barcelona', in Rugg and Hinchcliffe (2002), pp. 19–36.
de Forges, M.-T. (1978) biography, in Royal Academy of Arts (1978), pp. 22–50.
de Geus, M. (1999) *Ecological Utopia: Envisioning the Sustainable Society*, Utrecht, International Books.

d'Harnoncourt, A. and McShine, K. (1973) *Marcel DuChamp* [catalogue], New York, Museum of Modern Art.

Derrida, J. (2002) *On Cosmopolitanism and Forgiveness*, London, Routledge.

Descartes, R. [1637] (1960) *Discourse on Method*, trans. Wollaston, A., Harmondsworth, Penguin.

Deutsche, R. (1991) 'Uneven development: Public art in New York City', in Ghirardo (1991), pp. 157–219 [first published in *October*, Winter 1988].

Deutsche, R. (1992) 'Public art and its uses', in Senie and Webster (1992), pp. 158–70.

Deutsche, R. (1996) *Evictions*, Cambridge, Mass., MIT.

Deutscher, P. (2002) *A Politics of Impossible Difference: The Later Work of Luce Irigaray*, Ithaca, N.Y., Cornell University Press.

Diderot, D. (1963) *Salons*, eds Selznec, J. and Adhemar, J., Oxford, Clarendon Press.

Dinsdale, J. E. (1980) *Survivors, Victims, and Perpetrators: Essays in the Nazi Holocaust*, Washington, D.C., Hemisphere Publishing.

Dobson, A. (1995) *Green Political Thought*, 2nd edition, London, Routledge.

Dodd, D. and van Hemmel, A. (1999) *Planning Cultural Tourism in Europe: A Presentation of Theories and Cases*, Amsterdam, Boekman Stichting.

Donald, J. (1999) *Imagining the Modern City*, London, Athlone.

Douglas, M. (1987) *How Institutions Think*, London, Routledge & Kegan Paul.

Douglas, M. and Friedmann, J. (1998) *Cities for Citizens*, Chichester, Wiley.

Dovey, K. (1999) *Framing Places: Mediating Power in Built Form*, London, Routledge.

Drakakis-Smith, D. (1990) *Third World Cities*, London, Routledge.

Drakulic, S. (1993) *How We Survived Communism and Even Laughed*, New York, HarperCollins.

Duncan, C. (1993) *The Aesthetics of Power*, Cambridge, Cambridge University Press.

Duncan, C. (1995) *Civilizing Rituals: Inside Public Art Museums*, London, Routledge.

Eade, J., ed. (1997) *Living the Global City*, London, Routledge.

Edwards, S., ed. (1973) *The Communards of Paris, 1871*, London, Thames & Hudson.

Eitner, L., ed. (1971) *Neoclassicism and Romanticism, 1750–1850*, Sources and Documents in the History of Art Series, London, Prentice-Hall.

Elias, N. and Scotson, J. (1965) *The Established and the Outsiders*, London, F. Cass.

Ellin, N., ed. (1997) *Architecture of Fear*, New York, Princeton Architectural Press.

Elliott, J. (1999) *An Introduction to Sustainable Development*, 2nd edition, London, Routledge.

Engels, F. (n.d.) *Ludwig Feuerbach and the Outcome of Classical German Philosophy*, London, Martin Lawrence [Marxist Leninist Library].

Ernst, A.-S. and Klinger, G. (1997) 'Socialist Socrates: Ernst Bloch in the GDR', *Radical Philosophy*, 84, pp. 6–21.

Escobar, A. (1995) *Encountering Development: The Making and Unmaking of the Third World*, Princeton, N.J., Princeton University Press.

Escobar, A. (1996) 'Constructing Nature: Elements for a Poststructural Political Ecology', in Peet and Watts (1996), pp. 46–68.

Evans, G. (2001) *Cultural Planning – an Urban Renaissance?* London, Routledge.

Extra]muros[(2002) *Lisboa capital do nada, Marvila, 2001*, Lisbon, Extra]muros[.

Fanon, F. [1961] (1967) *The Wretched of the Earth*, Harmondsworth, Penguin.

Farrell, J. J. (1997) *The Spirit of the Sixties: The Making of Postwar Radicalism*, London, Routledge.

Fat (2001) 'It's not unusual: Projects and tactics', in Borden *et al.* (2001), pp. 340–55.

Fathy, H. (1973) *Architecture for the Poor*, Chicago, University of Chicago Press [reprint with identical text and pagination to [1969] (1989) *Gourna – A Tale of Two Villages*, Cairo, Egyptian Ministry of Culture].

Fathy, H. (1984) 'Palaces of mud', *Resurgence* 103, 16–17.
Fathy, H. (1986) *Natural Energy and Vernacular Architecture: Principles and Examples with Reference to Hot Arid Climates*, Chicago, University of Chicago Press.
Felman, S. and Laub, D. (1992) *Testimony – Crisis of Witnessing in Literature, Psychoanalysis and History*, London, Routledge.
Felshin, N. ed. (1995) *But Is It Art?*, Seattle, Wash., Bay Press.
Felstiner, J. (1995) *Paul Celan: Poet, Survivor, Jew*, New Haven, Conn., Yale University Press.
Ferguson, J. (1990) *The Anti-Politics Machine: 'Development', Depoliticisation, and Bureaucratic State Power in Lesotho*, Cambridge, Cambridge University Press.
Ferguson, S. (1999a) 'The death of Little Puerto Rico: NYC gardens are getting plowed by a new wave of urban development', in Wilson and Weinberg (1999), pp. 60–79.
Ferguson, S. (1999b) 'A brief history of grassroots greening on the Lower East Side', in Wilson and Weinberg (1999), pp. 80–90.
Fernandez, E. and Varley, A., eds (1998) *Illegal Cities: Law and Urban Change in Developing Countries*, London, Zed Books.
Field, P. (1999) 'The anti-roads movement: The struggle of memory against forgetting', in Jordan and Lent (1999), pp. 68–79.
Finkelstein, N. G. (2000) *The Holocaust Industry: Reflections on the Exploitation of Jewish Suffering*, London, Verso.
Finnegan, R. (1998) *Tales of the City: A Study of Narrative and Urban Life*, Cambridge, Cambridge University Press.
Fischer, E. (1970) *Marx in His Own Words*, Harmondsworth, Penguin.
Fitzpatrick, S. (1999) *Everyday Stalinism – Ordinary Life in Extraordinary Times: Soviet Russia in the 1930s*, Oxford, Oxford University Press.
Flaubert, G. (1983) *Flaubert in Egypt*, trans. and ed. Steegmuller, F., London, Michael Haag.
Foran, J., ed. (2003) *The Future of Revolutions: Rethinking Radical Change in the Age of Globalization*, London, Zed Books.
Forty, A. (2001) 'The Royal Festival Hall – a "Democratic" Space?', in Borden *et al.* (2001), pp. 200–11.
Foster, H., ed. (1983) *The Anti-Aesthetic: Essays on Postmodern Culture*, Seattle, Wash., Bay Press.
Foster, H. (1985) *Recodings: Art, Spectacle, Cultural Politics*, Seattle, Wash., Bay Press.
Foster, H., ed. (1988) *Vision and Visuality*, Seattle, Wash., Bay Press.
Foster, H. (1996) *The Return of the Real: The Avant-Garde at the End of the Century*, Cambridge, Mass., MIT.
Foster, S. C. (2000) 'Dada and the constitution of culture: (Re-) conceptualising the avant-garde', in Scheunemann (2000), pp. 49–68.
Foucault, M. [1961] (1967) *Madness and Civilization: A History of Madness in the Age of Reason*, London, Tavistock.
Foucault, M. [1975] (1991) *Discipline and Punish: The Birth of the Prison*, Harmondsworth, Penguin.
Fowkes, R. (2002) 'The role of monumental sculpture in the construction of socialist space in Stalinist Hungary', in Crowley and Reid (2002), pp. 65–84.
Franklin, S., Lury, C. and Stacey, J. (2000) *Global Nature, Global Culture*, London, Sage.
Frascina, F. (1999) *Art, Politics and Dissent: Aspects of the Art Left in Sixties America*, Manchester, Manchester University Press.
Fraser, N. (1993) 'Rethinking the public sphere: A contribution to the critique of actually existing democracy', in Robbins (1993), pp. 1–34.
Freire, P. (1972) *Pedagogy of the Oppressed*, Harmondsworth, Penguin.
Fremian, Y. (2002) *Orgasms of History: 3000 Years of Spontaneous Insurrection*, Edinburgh, A. K. Press.

Freshman, P., ed. (1992) *Public Address: Krzysztof Wodiczko*, Minneapolis, Minn., Walker Art Centre.
Freud, S. (1962) *Two Short Accounts of Psycho-Analysis*, Harmondsworth, Penguin.
Freud, S. (1991) *On Metapsychology*, Harmondsworth, Penguin.
Frisby, D. (1985) *Fragments of Modernity*, Cambridge, Polity Press.
Frisby, D. and Featherstone, M., eds (1997) *Simmel on Culture*, London, Sage.
Fromm, E. (1959) *Sigmund Freud's Mission*, London, George Allen & Unwin.
Frye, N. (1973) 'Varieties of literary utopias', in Manuel (1973), pp. 25–49.
Fuller, P. (1980) *Beyond the Crisis in Art*, London, Readers and Writers Cooperative.
Fuller, P. [1980] (1988) *Art and Psychoanalysis*, London, Hogarth Press.
Fyfe, N., ed. (1998) *Images of the Street*, London, Routledge.
Gablik, S. (1995) *Conversations Before the End of Time: Dialogue on Art, Life and Spiritual Renewal*, London, Thames and Hudson.
Galbraith, J. K. [1958] (1962) *The Affluent Society*, Harmondsworth, Penguin.
Gandy, M. (1997) 'Contradictory modernists: Conceptions of nature in the art of Joseph Beuys and Gerhard Richter', *Annals of the Association of American Geographers*, vol. 87, no.4, pp. 636–59.
Gandy, M. (2002) *Concrete and Clay: Reworking Nature in New York City*, Cambridge, Mass., MIT.
Garcia, D. and Lovink, G. (2001) 'The ABC of tactical media', *Sarai Reader* vol. 1, pp. 90–2.
Gardiner, M. E. (2000) *Critiques of Everyday Life*, London, Routledge.
Gayle, M. and Cohen, M., eds (1988) *Manhattan's Outdoor Sculpture*, Art Commission and Municipal Art Society Guide, New York, Prentice Hall.
Gentile, E. (1997) 'The myth of national regeneration in Italy: From modernist avant-garde to fascism', in Affron and Antliff (1997), pp. 25–45.
Geoghegan, V. (1987) *Utopianism & Marxism*, London, Methuen.
Geoghegan, V. (1996) *Ernst Bloch*, London, Routledge.
Gerz, J. (1999) *Das Berkeley Orakel: Fragen ohne Antwort*, Düsseldorf, Richter Verlag.
Gerz, J. (2001) unpublished conference paper, Coventry, Herbert Art Gallery.
Geuss, R. (1981) *The Idea of a Critical Theory*, Cambridge, Cambridge University Press.
Ghirardo, D. ed. (1991) *Out of Site: A Social Criticism of Architecture*, Seattle, Wash., Bay Press.
Ghirardo, D. (1996) *Architecture after Modernism*, London, Thames & Hudson.
Gilbert, A. and Gugler, J. (1992) *Cities, Poverty and Development: Urbanization in the Third World*, Oxford, Oxford University Press.
Gillis, J. R., ed. (1994) *Commemorations: The Politics of National Identity*, Princeton, N.J., Princeton University Press.
Gilloch, G. (1996) *Myth & Metropolis: Walter Benjamin and the City*, Cambridge, Polity Press.
Gilloch, G. (2002) *Walter Benjamin: Critical Constellations*, Cambridge, Polity Press.
Girardet, H. (1992) *The Gaia Atlas of Cities: New Directions for Sustainable Urban Living*, London, Gaia Books.
Goddard, D. (2001) 'Mierle Laderman Ukeles: Penetration and transparency: Morphed' [http://www.newyorkartworld.com]
Goldblatt, D. (1996) *Social Theory and the Environment*, Cambridge, Polity Press.
Goldhagen, D. J. (1996) *Hitler's Willing Executioners: Ordinary Germans and the Holocaust*, London, Little, Brown.
Goldman, R. and Papson, S. (1998) *Nike Culture*, London, Sage.
Gordon, D. (1968) *Women of Algeria: An Essay on Change*, Cambridge, Mass., MIT.
Goto, J. (1988) '*Terezin*, Oxford, remembering the future' [exhibition catalogue], University of Oxford, 1988; Cambridge Darkroom, 1988; John Hansard Gallery, 1989.

Goto, J. (1998) *The Commissar of Space*, Oxford, Museum of Modern Art.
Goto, J. (2002) *Loss of Face* [exhibition catalogue] London, Tate.
Goto, J. and Ortzan, D. (1993) *The Scar*, London, Benjamin Rhodes Gallery.
Goto, J. and Taylor, B. (1998) *The Commissar of Space*, Oxford, published by the artist to accompany an exhibition of the same title at the Museum of Modern Art, Oxford.
Gottdiener, M. (1985) *The Social Production of Urban Space* [2nd edition 1994], Austin, Tex., University of Texas Press.
Gottdiener, M., ed. (2000) *New Forms of Consumption: Consumers, Culture, and Commodification*, Lanham, Md., Rowman & Littlefield.
Gratz, R. B. (1989) *The Living City*, New York, Simon & Schuster.
Green, D. and Seddon, P., eds (2000) *History Painting Reassessed*, Manchester, Manchester University Press.
Greenberg, C., 1988, *Collected Essays and Criticism*, ed. O'Brien, J., vol. 1 1939–1944, Chicago, University of Chicago Press.
Gregson, N. and Crewe, L. (2003) *Second-Hand Cultures*, Oxford, Berg.
Griffin, R., ed. (1995) *Fascism*, Oxford, Oxford University Press.
Griffiths, J. and Kemp, P. (1999) *Quaking Houses – Art, Science and the Community: A Collaborative Approach to Water Pollution*, Charlbury, Jon Carpenter.
Griswold, C. L. (1992) 'The Vietnam Veterans Memorial and the Washington Mall: Philosophical thoughts on political iconography', in Mitchell (1992), pp. 79–112.
Grunenberg, C. (1994) 'The politics of presentation: The Museum of Modern Art, New York', in Pointon (1994), pp. 192–211.
Guerrilla Girls (1995) *Confessions of the Guerrilla Girls: How a Bunch of Masked Avengers Fight Sexism and Racism in the Art World with Facts, Humour, and Fake Fur*, New York, Pandora.
Guha, R. and Martinez-Alier, J., eds (1997) *Varieties of Environmentalism: Essays North and South*, London, Earthscan.
Gunn, W. and Renwick, G. (1998) 'Whaur extremes meet', *Architectural Design* profile 135 'Ephemeral/Portable Architecture', ed. Kronenburg, R., London, Academy Editions.
Gupta, S. (2002) *The Replication of Violence: Thoughts on International Terrorism after September 11th*, London, Pluto.
Habermas, J. (1989) *The New Conservatism: Cultural Criticism and the Historians' Debate*, ed. and trans. Nicholson, S. W., Cambridge, Mass., MIT.
Hadjinicolaou, N. (1978) *Art History and Class Struggle*, London, Pluto.
Hall, P. [1988] (1996) *Cities of Tomorrow*, updated edition, Oxford, Blackwell.
Hall, T. (2003) 'Birmingham as a Cultural City', in Mile and Kirkham (2003), pp. 49–57.
Hall, T. and Hubbard, P., eds (1998) *The Entrepreneurial City: Geographies of Politics, Regime and Representation*, Chichester, Wiley.
Hamburger, M. (1972) *The Truth of Poetry: Tensions in Modern Poetry from Baudelaire to the 1960s*, Harmondsworth, Penguin.
Hamdi, N. (1995) *Housing without Houses*, London, Intermediate Technology Publications.
Hamdi, N. with El-Sherif, A., eds (1996) *Educating for Real: The Training of Professionals for Development Practice*, London, Intermediate Technology Publications.
Hamdi, N. and Goethert, R. (1998) 'Urban development and urban design: deciding the parameter', *Urban Design International*, vol. 3, no.1, pp. 23–31.
Hamm, B. and Muttagi, P., eds (1998) *Sustainable Development and the Future of Cities*, London, Intermediate Technology Development Group.
Hansen, M. B. (2001) '*Schindler's List* is not *Shoah*: Second commandment, popular modernism, and public memory', in Zelizer (2001), pp. 127–51.
Haraway, D. (1991) *Simians, Cyborgs and Women: The Reinvention of Nature*, London, Routledge.

Harbison, R. (1991) *The Built, the Unbuilt, & the Unbuildable*, London, Thames & Hudson.
Harding, J. M. (1997) *Adorno and 'A Writing of the Ruins': Essays on Modern Aesthetics and Anglo-American Literature and Culture*, Albany, N.Y., State University of New York Press.
Harper, G., ed. (1998) *Interventions and Provocations: Conversations on Art, Culture, and Resistance*, Albany, N.Y., State University of New York Press.
Harris, S. and Berke, D., eds (1997) *Architecture of the Everyday*, New York, Princeton Architectural Press.
Harrison, C. and Wood, P., eds (1992) *Art in Theory 1900–1990: An Anthology of Changing Ideas*, Oxford, Blackwell.
Harrison, C. and Wood, P., with Gaiger, J., eds (1998) *Art in Theory 1815–1900: An Anthology of Changing Ideas*, Oxford, Blackwell.
Harvey, D. (1989) *The Urban Experience*, Baltimore, Md., Johns Hopkins University Press [abridged version of *The Urbanization of Capital: Studies in the History and Theory of Capitalist Urbanization* and *Consciousness and the Urban Experience: Studies in the History and Theory of Capitalist Urbanization*, originally published 1985].
Harvey, D. (1991) 'Afterword', in Lefebvre (1991), pp. 425–32.
Hauerwas, S. and Lentricchia, F., eds (2002) *Dissent from the Homeland: Essays after September 11*, [special issue, no.101–2, *The South Atlantic Quarterly*] Durham, N.C., Duke University Press.
Hattenstone, S. (1997) 'From house to holocaust' [interview with Rachel Whiteread], *Guardian*, May 13th, p. 3.
Hayden, D. (1995) *The Power of Place: Urban Landscapes as Public History*, Cambridge Mass., MIT.
Hayden, D. (2001) 'Claiming women's history in the urban landscape: Projects from Los Angeles', in Borden *et al.* (2001), pp. 356–69.
Hayward, T. (1998) *Political Theory and Ecology*, Cambridge, Polity Press.
Heartney, E. (2001) 'Marjetica Potrč at the Guggenheim Museum', *Art in America*, July, p. 10.
Hebdige, D. (1993) 'The machine is *Unheimlich*: Wodiczko's homeless vehicle project', in Walker Art Centre (1993), pp. 55–73.
Heidegger, M. (1959) *An Introduction to Metaphysics*, New Haven, Conn., Yale University Press.
Heller, C. (1999) *Ecology of Everyday Life: Rethinking the Desire for Nature*, Montreal, Black Rose Books.
Herr, M. (1978) *Dispatches*, New York, Avon.
Hertmans, S. (2001) *Intercities*, London, Reaktion.
Hess, T. B. and Ashbery, J., eds, 1968, *Avant-Garde Art*, New York, Collier-Macmillan.
Highmore, B. (2000) 'Awkward moments: Avant-gardism and the dialectics of everyday life' in Scheunemann (2000), pp. 245–65.
Highmore, B. (2002) *Everyday Life and Cultural Theory: An Introduction*, London, Routledge.
Hildebrandt, R. (1988) *Die Mauer Spricht / The Wall Speaks*, Berlin, Verlag. Haus am Checkpoint Charlie [bilingual text, not paginated].
Hill, J., ed. (1998) *Occupying Architecture: Between the Architect and the User*, London, Routledge.
Hiller, S., ed. (1991) *The Myth of Primitivism: Perspectives on Art*, London, Routledge.
Hirsch, M. (2001) 'Surviving images: Holocaust photographs and the work of photography', in Zelizer (2001), pp. 215–46.
Hoberman, J. (1993) 'Spielberg's Oskar', *Village Voice*, December 21st, p. 65.

Hoffman, J. (1999) 'The idea of the art strike and its astonishing effects', in Cole (1999), pp. 26–8.
hooks, b. (1984) *Feminist Theory: From Margin to Center*, Boston, South End Press.
hooks, b. (1990) *Yearning: Race, Gender, and Cultural Politics*, Boston, South End Press.
Hooper, B. (1995) 'The poem of male desires: female bodies, modernity, and "Paris: Capital of the Nineteenth Century"', *Planning Theory*, no.13, pp. 105–29 [reprinted in Sandercook (1998b), pp. 227–54].
Horkheimer, M. (1972) *Critical Theory*, New York, Continuum.
Horkheimer, M. (1993) *Between Philosophy and Social Science*, trans. and ed. Hunter, G. F., Kramer, M. S. and Torpey, J., Cambridge, Mass., MIT.
Horsfield, C. (1988) '2,5.45 / april 88', in Goto (1988), pp. 53–5.
Houtard, F. (2001) 'A summary of the process', in Houtart and Polet (2001), pp. 113–16.
Houtard, F. and Polet, F., eds (2001) *The Other Davos: The Globalization of Resistance to the World Economic System*, London, Zed Books.
Hoy, D. C. and McCarthy, T. (1994) *Critical Theory*, Oxford, Blackwell.
Hughes, J. (2000) 'After non-plan: Retrenchment and reassertion', in Hughes and Sadler (2000), pp. 166–83.
Hughes, J. and Sadler, S., eds (2000) *Non-Plan: Essays on Freedom, Participation and Change in Modern Architecture and Urbanism*, Oxford, Architectural Press.
Huhn, T. (1990) 'The sublimation of culture in Adorno's aesthetics', in Roblin (1990), pp. 290–307.
Huhn, T. (1997) 'Kant, Adorno, and the social opacity of the aesthetic', in Huhn and Zuidervaart (1997), pp. 237–58.
Huhn, T. and Zuidervaart, L., eds (1997) *The Semblance of Subjectivity: essays in Adorno's Aesthetic Theory*, Cambridge, Mass., MIT.
Hunt, L., ed. (1989) *The New Cultural History*, Berkeley, Calif., University of California Press.
Huysmans, J.-K. [1884] (1959) *Against Nature* [*A Rebours*], trans. Baldick, R., Harmondsworth, Penguin.
Illich, I. (1986) *H_2O and the Waters of Forgetfulness*, London, Marion Boyars.
Independent Commission on International Development Issues (1982) *North–South: A Programme for Survival*, London, Pan Books.
Irigaray, L. (1978) interview, in Hans, M. F. and Lapouge, G., eds (1978), *Les Femmes, La Pornographie, L'Erotisme*.
Irigaray, L. (1994) *Thinking the Difference: For a Peaceful Revolution*, London, Athlone.
Isaak, J. A. (1996) *Feminism & Contemporary Art*, London, Routledge.
Jackson, H. [1913] (1950) *The Eighteen Nineties*, Harmondsworth, Penguin.
Jacobs, J. (1961) *The Death and Life of Great American Cities*, New York, Random House.
Jacobs, J. M. (1996) *Edge of Empire: Postcolonialism and the City*, London, Routledge.
Jacobus, J. (1973) *Matisse*, London, Thames and Hudson.
Jagger, C. S. (1933) *Modelling and Sculpture in the Making*, London, Studio.
Jameson, F. (1991) *Postmodernism, or the Cultural Logic of Late Capitalism*, London, Verso.
Jamison, A. (2001) *The Making of Green Knowledge: Environmental Politics and Cultural Information*, Cambridge, Cambridge University Press.
Jarosz, L. (1996) 'Defining deforestation in Madagascar', in Peet and Watts (1996), pp. 148–64.
Jarvis, S. (1998) *Adorno: A Critical Introduction*, Cambridge, Polity Press.
Jay, M. (1984) *Adorno*, Cambridge, Mass., Harvard.
Johnson, H. (1999) 'Local forms of resistance: Weapons of the weak', in Skelton and Allen (1999), pp. 159–66.
Jones, B. (2002) *Building with Straw Bale: A Practical Guide for the UK and Ireland*, Dartington, Green Books.
Jones, S. (2000) 'Fertile minds', *Guardian*, April 26th, Society section, p. 4.

Jordan, T. (2002) *Activism: Direct Action, Hacktivism and the Future of Society*, London, Reaktion.

Jordan, T. and Lent, A., eds (1999) *Storming the Millennium: The New Politics of Change*, London, Lawrence and Wishart.

Julian, P. (1977) *The Orientalists*, Oxford, Phaidon.

Kabeer, N. (1994) *Reversed Realities: Gender Hierarchies in Development Thought*, London, Verso.

Kagarlitsky, B. (2001) 'The road to consumption', in Katsiaficas (2001), pp. 52–66.

Kahnweiler, D.-H., with Crémieux, F. (1961) *Mes galleries et mes peintres*, Paris, Gallimard [English translation: *My Galleries and Painters* (1971) trans. Weaver, H., London, Thames & Hudson].

Kaminski, A. (1993) 'Art in the twilight of totalitarianism', in Bown and Taylor (1993), pp. 140–53.

Kandinsky, W. (1912) *Über das Geistige in der Kunst, Insbesondere in der Malerei*, Munich, Piper Verlag.

Kandinsky, W. (1947) *Concerning the Spiritual in Art, and Painting in Particular, 1912*, New York, Wittenborn, Schultz.

Kandinsky, W. (1977) *Concerning the Spiritual in Art*, trans. Sadler, M. T. H., New York, Dover [first published in Munich in 1911 (dated 1912) as *Über das Geistige in der Kunst*, and in English in 1914 as *The Art of Spiritual Harmony*, London, Constable].

Kaprow, A. (1993) *Essays on the Blurring of Art and Life*, ed. Kelley, J., Berkeley, Calif., University of California Press.

Karnouk, L. (1988) *Modern Egyptian Art: The Emergence of a National Style*, Cairo, American University in Cairo Press.

Karp, I., Kreamer, C. M. and Lavine, S. D., eds (1992) *Museums and Communities: The Politics of Public Culture*, Washington, Smithsonian Institution.

Katsiaficas, G., ed. (2001) *After the Fall: 1989 and the Future of Freedom*, London, Routledge.

Katz, B. (1982) *Herbert Marcuse: Art of Liberation*, London, Verso.

Katz, B. (1990) 'The liberation of art and the art of liberation: The aesthetics of Herbert Marcuse', in Roblin (1990), pp. 152–85.

Katz, S. (1999) 'The idea of civil society', in *Civil Society: A New Agenda for US–Japan Intellectual Exchange*, Tokyo, Japan Foundation.

Keil, R., Bell, D. V. J., Penz, P., and Fawcett, L., eds (1998) *Political Ecology: Global and Local*, London, Routledge.

Keller, J. R. (2002) *Samuel Beckett and the Primacy of Love*, Manchester, Manchester University Press.

Kellner, D. (2003) 'Globalization, technopolitics and revolution', in Foran (2003), pp. 180–94.

Kennedy, J. F., Smith, M. G. and Wanek, C., eds (2002) *The Art of Natural Building: Design, Construction and Resource*, Gabriola Island, British Columbia, New Society Publishers.

Kenny, M. and Meadowcroft, J., eds (1999) *Planning Sustainability*, London, Routledge.

Kerr, J. (2001) 'The uncompleted monument: London, war, and the architecture of remembrance', in Borden *et al.* (2001), pp. 68–89.

King, A., ed. (1980) *Buildings and Society*, London, Routledge & Kegan Paul.

King, A., ed. (1996) *Re-Presenting the City: Ethnicity, Capital and Culture in the 21st Century Metropolis*, Basingstoke, Macmillan.

Kinney, L. and Çelik, Z. (1990) 'Ethnography and exhibitionism at the Expositions Universelles', *Assemblages*, no.13, pp. 35–59.

Kitching, G. (1988) *Karl Marx and the Philosophy of Praxis*, London, Routledge.

Kofman, E. and Lebas, E. (1996) 'Last in transposition – time, space and the city', in Lefebvre (1996), pp. 3–60.

Krauss, R. (1972) 'A view of modernism', *Artforum*, September, pp. 48–51.
Krauss, R. [1979] (1983) 'Sculpture in the expanded field', in Foster (1983), pp. 31–42.
Krauss, R. (1985) *The Originality of the Avant-Garde and Other Modernist Myths*, Cambridge, Mass., MIT.
Kristeva, J. (2002) *Revolt, She Said*, Los Angeles, Semiotext(e).
Kruft, H.-W. (1994) *A History of Architectural Theory, from Vitruvius to the Present*, New York, Princeton Architectural Press.
Kugelmass, J. (1992) 'The rites of the tribe: American Jewish tourism in Poland', in Karp, Kreamer and Lavine (1992), pp. 382–427.
Kumar, K. and Bann, S., eds (1993) *Utopias and the Millennium*, London, Reaktion.
Kuspit, D. (1993) *The Cult of the Avant-Garde Artist*, Cambridge, Cambridge University Press.
Labowitz-Strauss, L. and Lacy, S. [1978] (2001) 'In mourning and in rage', in Robinson (2001), pp. 102–6.
Laclau, E. (1996) *Emancipation(s)*, London, Verso.
Laclau, E. and Mouffe, C. (1985) *Hegemony and Socialist Strategy: Towards a Radical Democratic Politics*, London, Verso.
Lacour, C. B. (1996) *Lines of Thought: Discourse, Architectonics, and the Origin of Modern Philosophy*, Durham, N.C., Duke University Press.
Lacy, S., ed. (1995) *Mapping the Terrain*, Seattle, Wash., Bay Press.
Landes, J. (1988) *Women and the Public Sphere in the Age of the French Revolution*, Ithaca, N.Y., Cornell University Press.
Lash, S. (1999) *Another Modernity, A Different Rationality*, Oxford, Blackwell.
Lasky, M. J., 1976, *Utopia and Revolution*, Chicago, Chicago University Press.
Lawson, H. (2001) *Closure: A Story of Everything*, London, Routledge.
Leach, N., ed. (1997) *Rethinking Architecture: A Reader in Cultural Theory*, London, Routledge.
Leach, N., ed. (1999) *Architecture and Revolution: Contemporary Perspectives on Central and Eastern Europe*, London, Routledge.
Leadbeater, B. J. R. and Way, N., eds (1996) *Urban Girls: Resisting Stereotypes, Creating Identities*, New York, New York University Press.
Le Corbusier (1960) *Creation is a Patient Search*, New York, Praeger [from *L'Atelier de la recherche patiente*].
Le Corbusier [1929] (1987a) *The City of Tomorrow and Its Planning*, trans. Etchells, F., New York, Dover [from 8th edition, *Urbanisme*].
Le Corbusier [1935] (1987b) *Aircraft*, London, Trefoil Press.
Le Corbusier [1923] (1987c) *Towards a New Architecture*, trans. Etchells, F., enlarged edition, New York, Dover [from *Vers une architecture*, 13th edition].
Lee, M. L. (1995) *Earth First!: Environmental Apocalypse*, Syracuse, N.Y., Syracuse University Press.
Lefebvre, H. (1991) *The Production of Space*, Oxford, Blackwell.
Lefebvre, H. (1996) *Writings on Cities*, Oxford, Blackwell.
Lefebvre, H. (2000) *Everyday Life in the Modern World*, trans. Rabinovitch, S., London, Athlone.
Lefebvre, H. (2003) *The Urban Revolution*, foreword Smith, N., trans. Bononno, R., Minneapolis, University of Minnesota Press.
LeGates, R. T. and Stout, F., eds (2000) *The City Reader*, 2nd edition, London, Routledge.
Lenin, V. I. [1952] (1975) *What Is To Be Done?*, Peking, Foreign Language Press.
Lentricchia, F. and McAuliffe, J. (2002) 'Groundzeroland', in Hauerwas and Lentricchia (2002), pp. 349–60.
Leslie, E. (1999) 'Telescoping the microscopic object: Benjamin the collector', *de-, dis-, ex-*, 3, pp. 58–91.

Leslie, E. (1999) 'Space and west end girls: Walter Benjamin versus cultural studies', *New Formations*, no.38, pp. 110–24.
Leslie, E. (2000) *Walter Benjamin: Overpowering Conformism*, London, Pluto.
Leslie, E. (2001) 'Tate Modern: A year of sweet success', *Radical Philosophy*, no.109, pp. 2–5.
Levi, P. (2001a) *The Search for Roots: A Personal Anthology*, Harmondsworth, Penguin.
Levi, P. (2001b) *The Voice of Memory*, ed. Belpoliti, M. and Gordon, R., Cambridge, Polity Press.
Levinson, S. (1998) *Written in Stone: Public Monuments in Changing Societies*, Durham, N.C., Duke University Press.
Levy, Z. (1997) 'Utopia and reality in the philosophy of Ernst Bloch', in Daniel and Moylan (1997), pp. 175–85.
Libeskind, D. (1992) *Countersign*, New York, Rizzoli.
Light, A., ed (1998) *Social Ecology After Bookchin*, New York, Guilford Press.
Lindblom, C. E. (1999) 'A century of planning', in Kenny and Meadowcroft, (1999), pp. 39–65.
Lingwood, J. (1990) *New Works for Different Places: TSWA Four Cities Project*, Bristol, TSWA.
Lingwood, J., ed. (1995) *House*, London, Phaidon.
Lippard, L. (1973) *Six Years: The Dematerialization of the Art Object*, London, Studio Vista.
Lippard, L. (1981) 'Hot potatoes: Art and politics in 1980', *Block*, no.4, pp. 2–9 [reprinted in Robinson (2001), pp. 107–18].
Lippard, L. (1995) 'Looking around: Where we are, where we could be', in Lacy (1995), pp. 114–30.
Lippard, L. (1997) *Lure of the Local*, New York, The New Press.
Lloyd, J. (1991) 'Emil Nolde's "ethnographic" still lifes: primitivism, tradition, and modernity', in Hiller (1991), pp. 90–112.
Lodziak, C. (2002) *The Myth of Consumption*, London, Pluto.
Loftman, P. and Nevin, B. (1998) 'Pro-growth local economic development strategies: Civic promotion and local needs in Britain's second city', in Hall and Hubbard (1998), pp. 129–48.
Loftman, P. and Nevin, B. (2003) 'Prestige projects, city centre restructuring and social exclusion: Taking the long-term view', in Miles and Hall (2003), pp. 76–91.
Long, R.-C. W. (1972) 'Kandinsky and abstraction: the role of the hidden image', *Artforum*, June, pp. 42–9.
Long, R.-C. W. (1975) 'Kandinsky's abstract style: The veiling of apocalyptic folk imagery', *Art Journal*, vol. XXXIV, Spring, pp. 217–27.
Longo, P. J. (1998) 'Environmental injustices and traditional environmental organizations: Potential for coalition building', in Camacho (1998), pp. 165–76.
Lopes, S. (1987) *The Wall: Images and Offerings from the Vietnam Veterans Memorial*, New York, Collins.
Lorde, A. [1979] (2000) 'The master's tools will never dismantle the mater's house', in Rendell, Penner, and Borden (2000), pp. 53–5.
Low, S. M. (2000) *On the Plaza: The Politics of Public Space and Culture*, Austin, Tex., University of Texas.
Lukács, G. (1965) 'The sociology of modern drama', in Bentley (1965), pp. 425–50.
Lukács, G. (1980) 'Realism in the balance', in Adorno *et al.* (1980), pp. 28–59.
Lunn, E. (1984) *Marxism & Modernism: An Historical Study of Lukács, Brecht, Benjamin and Adorno*, Berkeley, University of California Press.
Lynch, K. (1981) *Good City Form*, Cambridge, Mass., MIT.
Lynch, K. (1972) *What Time is this Place?*, Cambridge, Mass., MIT.

Lyotard, J.-F. (1984) *The Postmodern Condition: A Report on Knowledge*, Manchester, Manchester University Press.
MacAvera, B. (n.d. but 1990) *Art, Politics and Ireland*, Dublin, Open Air.
McCreery, S. (2001) 'The Claremont Road situation', in Borden *et al.* (2001), pp. 227–45.
McEwen, I. K. (1993) *Socrates' Ancestor: An Essay on Architectural Beginnings*, Cambridge, Mass., MIT.
McGuigan, J. (1996) *Culture and the Public Sphere*, London, Routledge.
McKay, G. (1996) *Senseless Act of Beauty: Cultures of Resistance since the Sixties*, London, Verso.
McLeod, M. (1980) 'Le Corbusier and Algiers', *Oppositions*, no.19–20, pp. 53–85.
McLeod, M. (1983) 'Architecture or revolution: Taylorism, technocracy, and social change', *Art Journal*, vol. 43, no.2, pp. 132–47.
McLeod, M. (1985) 'Urbanism and utopia: Le Corbusier from regional syndicalism to Vichy', doctoral thesis, Princeton University, N.J.
McLeod, M. (1997) 'Henri Lefebvre's critique of everyday life: An introduction', in Harris and Berke (1997), pp. 9–29.
Magwood, C. and Mark, P. (2000) *Straw Bale Building*, Gabriola Island, British Columbia, New Society Publishers.
Madsun, P. and Plunz, R., eds (2002) *The Urban Life World: Formation, Perception, Representation*, London, Routledge.
Mahler, A. [1940] (1973) *Gustav Mahler: Memoirs and Letters*, ed. Mitchell, D., London, John Murray.
Manco, T. (2002) *Stencil Graffiti*, London, Thames & Hudson.
Mandelker, A., ed. (1995) *Bakhtin in Contexts*, Evanston, Ill., Northwestern University Press.
Manuel, F. E. (1973) *Utopias and Utopian Thought*, London, Souvenir Press.
Marcus, G. (1989) *Lipstick Traces*, Cambridge, Mass., Harvard.
Marcuse, H. [1956] (1987) *Eros and Civilisation*, London, Routledge & Kegan Paul.
Marcuse, H. (1964) *One-Dimensional Man*, Boston, Mass., Beacon Press.
Marcuse, H. (1968a) *Negations*, Harmondsworth, Penguin.
Marcuse, H. (1968b) 'Liberation from the affluent society', in Cooper (1968), pp. 175–92.
Marcuse, H. (1969) *An Essay on Liberation*, Harmondsworth, Penguin.
Marcuse, H. (1970) *Five Lectures*, Harmondsworth, Penguin.
Marcuse, H. (1972) *Counter-Revolution and Revolt*, Boston, Beacon Press.
Marcuse, H. (1978) *The Aesthetic Dimension: Toward a Critique of Marxist Aesthetics*, Boston, Beacon Press.
Marcuse, H. (1998) *Herbert Marcuse: Technology, War and Fascism*, ed. and intro. Kellner, D., vol. 1, Collected Papers, London, Routledge.
Marcuse, H. (2001) *Herbert Marcuse: Towards a Critical Theory of Society*, ed. and intro. Kellner, D., vol. 2, Collected Papers, London, Routledge.
Marcuse, P. (2002) 'The layered city', in Madsen and Plunz (2002), pp. 94–114.
Markus, T. A. (1993) *Buildings and Power*, London, Routledge.
Marra, A., ed. (1999) *Eco-Tec, Architecture of the In-Between*, New York, Princeton Architectural Press.
Martin, J. L., Nicholson, B., Gabo, N., eds (1937) *Circle*, London, Faber and Faber [1971 reprint].
Marx, K. and Engels, F. (1980) *On the Paris Commune*, Moscow, Progress.
Massey, D. (1994) *Space, Place and Gender*, Cambridge, Polity Press.
Massey, D. (2001) 'Living in Wythenshaw', in Borden *et al.* (2001), pp. 458–75.
Maxwell, K. (2002) 'Lisbon: The earthquake of 1755 and the recovery under the Marquês de Pombal', in Ockman (2002), pp. 20–45.
Mayall, D. (1988) *Gypsy-Travellers in Nineteenth-Century Society*, Cambridge, Cambridge University Press.

Mayer, G., [1936] 1969, *Friedrich Engels*, trans. Highet, G. and Highet, H., New York, Howard Fertig.
Meadowcroft, J. (1999) 'Planning for sustainable development: What can be learned from the critics?', in Kenny and Meadowcroft (1999), pp. 12–38.
Megill, A. (1985) *Prophets of Extremity*, Berkeley, University of California Press.
Meikle, G. (2002) *Future Active: Media Activism and the Internet*, London, Routledge.
Melehy, H. (1997) *Writing Cogito: Montaigne, Descartes, and the Institution of the Modern Subject*, Albany, N.Y., State University of New York Press.
Meller, H. (2001) *European Cities 1890–1930s: History, Culture and the Built Environment*, Chichester, Wiley.
Merquior, J. G. (1983) *Foucault*, London, Fontana.
Merrifield, A. and Swyngedouw, E., eds (1996) *The Urbanization of Injustice*, London, Lawrence and Wishart.
Meskimmon, M. (1997) *Engendering the City: Women Artists and Urban Space*, London, Scarlet Press [*Nexus*, vol. 1].
Meštrovič, S. (1994) *The Balkanization of the West: The Confluence of Postmodernism and Postcommunism*, London, Routledge.
Metzger, G. (1999) 'Earth to galaxies: On destruction and destructivity', in Cole (1999), pp. 44–7.
Michalski, S. (1998) *Public Monuments: Art in Political Bondage 1870–1997*, London, Reaktion.
Midnight Notes Collective (1992) *Midnight Oil: Work, Energy, War 1973–92*, New York, Autonomedia.
Miles, M. (1997) *Art, Space & the City*, London, Routledge.
Miles, M. (1998) 'A game of appearance: Public art and urban development – complicity or sustainability?', in Hall and Hubbard (1998), pp. 203–24.
Miles, M. (2000) *The Uses of Decoration: Essays in the Architectural Everyday*, Chichester, Wiley.
Miles, M. (2002) 'Seeing through place: Local approaches to global problems', in Rugg and Hinchcliffe (2002), pp. 77–89.
Miles, M. (2003) 'Strange days', in Miles and Hall (2003), pp. 44–61.
Miles, M., Hall, T., and Borden, I., eds (2000) *The City Cultures Reader*, London, Routledge.
Miles, M. and Hall, T. eds (2003) *Urban Futures: Critical Essays on Shaping the City*, London, Routledge.
Miles, M. and Kirkham, N., eds (2003) *Culture and Settlement*, Bristol, Intellect Books.
Milton, S. (1991) *In Fitting Memory: The Art and Politics of Holocaust Memorials*, Detroit, Wayne State University Press.
Mirzoeff, N., ed. (1998) *The Visual Culture Reader*, London, Routledge.
Mitchell, W. J. T. (1992) 'The violence of public art: *Do the Right Thing*', in Mitchell (1992), pp. 29–48.
Mitchell, W. J. T., ed. (1992) *Art and the Public Sphere*, Chicago, University of Chicago Press.
Morgan, C. (2000) 'Less is More', *Eco Design*, vol. VIII, no.1, Spring/Summer, pp. 12–13.
Morley, H. (1887) *Ideal Commonwealths*, London, George Routledge and Sons.
Mosbacher, M. (2002) *Marketing the Revolution: The New Anti-Capitalism and the Attack upon Corporate Brands*, London, The Social Affairs Unit.
Mozingo, L. (1989) 'Women and downtown open spaces', *Places*, vol. 6, no.1, pp. 38–47.
Mulhern, F. (2000) *Culture/Metaculture*, London, Routledge.
Mulvey, L. (1999) 'Reflections on disgraced monuments', in Leach (1999), pp. 219–27.
Mumford, L. (1956) *The Human Prospect*, London, Secker & Warburg.
Mumford, L. (1971) *The Myth of the Machine*, London, Secker & Warburg.
Mumford, L. (1973) 'Utopia, the city and the machine', in Manuel (1973), pp. 3–24.
Munt, S. (2001) 'The Lesbian Flâneur', in Borden *et al.* (2001), pp. 246–61.

Museum for Modern Art, Bolzano (1999) *Jochen Gerz: Res Publics, The Public Works 1968–1999*, Ostfildern, Hatje Cantz Verlag.

Muttitt, G. and Marriott, J. (2002) *Some Common Concerns: Imagining BP's Azerbaijan-Georgia-Turkey Pipelines System*, London, published jointly by PLATFORM, The Corner House, Friends of the Earth International, Campagna per la Riforma della Banca Mondiale, CEE Bankwatch Network, and the Kurdish Human Rights Project.

Myrhrman, M. and MacDonald, J. O. (1999) *Building with Bales: A Step by Step Guide for Straw-Bale Construction*, Tucson, Ariz., Out on Bale.

Myrvoll, S. (1999) 'Cultural heritage tourism in Norway, with the focus on Bergen', in Dodd and van Hemmel (1999), pp. 34–43.

Naess, A. (1989) *Ecology, Community and Lifestyle*, trans. and ed. Rothenberg, D., Cambridge, Cambridge University Press.

Nagel, M. (2002) *Masking the Abject: A Genealogy of Play*, Lanham, Md., Lexington Books.

Nicolis, G. (1995) *Introduction to Nonlinear Science*, Cambridge, Cambridge University Press.

Nietzsche, F. (1969) *Thus Spoke Zarathustra*, trans. Hollingdale, R. J., Harmondsworth, Penguin.

Nietzsche, F. (1977) *A Nietzsche Reader*, ed. and trans. Hollingdale, R. J., Harmondsworth, Penguin.

Nochlin, L., (1968) 'The invention of the avant-garde: France 1830–80', in Hess and Ashbery, 1968, pp. 1–24.

Nochlin (1991) *The Politics of Vision: Essays on Nineteenth-Century Art and Society*, London, Thames & Hudson.

North, M. (1992) 'The public a sculpture: From heavenly city to mass ornament', in Mitchell (1992), pp. 9–29.

O'Connor, J. and Wynne, D., eds (1996) *From the Margins to the Centre: Cultural Production and Consumption in the Post-Industrial City*, Aldershot, Ashgate.

O'Donoghue, B. (1982) *The Courtly Love Tradition*, Manchester, Manchester University Press.

Ockman, J., ed. (2002) *Out of Ground Zero: Case Studies in Urban Reinvention*, Munich, Prestel Verlag.

Oh, M. and Arditi, J. (2000) 'Shopping and postmodernism: Consumption, production, identity, and the Internet', in Gottdiener (2000), pp. 71–89.

Okome, O., ed (2000) *Before I am Hanged: Ken Sara-Wiwa: Literature, Politics and Dissent*, Trenton, N.J., Africa World Press.

Orton, F. (1996) 'Present, the scene of . . . selves, the occasion of . . . ruses', in Bird *et al.* (1996), pp. 87–114.

Osborne, J. (1989) *Strindberg's The Father & Ibsen's Hedda Gabler* [adaptations], London, Faber and Faber.

Osborne, P. (2000) *Philosophy in Cultural Theory*, London, Routledge.

Owens, C. (1985) 'The discourse of others: Feminists and postmodernism', in Foster (1983), pp. 57–82.

Paetzold, H. ed. (1997) *City Life: Essays on Urban Culture*, Maastricht, Akademie Jan van Eyck.

Page, M. (2002) 'New York: Creatively destroying New York: fantasies, premonitions, and realities in the provisional city', in Ockman (2002), pp. 166–83.

Papanek, V. (1984) *Design for the Real World: Human Ecology and Social Change*, London, Thames and Hudson.

Papanek, V. (1995) *The Green Imperative: Ecology and Ethics in Design and Architecture*, London, Thames and Hudson.

Paschich, E. and Zimmerman, J. (2001) *Mainstreaming Sustainable Architecture: Casa de Paja, a Demonstration*, Corrales, N.Mex., High Desert Press.

Pearson, D. (1989) *The New Natural House Book: Creating a Healthy, Harmonious, and Ecologically Sound Home*, New York, Simon and Schuster [2nd edition, 1998].
Peet, R. and Watts, M., eds (1996) *Liberation Ecologies: Environment, Development, Social Movements*, London, Routledge.
Pepper, D. (1996) *Modern Environmentalism: An Introduction*, London, Routledge.
Petegorsky, D. W. [1940] (1999) *Left-Wing Democracy in the English Civil War*, Trowbridge, Redwood Books.
Petropoulos, J. (1996) *Art as Politics in the Third Reich*, Chapel Hill, University of North Carolina Press.
Phillips, P. (1988) 'Out of order: The public art machine', *Artforum*, Dec., pp. 92–6.
Phillips, P. (1992) 'Temporality and public art', in Senie and Webster (1992), pp. 295–304.
Phillips, P. (1993) 'Images of repossession', in Walker Art Centre (1993), pp. 43–53.
Phillips, P. (1994) 'The private is public: Peggy Diggs and the system', *Public Art Review*, vol. 5, no.2, Spring/Summer, pp. 13–15.
Phillips, P. (1995a) 'Maintenance activity: creating a climate for change', in Felshin (1995), pp. 165–94.
Phillips, P. (1995b) ''Peggy Diggs: Private acts and public art', in Felshin (1995), pp. 283–308.
Pile, S., Brook, C. and Mooney, G. (1999) *Unruly Cities?*, London, Routledge.
Plant, S. (1992) *The Most Radical Gesture: The Situationist International in a Postmodern Age*, London, Routledge.
Poggioli, R. (1968) *The Theory of the Avant-Garde*, Cambridge, Mass., Harvard.
Pointon, M., ed. (1994) *Art Apart: Art Institutions and Ideology across England and North America*, Manchester, Manchester University Press.
Pollan, M. (1998) *A Place of My Own: The Education of an Amateur Builder*, London, Bloomsbury.
Pollock, G. (1988) *Vision & Difference: Femininity, Feminism and the Histories of Art*, London, Routledge.
Pollock, G. (1994) 'Territories of desire: reconsiderations of an African childhood', in Robertson *et al.* (1994), pp. 63–89.
Poggioli, R. (1968) *The Theory of the Avant-Garde*, Cambridge, Mass., MIT.
Pointon, M., ed. (1994) *Art Apart: Art Institutions and Ideology Across England and North America*, Manchester, Manchester University Press.
Potrč, M. (2001) interview with Ulrike Groos, in exhibition catalogue, Berlin, Künstlerhaus Bethanien.
Pradervand, P. (1989) *Listening to Africa: Developing Africa from the Grassroots*, New York, Praeger.
Prigann, H. (1984) *Der Wald: Ein Zyklus*, Vienna, Medusa Verlag.
Prigann, H. ed. (1993) *Ring der Erinnerung*, Berlin, Verlag Dirk Nishen [bi-lingual German/English].
Proudhon, P.-J. (1969) ed. Edwards, S., trans. Fraser, E., *Selected Writings*, London, Macmillan.
Pugh, S. (1988) *Garden Nature Language*, Manchester, Manchester University Press.
Rabinbach, A. (1997) *In the Shadow of Catastrophe: German Intellectuals between Apocalypse and Enlightenment*, Berkeley, University of California Press.
Rabinow, P. (1989) *French Modern: Norms and Forms of the Social Environment*, Cambridge, Mass., MIT.
Rahnema, M. and Bawtree, V., eds (1997) *The Post-Development Reader*, London, Zed Books.
Rangan, H. (1996) 'From Chipko to Uttaranchai: Development, environment, and social protest in the Garhwal Himalayas, India', in Peet and Watts (1996), pp. 205–24.
Rapoport, A. (1980) 'Vernacular architecture and the cultural determinants of form', in King (1980), pp. 283–305.

Raven, A., ed. (1993) *Art in the Public Interest*, New York, da Capo Press.
Raymond, M. (1970) *From Baudelaire to Surrealism*, London, Methuen.
Reij, C. (1988) 'Soil and water conservation in Yatenga, Burkina Faso', in Conroy and Litvinoff (1988), pp. 74–7.
Rendell, J. (1998) 'Doing It, (un)doing It, (over)doing it yourself: Rhetorics of architectural abuse', in Hill (1998), pp. 229–45.
Rendell, J. (1999) 'Thresholds, passages and surfaces: Touching, passing and seeing in the Burlington Arcade', in Coles (1999), pp. 168–91.
Rendell, J. (2002) *The Pursuit of Pleasure: Gender, Space and Architecture in Regency London*, London, Continuum.
Rendell, J., Penner, B. and Borden, I., eds (2000) *Gender Space Architecture: An Interdisciplinary Introduction*, London, Routledge.
Rich, B. R. (1994) 'Dissed and disconnected: Notes on present ills and future dreams', in Becker (1994), pp. 223–48.
Richon, O. [1985] (1996) 'Representation, the harem and the despot', *Block*, no.10, in Bird *et al.* (1996), pp. 242–57.
Rifkin, A. (1979) 'Cultural movement and the Paris Commune', *Art History*, vol. 2, no.2, pp. 201–20.
Rinder, L. (1999) 'Where is my future?' in Gerz (1999), pp. 20–5.
Robbins, E. (1996) 'Thinking space/seeing space: Thamesmead revisited', *Urban Design International*, vol. 1, no. 3, pp. 283–91.
Roberts, J. (2001) 'Art, politics, and provincialism', *Radical Philosophy*, no.106, pp. 2–6.
Roberts, J. (2002) 'The labour of subjectivity, The subjectivity of labour: Reflections on contemporary political theory and culture', *Third Text*, vol. 16, no.4, pp. 367–85.
Robertson, G., Mash, M., Tickner, L., Bird, J., Curtis, B. and Putnam, T., eds (1994) *Travellers Tales: Narratives of Home and Displacement*, London, Routledge.
Robinson, D. (1990) *SoHo Walls: Beyond Graffiti*, London, Thames & Hudson.
Robinson, H., ed. (2001) *Feminism – Art – Theory*, Oxford, Blackwell.
Robbins, B. (1993) *The Phantom Public Sphere*, Minneapolis, University of Minnesota Press.
Robbins, E. (1996) 'Thinking space / seeing space: Thamesmead revisited' *Urban Design International*, vol. 1, no.3, pp. 283–91.
Roblin, R. (1990) *The Aesthetics of the Critical Theorists*, Lampeter, Edwin Mellen Press.
Roessler, B. (2002) 'Revelation and concealment: Staging private life in public', in Bangma (2002), pp. 36–45.
Rogoff, I. (1998) 'Studying visual culture', in Mirzoeff (1998), pp. 14–26.
Rogoff, I. (2000) *Terra Infirma: Geography's Visual Culture*, London, Routledge.
Rolston, B. (1992) *Drawing Support: Murals in the North of Ireland*, Belfast, Beyond the Pale Publications.
Roos, J. M. (1996) *Early Impressionism and the French State (1866–1874)*, Cambridge, Cambridge University Press.
Rose, M. A. (1988) *Marx's Lost Aesthetic: Karl Marx and the visual arts*, Cambridge, Cambridge University Press.
Rose, N. (1999) *Powers of Freedom: Reframing Political Thought*, Cambridge, Cambridge University Press.
Rosen, J. (1991) 'America's holocaust', *Forward*, April 12th.
Rosenblum, R. (1975) *Modern Painting and the Northern Romantic Tradition: Friedrich to Rothko*, London, Thames & Hudson.
Rosenthal, N. (1979) 'Idealism and naturalism in painting', *Post-impressionism* [catalogue], London, Royal Academy of Arts, pp. 150–3.
Rosler, M. (1991) 'Fragments of a metropolitan viewpoint', in Wallis (1991), pp. 15–44.
Rosler, M. (1994) 'Place, position, power, politics', in Becker (1994), pp. 55–76.

Rosler, M. (1996) 'Video: Shedding the utopian moment', in Bird *et al.* (1996), pp. 258–78.
Ross, K. (1988) *The Emergence of Social Space: Rimbaud and the Paris Commune*, Minneapolis, University of Minnesota Press.
Roth, M. (1993) 'Suzanne Lacy: Social reformer and witch', in Raven (1993), pp. 155–74.
Roth, M. S. with Lyons, C. and Merewether, C. (1997) *Irresistible Decay*, Los Angeles, Getty Research Institute for the History of Art and the Humanities.
Rowe, C. and Koetter, F. (1982) *Collage City*, Cambridge, Mass., MIT.
Royal Academy of Arts (1978) *Courbet* [exhibition catalogue], London, Royal Academy of Arts.
Royal Academy of Arts (1999) *Charlotte Salomon: 'Life or Theatre?'* [exhibition catalogue], London, Royal Academy of Arts.
Rubin, J. H. (1980) *Realism and Social Vision in Courbet and Proudhon*, Princeton, N.J., Princeton University Press.
Rugg, J. and Sedgewick, M. (2001) 'Budapest's statue park: memorial or counter-monument?' *Soundings*, no.17, pp. 94–112.
Rugg, J. (2002) 'Budapest's statue park: collective memory or collective amnesia?', unpublished conference paper.
Rugg, J. and Hinchcliffe, D., eds (2002) *Recoveries and Reclamations*, Bristol, Intellect Books.
Rupp, J. M. (1992) *Art in Seattle's Public Places: An Illustrated Guide*, Seattle, University of Washington Press.
Ryan, M. P. (1990) *Women in Public: Between Banners and Ballots, 1825–1880*, Baltimore, Md., Johns Hopkins University Press.
Ryan, J. and Fitzpatrick, H. (1996) 'The space that difference makes: negotiation and urban identities through consumption practices', in O'Connor and Wynne (1996), pp. 169–202.
Sabini, J. P. and Silver, M. (1980), 'Destroying the innocent with a clear conscience: A sociopsychology of the Holocaust', in Dinsdale (1980), pp. 329–30.
Sachs, W. (1992) *The Development Dictionary*, London, Zed Books.
Sadler, S. (1998) *The Situationist City*, Cambridge, Mass., MIT.
Saint-Simon, C.-H. de Rouvroy, Comte de, 1825, *Opinions litteraires, philosophiques et industrielles*, Paris, Gallerie de Bossange père.
Said, E. (1990) 'Narrative, geography, and interpretation', *New Left Review*, no.180, p. 88.
Said, E. W. (1991) *Orientalism: Western Conceptions of the Orient*, Harmondsworth, Penguin.
Said, E. W. (1994) *Culture and Imperialism*, London, Verso.
Salecl, R. (1999) 'The state as a work of art: the trauma of Ceausescu's Disneyland', in Leach (1999), pp. 92–111.
Saltzman, L. (2001) 'Lost in Translation: Clement Greenberg, Anselm Kiefer, and the Subject of History', in Zelizer (2001), pp. 74–90.
Sandercock, L. (1998a) *Towards Cosmopolis*, Chichester, Wiley.
Sandercock, L., ed. (1998b) *Making the Invisible Visible: A Multicultural Planning History*, Berkeley, University of California Press.
Sandercock, L. and Forsyth, A. [1992] (2000) 'A gender agenda: New directions for planning theory', in LeGates and Stout (2000), pp. 446–59.
Sangregorio, I.-L. (1998) 'Having it all? A question of collaborative housing', in Ainley (1998), pp. 101–11.
Santos, D. (2002) 'Singularity in the post-modern condition, or the possible mediation of contemporary art', in Extra]muros[(2002), pp. 160–9.
Sassower, R. and Cicotello, L. (2000) *The Golden Avant-Garde: Idolatry, Commercialism, and Art*, Charlottesville, Va., University Press of Virginia.
Satterthwaite, D., ed. (1999) *The Earthscan Reader in Sustainable Cities*, London, Earthscan.

Savage, K. (1994) 'The politics of memory: Black emancipation and the Civil War Monument', in Gillis (1994), pp. 127–49.

Schelling, V. (1999) 'The people's radio of Vila Nossa Senhora Aparecida: alternative communication and cultures of resistance in Brazil', in Skelton and Allen (1999), pp. 167–79.

Scheunemann, D., ed. (2000) *European Avant-Garde: New Perspectives*, Amsterdam, Rodopi.

Schirmacher, W., ed. (2000) *The Frankfurt School* (German 20th Century Philosophy series), New York, Continuum.

Schmidt, A. P. and de Graaf, J. (1982) *Violence as Communication: Insurgent Terrorism and the Western News Media*, London, Sage.

Schnädelbach, H. (1999) 'The cultural legacy of critical theory'. *New Formations*, no.38, pp. 64–77.

Schwartz, W. and Schwartz, D. (1998) *Living Lightly: Travels in Post-Consumer Society*, Charlbury, Jon Carpenter.

Scott, J. C. (1985) *Weapons of the Weak: Everyday Forms of Peasant Resistance*, New Haven, Conn., Yale.

Scruton, R. (2001) *Kant – A Very Short Introduction* [revised from 1982 version], Oxford, Oxford University Press.

Seabrook, J. (1993) *Pioneers of Change: Experiments in Creating a Humane Society*, London, Zed Books.

Seabrook, J. (1996) *In the Cities of the South: Scenes from a Developing World*, London, Verso.

Seddon, P. (2000) 'From eschatology to ecology: the ends of history and nature', in Green and Seddon (2000), pp. 82–96.

Segal, W. (1980) 'The housing crisis in Western Europe: Britain – assessment and options', in Mikellides (1980), pp. 171–5.

Senie, H. and Webster, S. (1992) *Critical Issues in Public Art*, Washington, D.C., Smithsonian Institution Press.

Sennett, R. (1970) *The Uses of Disorder: Personal Identity and City Life*, New York, Norton.

Sennett, R. [1977] (1986) *The Fall of Public Man*, New York, Norton.

Sennett, R. (1980) *Authority*, New York, Norton.

Sennett, R. (1990) *The Conscience of the Eye*, New York, Norton.

Sennett, R. (1995) *Flesh and Stone: The Body and the City in Western Civilization*, London, Faber and Faber.

Serpentine Gallery (1988) *Kevin Atherton, a Body of Work, 1982–1988*, London, Serpentine Gallery.

Shattuck, R. (1969) *The Banquet Years: the Origins of the Avant-Garde in France: 1885 to World War I*, London, Cape.

Shields, R. (1999) *Lefebvre, Love & Struggle*, London, Routledge.

Shor, I. and Freire, P. (1987) *A Pedagogy for Liberation: Dialogues on Transforming Education*, New York, Bergin and Garvey.

Short, J. R. (1996) *The Urban Order: An Introduction to Cities, Culture and Power*, Oxford, Blackwell.

Sibley, D. (1995) *Geographies of Exclusion*, London, Routledge.

Siegert, H. and Stern, M. (2002) 'Berlin: Film and the representation of urban reconstruction since the fall of the Wall', in Ockman (2002), pp. 116–37.

Simmel, G. (1990) ed. Frisby, D., *The Philosophy of Money*, 2nd edition, London, Routledge.

Simony, C., Brodt, J. and Pryor, K., eds (1998) *Ample Opportunity: A Community Dialogue*, Pittsburgh, Studio for Creative Inquiry, Carnegie Mellon University.

Singer, D. (2001) '1989: The end of communism', in Kastiafica (2001), pp. 11–19.

Skelton, T. and Allen, T., eds (1999) *Culture and Global Change*, London, Routledge.

Smith, G., ed. (1986) *Walter Benjamin: Moscow Diary*, Cambridge Mass., Harvard.

Smith, N. (1997) *The New Urban Frontier: Gentrification and the Revanchist City*, London, Routledge.
Soja, E. (1996) *Thirdspace: Journeys to Los Angeles and Real-And-Imagined Places*, Oxford, Blackwell.
Soja, E. (2000) *Postmetropolis: Critical Studies of Cities and Regions*, Oxford, Blackwell.
Sontag, S. (1979) *On Photography*, Harmondsworth, Penguin.
Sorkin, M., ed. (1992) *Variations on a Theme Park: The New American City and the End of Public Space*, New York, Hill & Wang.
Sorkin, M. (2002) *Some Assembly Required*, Minneapolis, University of Minnesota Press.
Speer, A. (1970) *Inside the Third Reich*, New York, Weidenfeld and Nicolson.
Spender, D. (1985) *Man Made Language*, 2nd edition, London, Routledge.
Spivak, G. C. (1988) *In Other Worlds: Essays in Cultural Politics*, London, Routledge.
Starhawk (2002) *Webs of Power: Note from the Global Uprising*, Gabriola Island, British Columbia, New Society Publishers.
Starr, P. (1995) *Logics of Failed Revolt: French Theory After May '68*, Stanford Calif., Stanford University Press.
Steegmuller, F., ed. (1983) *Flaubert in Egypt*, London, Michael Haag.
Steele, J. (1988) 'Hassan Fathy' *Architectural Monograph*, no.13, London, Academy Editions.
Steele, J. (1997) *An Architecture for People: The Complete Works of Hassan Fathy*, London, Thames and Hudson.
Sternhell, Z. (1986) *Neither Right nor Left: Fascist Ideology in France*, trans. Maisel, D., Princeton, N.J., Princeton University Press.
Sternhell, Z. (1987) 'The anti-materialist revision of Marxism as an aspect of the rise of fascist ideology', *Journal of Contemporary History*, no.22, pp. 379–400.
Stirk, P. M. (2000) *Critical Theory, Politics and Society*, London, Pinter.
Storr, A. (1976) *The Dynamics of Creation*, Harmondsworth, Penguin.
Stout, F. (1999) 'Visions of a new reality: The city and the emergence of modern visual culture', in LeGates and Stout (2000), pp. 143–6.
Strelow, H., ed. (1999) *Natural Reality* [exhibition catalogue], Stuttgart, DACO-Verlag.
Suttie, I. A. [1935] (1960) *The Origins of Love and Hate*, Harmondsworth, Penguin.
Tafuri, M. (1976) *Architecture and Utopia: Design and Capitalist Development*, Cambridge, Mass., MIT.
Tagg, J. (1992) *Grounds of Dispute: Art History, Cultural Politics and the Discursive Field*, Basingstoke, Macmillan.
Tajbakhsh, K., 2001, *The Promise of the City: Space, Identity, and Politics in Contemporary Social Thought*, Berkeley, University of California Press.
Tate Gallery (1983) *The Essential Cubism 1907–1920* [catalogue], London, Tate Publishing.
Taylor, B. (1994) 'From penitentiary to temple of art: Early metaphors of improvement at the Millbank Tate', in Pointon (1994), pp. 9–32.
Taylor, B. (1998) 'Commissar of space', in Goto and Taylor (1998), pp. 5–21.
Taylor, B. (2000) 'History painting west and east', in Green and Seddon (2000), pp. 66–81.
Taylor, L. (1990) *Housing: Symbol, Structure, Site*, New York, Rizzoli.
Thomson, I. (2001) 'Primo Levi in conversation', in Belpoliti and Gordon (2001), pp. 34–44 [first published in 1987, *PN Review*, vol. 14, no.2, pp. 15–19].
Thoré, T., 1868, *Salons de T. Thoré*, Paris, Librairie Internationale.
Till, J. (1998) 'Architecture of the impure community', in Hill (1998), pp. 61–75.
Tillet, W. (2002) 'Encompassing the movement. Containment. Filtering. Mediation' (http://www.amsterdam.nettime.org/Lists-Archive/nettime-bold-0203/msg00304.html).
Tisdall, C. (1974) *Art into Society, Society into Art* [exhibition catalogue] London, Institute of Contemporary Arts.

Tolstoy, V., Bibikova, I. and Cooke, C. (1990) *Street Art of the Russian Revolution*, New York, Vendome Press.
Toussaint, H. (1978) catalogue notes, Royal Academy of Arts, 1978, pp. 75–235; and 'The dossier on "The Studio" by Courbet', pp. 249–85.
Toulmin, S. (1990) *Cosmopolis: The Hidden Agenda of Modernity*, Chicago, University of Chicago Press.
Towers, G. (1995) *Building Democracy: Community Architecture in the Inner Cities*, London, UCL Press.
Trowell, J. (2000) 'The snowflake in hell and the baked alaska: Improbability, intimacy and change in the public realm', in Bennett and Butler (2000), pp. 99–109.
Tucker, V., ed (1997) *Cultural Perspectives on Development*, London, Frank Cass.
Turner, J. F. C. (1976) *Housing By People: Towards Autonomy in Building Environments*, London, Marion Boyars.
Turner, R. K. (1988) *Sustainable Environmental Management*, London, Belhaven.
Ukeles, M. L. [1980] (2001) 'Touch sanitation', in Robinson (2001), pp. 196–7.
Umenyilora, C. (2000) 'Empowering the self-builder', in Hughes and Sadler (2000), pp. 210–21.
Urry, J. (1990) *The Tourist Gaze: Leisure and Travel in Contemporary Societies*, London, Sage.
Urry, J. (1995) *Consuming Places*, London, Routledge.
Vale, B. and Vale, R. (1991) *Green Architecture: Design for a Sustainable Future*, London, Thames and Hudson.
van Alphen, E. (2001) 'Deadly historians: Boltanski's intervention in Holocaust historiography', in Zelizer (2001), pp. 45–73.
van Moos, S. (1979) *Le Corbusier: Elements of a Synthesis*, Cambridge, Mass., MIT.
Vaneigem, R. [1967] (1994) *The Revolution of Everyday Life*, trans. Nicholson-Smith, D., revised, London, Rebel Press and Left Bank Books.
Vergo, P. (1975) *Art in Vienna 1898–1918*, London, Phaidon.
Verrier, M. (1979) *The Orientalists*, New York, Rizzoli.
Wackernagel, M. and Rees, W. (1996) *Our Ecological Footprint: Reducing Human Impact on the Earth*, Gabriola Island, British Columbia, New Society Publishers.
Wall, D. (1999) *Earth First! and the Anti-Roads Movement: Radical Environmentalism and Comparative Social Movements*, London, Routledge.
Walker Art Centre (1993) *Public Address: Krzysztof Wodiczko*, Minneapolis, Minn., Walker Art Centre.
Waltzer, M., ed. (1995) *Toward a Global Civil Society*, Providence, Mass., Berghahn Books.
Warner, M. (1981) *Joan of Arc: The Image of Female Heroism*, London, Weidenfeld and Nicolson.
Warner, M. (1987) *Monuments and Maidens: The Allegory of the Female Form*, London, Pan Books.
Warnke, M. (1994) *Political Landscape: The Art History of Nature*, London, Reaktion.
Washton Long, R.-C. (1976) 'Kandinsky's abstract style: The veiling of apocalyptic folk imagery', *Art Journal*, no. XXXIV, Spring, pp. 217.
Wates, N. and Knevitt, C. (1987) *Community Architecture: How People Are Creating Their Own Environment*, Harmondsworth, Penguin.
Weinberg, B. (1999) '¡Viva Loisada Libre!', in Wilson and Weinberg (1999), pp. 38–56.
Weissberg, L. (2001) 'In plain sight', in Zelizer (2001), pp. 13–27.
Welsch, W. (1997) *Undoing Aesthetics*, London, Sage.
Werckmeister, O. K. (1989) *The Making of Paul Klee's Career, 1914–20*, Chicago, University of Chicago Press.
West, C. (1993) *Race Matters*, New York, Random House.

Westwood, S. and Williams, J., eds (1997) *Imagining Cities: Scripts, Signs, Memory*, London, Routledge.
Weyergraf Serra, C. and Buskirk, M., eds (1991) *The Destruction of Tilted Arc: Documents*, Cambridge, Mass., MIT.
Whyte, W. H. (1980) *The Social Life of Small Urban Spaces*, Washington, D.C., Conservation Foundation.
Whyte, W. H. (1988) *City: Rediscovering the Centre*, New York, Doubleday.
Wigglesworth, S. and Till, J. (1998a) 'The Everyday and Architecture', *Architectural Design*, vol. 68, no.7/8, July/August, profile 134, pp. 7–9.
Wigglesworth, S. and Till, J. (1998b) 'Table manners', *Architectural Design*, vol. 68, no.7/8, July/August, profile 134, pp. 31–5.
Wigglesworth, S. and Till, J. (2001) *9/10 Stock Orchard Street: A Guidebook*, London, Bank of Ideas.
Wigley, M. (1992) 'Untitled: The housing of gender', in Colomina (1992), pp. 327–89.
Wilbert, C. (2003) 'No to Kyoto', *Radical Philosophy* no.110, pp. 2–7.
Willett, J. (1967) *Art in a City*, London, Methuen.
Willett, J. (1982) *The New Sobriety: Art and Politics in the Weimar Period 1917–33*, London, Thames and Hudson.
Williams, R. (1989) *The Politics of Modernism*, ed. Pinkney, T., London, Verso.
Willis, S. (2002) 'Old Glory', in Hauerwas and Lentricchia (2002), pp. 375–84.
Wilson, E. [1931] (1961) *Axel's Castle: Essays on Yeats, Valéry, T. S. Eliot, Proust, James Joyce, Gertrude Stein*, London, Fontana.
Wilson, E. (2003) *Bohemians: The Glamorous Outcasts*, London, I. B. Taurus.
Wilson, P. L. and Weinberg, B., eds (1999) *Avant Gardening: Ecological Struggle in The City and the World*, New York, Autonomedia.
Wilson, S. (1993) 'The Soviet Pavilion in Paris', in Bown and Taylor (1993) pp. 106–20.
Williams, R. [1989] (1996) *The politics of modernism*, London, Verso.
Wines, J. (1987) *De-Architecture*, New York, Rizzoli.
Winnicott, D. W. (1986) *Home Is Where We Start From: Essay by a Psychoanalyst*, Harmondsworth, Penguin.
Witkin, R. W. (2003) *Adorno on Popular Culture*, London, Routledge.
Wolff, J. (1989) 'The invisible flâneuse: Women and the literature of modernity', in Benjamin (1989), pp. 141–56.
Wolin, R. (1994) *Walter Benjamin: An Aesthetic of Redemption*, Berkeley, Calif., University of California Press.
Wood, P., ed., 1999, *The Challenge of the Avant-Garde*, New Haven, Conn., Yale University Press.
World Commission on Environment and Development (WCED) (1987) *Our Common Future*, Oxford, Oxford University Press.
Worsley, P. (1999) 'Culture and development theory', in Skelton and Allen (1999), pp. 30–41.
Wright, G. (1991) *The Politics of Urban Design in French Colonial Urbanism*, Chicago, University of Chicago Press.
Young, I. M. (1990) *Justice and the Politics of Difference*, Princeton, N.J., Princeton University Press.
Young, I. M. (2000) *Inclusion and Democracy*, Oxford, Oxford University Press.
Young, J. E. (1993) *The Texture of Memory: Holocaust Memorials and Meaning*, New Haven, Conn., Yale University Press.
Young, J. E. (1992) 'The counter-monument: Memory against itself in Germany today', in Mitchell (1992), pp. 49–78.
Young, J. E. (2001) 'Daniel Libeskind's Jewish Museum in Berlin: The uncanny arts of memorial architecture', in Zelizer (2001), pp. 179–97.

Younge, G. (1988) *Art of the South African Townships*, London, Thames & Hudson.
Zelizer, B., ed. (2001) *Visual Culture and the Holocaust*, London, Athlone.
Zimmerer, K. S. (1996) 'Discourses on Soil Loss in Bolivia: sustainability and the search for socioenvironmental "middle ground"', in Peet and Watts (1996), pp. 110–24.
Zola, E. (1991) *Ecrit sur l'art*, ed. Leduc-Adine, J.-P., Paris, Gallimard.
Zukin, S. (1989) *Loft Living: Culture and Capital in Urban Change*, 2nd edition, New Brunswick, N.J., Rutgers University Press.
Zukin, S. (1995) *The Cultures of Cities*, Oxford, Blackwell.
Zukin, S. (1996a) 'Space and symbols in an age of decline', in King (1996), pp. 43–59.
Zukin, S. (1996b) 'Cultural strategies of economic development and the hegemony of vision', in Merrifield and Swyngedouw (1996), pp. 223–43.

INDEX

•

Page numbers in *italics* refer to plates; n means note.